Fish Bioenergetics

CHAPMAN & HALL FISH AND FISHERIES SERIES

Amongst the fishes, a remarkably wide range of fascinating biological adaptations to diverse habitats has evolved. Moreover, fisheries are of considerable importance in providing human food and economic benefits. Rational exploitation and management of our global stocks of fishes must rely upon a detailed and precise insight of the interaction of fish biology with human activities.

The *Chapman & Hall Fish and Fisheries Series* aims to present authoritative and timely reviews which focus on important and specific aspects of the biology, ecology, taxonomy, physiology, behaviour, management and conservation of fish and fisheries. Each volume will cover a wide but unified field with themes in both pure and applied fish biology. Although volumes will outline and put in perspective current research frontiers, the intention is to provide a synthesis accessible and useful to both experts and non-specialists alike. Consequently, most volumes will be of interest to a broad spectrum of research workers in biology, zoology, ecology and physiology, with an additional aim of the books encompassing themes accessible to non-specialist readers, ranging from undergraduates and postgraduates to those with an interest in industrial and commercial aspects of fish and fisheries.

Applied topics will embrace synopses of fishery issues which will appeal to a wide audience of fishery scientists, aquaculturists, economists, geographers and managers in the fishing industry. The series will also contain practical guides to fishery and analysis methods and global reviews of particular types of fisheries.

Books already published and forthcoming are listed below. The Publisher and Series Editor would be glad to discuss ideas for new volumes in the series . . .

Available titles

1. **Ecology of Teleost Fishes**
   Robert J. Wootton
2. **Cichlid Fishes**
   Behaviour, ecology and evolution
   Edited by Miles A. Keenlyside
3. **Cyprinid Fishes**
   Systematics, biology and exploitation
   Edited by Ian J. Winfield and Joseph S. Nelson
4. **Early Life History of Fish**
   An energetics approach
   Ewa Kamler

5. **Fisheries Acoustics**
   David N. MacLennan and E. John Simmonds
6. **Fish Chemoreception**
   Edited by Toshiaki J. Hara
7. **Behaviour of Teleost Fishes**
   Second edition
   Edited by Tony J. Pitcher
8. **Genetics and Fish Breeding**
   Colin R. Purdom
9. **Fish Ecophysiology**
   J. Cliff Rankin and Frank B. Jensen
10. **Fish Swimming**
    John J. Videler
11. **On the Dynamics of Exploited Fish Populations**
    Raymond J.H. Beverton and Sidney J. Holt
    (Facsimile reprint)
12. **Sea Bass**
    Graham D. Pickett and Michael G. Pawson
13. **Fish Bioenergetics**
    Malcolm Jobling
14. **On the Sex of Fish and Gender of Scientists**
    Daniel Pauly

Forthcoming titles

Fisheries Ecology
Second edition
Edited by T.J. Pitcher and P. Hart

Hake
Fisheries, products and markets
J. Alheit and T.J. Pitcher

Impact of Species Change in the African Lakes
Edited by T.J. Pitcher

Electric Fishes
P. Moller

Fish Bioenergetics

Malcolm Jobling

The Norwegian College of Fishery Science
University of Tromsø
Norway

CHAPMAN & HALL

London · Glasgow · Weinheim · New York · Tokyo · Melbourne · Madras

Published by Chapman & Hall, 2–6 Boundary Row, London SE1 8HN

Chapman & Hall, 2–8 Boundary Row, London SE1 8HN, UK

Blackie Academic & Professional, Wester Cleddens Road, Bishopbriggs, Glasgow G64 2NZ, UK

Chapman & Hall GmbH, Pappelallee 3, 69469 Weinheim, Germany

Chapman & Hall Inc., One Penn Plaza, 41st Floor, New York NY10119, USA

Chapman & Hall Japan, Thomson Publishing Japan, Hirakawacho Nemoto Building, 6F, 1-7-11 Hirakawa-cho, Chiyoda-ku, Tokyo 102, Japan

Chapman & Hall Australia, Thomas Nelson Australia, 102 Dodds Street, South Melbourne, Victoria 3205, Australia

Chapman & Hall India, R. Seshadri, 32 Second Main Road, CIT East, Madras 600 035, India

First edition 1994

Typeset in 10/12 Photina by Acorn Bookwork, Salisbury, Wilts

Printed in Great Britain by St. Edmundsbury Press, Bury St. Edmunds, Suffolk

ISBN 0 412 58090 X

A catalogue record for this book is available from the British Library

Library of Congress Catalog Card Number: 93-74904

∞ Printed on permanent acid-free text paper, manufactured in accordance with the proposed ANSI/NISO Z 39.48-199X and ANSI Z 39.48-1984

Contents

Preface		xi
Series foreword *T.J. Pitcher*		xiii
Part One Nutritional Requirements and Dietary Formulation		1
1	Nutrient classes and energy metabolism	3
	1.1 Introduction	3
	1.2 Nutrient classes	3
	1.3 The basics of energy metabolism	23
2	Nutritional requirements	31
	2.1 Introduction	31
	2.2 Protein and amino acid requirements	31
	2.3 Lipid and fatty acid requirements	37
	2.4 Vitamin requirements	41
	2.5 Mineral requirements	53
3	Dietary ingredients and feed formulation	57
	3.1 Introduction	57
	3.2 Binding agents and feed production	58
	3.3 Protein sources	61
	3.4 Lipid sources	66
	3.5 Vitamin destruction and dietary additions	69
	3.6 Miscellaneous feed components	72
4	Nutrient interactions and fish health	77
	4.1 Introduction	77
	4.2 Interactions between nutrients	77
	4.3 Nutrition and disease	82
5	Fish nutrition – concluding summary	87
	Part One: Further reading	88

Part Two Physiological Energetics: Feeding, Metabolism and Growth 91

6 Introduction to energetics 93
6.1 Introduction 93
6.2 The energy budget equation 93

7 Ingestion, absorption and excretion 99
7.1 Introduction 99
7.2 Factors influencing ingestion (R) 99
7.3 Losses in faecal material (F) 104
7.4 Products of nitrogenous excretion (U) 109

8 Respiration and metabolism 121
8.1 Introduction 121
8.2 Costs of maintaining bodily functions, M_M 121
8.3 Energy expenditure associated with swimming, M_A 128
8.4 Activity levels and 'field metabolic rates' 137
8.5 Effects of feeding on metabolic rate, M_F 139
8.6 Effects of previous nutritional history on metabolism 142

9 Production and growth 147
9.1 Introduction 147
9.2 Basic relationships between feeding and growth 148
9.3 Laboratory methods for growth studies 150

10 Environmental factors and growth 155
10.1 Introduction 155
10.2 Effects of temperature on growth 156
10.3 Effects of salinity on growth 160
10.4 Photoperiodic effects on growth 163
10.5 Oxygen availability and growth 166

11 Biotic factors and growth performance 169
11.1 Introduction 169
11.2 Fish size and growth 169
11.3 Behavioural factors and growth 171
11.4 Sustained exercise and growth 176
11.5 Hormonal manipulations and growth 179
11.6 Selection programmes for growth 189
11.7 Somatic (P_S) versus reproductive (P_R) growth 190
11.8 Compensatory or 'catch-up' growth 194

12 Bioenergetics – concluding summary 203
Part Two: Further reading 205

Part Three Environmental Tolerances and Preferences 207

13 Environmental tolerances – general introduction 209

14 Temperature 213
14.1 Thermal relations of fishes 213
14.2 Thermal acclimation and compensatory responses 219

15 Dissolved gases 231
15.1 Introduction 231
15.2 Dissolved oxygen 233
15.3 Aquatic surface respiration (ASR) 236
15.4 Air breathing – bimodally breathing fish 238
15.5 Physiological responses to hypoxia in unimodal fish 248
15.6 Dissolved oxygen and metabolic rates 255
15.7 Carbon dioxide and hypercapnia 263
15.8 Gas supersaturation 264

16 Pollutants and toxic chemicals 267
16.1 Introduction 267
16.2 Toxicity and bioaccumulation 268
16.3 Detoxification mechanisms 271
16.4 Freshwater acidification 274
16.5 Ammonia and other nitrogenous wastes 278

17 Environmental stressors 285
17.1 Introduction 285
17.2 Primary stress response – the endocrine system 286
17.3 Secondary responses – tissue alteration 287
17.3 Tertiary responses – longer-term effects 288

18 Adaptations to the environment – concluding summary 291
Part Three: Further reading 292

Author index 295

Species index 297

Subject index 300

Preface

In the years since the mid 1960s there has been an explosive increase in studies designed to examine various aspects of fish growth and nutrition. Much of this increase has been sparked by the rapid development of the aquaculture industry in many countries of the world, accompanied by the search for species which could have a commercial potential in fish farming.

Consequently, most authors of books on aquaculture devote a considerable proportion of the text to considerations of nutritional requirements and feeding practices. There are also monographs and texts devoted entirely to the topic of fish nutrition. Whilst aquaculture texts may provide information about the practical problems related to the feeding of fish in a farm setting, and nutritional requirements and biochemistry are treated at some length in monographs devoted to nutrition, both tend to give only passing reference to the principles of growth energetics.

In this book, which comprises three main sections, the nutrition and physiology of fish are examined from an energetic standpoint, with the environmental influences upon feeding and growth being emphasized. The opening section is devoted to a consideration of the nutrition of fish species, both with respect to the quantitative requirements for specific nutrients and from the viewpoint of practical feed formulation. This is followed by the section on physiological energetics, in which each component of the energy balance equation is described in some detail. The book closes with a discussion of the environmental tolerances and preferences of fish species. Greatest emphasis is placed on the description of general principles, and examples are selected from different experimental studies in order to illustrate specific points.

The intended readership is, primarily, senior undergraduate and postgraduate students taking courses in fish biology, aquaculture, resource management and aquatic biosciences, although it is also hoped that aquaculturists and fisheries managers will find the book to be of some interest.

Since this is a book for advanced students, descriptions of some of the more important experimental and analytical techniques used in studies of fish growth and nutrition have been included. Some discussion of the merits and weaknesses of the various methods is provided, and indications are given as to problem areas in which the unwary newcomer can be led astray.

To disrupt the flow of writing as little as possible, literature citations have been excluded from the main body of the text. A short list of major sources and additional reading is included at the end of each of the main sections. In adopting this approach, the view has been taken that the presentation of material in book form – albeit for a scientific audience – gains little and loses much if written in the style of an original research paper or review article. Whilst this decision may irritate some readers, it is my experience that students – including postgraduates – prefer to read a text uncluttered by literature citations, and because many, otherwise admirable, texts have been rendered almost unreadable by lists of citations, I have opted for the opposite course of action.

Many colleagues and students have contributed to the development of this book, either by commenting on the style or by suggesting ways in which the overall presentation could be improved, but especial thanks must be reserved for Hilka Falkseth, who managed to turn my crude drawings into clear figures and diagrams.

Malcolm Jobling
Tromsø, Norway

Series foreword

Among the fishes, a remarkably wide range of biological adaptations to diverse habitats has evolved. As well as living in the conventional habitats of lakes, ponds, rivers, rock pools and the open sea, fish have solved the problems of life in deserts, in the deep sea, in the cold antarctic, and in warm waters of high alkalinity or of low oxygen. Along with these adaptations we find the most impressive specializations of morphology, physiology and behaviour. For example, we can marvel at the high speed swimming of the marlins, sailfish and warm-blooded tunas, air-breathing in catfish and lungfish, parental care in the mouth-brooding cichlids and viviparity in many sharks and toothcarps.

Moreover, fish are of considerable importance to the survival of the human species in the form of nutritious, delicious and diverse food. Rational exploitation and management of our global stocks of fishes must rely upon a detailed and precise insight of their biology.

The *Chapman & Hall Fish and Fisheries Series* aims to present timely volumes reviewing important aspects of fish biology. Most volumes will be of interest to research workers in biology, zoology, ecology and physiology but an additional aim is for the books to be accessible to a wide spectrum of non-specialist readers ranging from undergraduates and postgraduates to those with an interest in industrial and commercial aspects of fish and fisheries.

Fish Bioenergetics by Malcolm Jobling comprises a further volume in the *Chapman & Hall Fish and Fisheries Series*. The author's original and extensive contribution to research in the field of fish bioenergetics allows him to provide a unique insight of the nutritional implications of fish diets, and allows him to give sound practical recommendations about the formulation of feeds in aquaculture.

Further advances in aquaculture depend not only upon greater insight of nutritional physiology, but also the laws governing the internal allocation of resources gained from food: diverse physiological processes such as swimming, reproduction compete with tissue growth for these nutritional resources. Part One of Jobling's book examines the nutritional physiology of fish and its implications for diet. Part Two examines each component of the energy balance equation in detail. Part Three surveys the environmental tolerances and preferences of fish.

Malcolm Jobling's book presents us with a scholarly review, digest and synthesis giving a stimulating and unique perspective of fish bioenergetics. I have every expectation that it will assume the role of a valuable reference work for a considerable time.

Professor Tony J. Pitcher
Editor, Chapman & Hall Fish and Fisheries Series
Director, Fisheries Centre, University of British Columbia,
Vancouver, Canada

Part One

Nutritional Requirements and Dietary Formulation

Chapter one

Nutrient classes and energy metabolism

1.1 INTRODUCTION

The performance of any activity, such as swimming, requires the expenditure of energy, and the fuelling of energy production requires the ingestion of food. The energy sources in the food are organic materials, consisting of carbohydrates, lipids and proteins. These organic molecules consist mostly of carbon, hydrogen and oxygen in varying proportions with minor quantities of other elements such as nitrogen, sulphur and phosphorus.

Potential foodstuffs come in many shapes and sizes and different fish species have become adapted to utilize many different food types ranging from other fish species to algae and aquatic macrophytes. These foodstuffs differ widely in biochemical composition and in the amounts that must be processed and ingested in order for the fish to obtain a given amount of any specific nutrient (Table 1.1). Thus, fish that feed on detritus may have to process 200 g or more of sedimented material in order to obtain 1 g protein, whereas the consumption of 6–7 g fish prey may supply this amount of protein.

1.2 NUTRIENT CLASSES

Carbohydrates, lipids and proteins make up the major portion of the dry weight of the ingested food, and these macronutrients can either be used directly as fuels (respiratory substrates) or be stored within the body for use at a later date. In addition to the macronutrients, the food will also contain a range of micronutrients. These micronutrients are the vitamins and minerals, which are required to be consumed in small doses if the fish is to survive and grow well.

Table 1.1 Chemical compositions of a range of foodstuffs and natural prey organisms consumed by fish species

Prey type	Percentage composition					Energy content ($kJ\ g^{-1}$)	Weight required per	
	Water	Ash	Protein	Lipid	Carbohydrate		g protein	kJ energy
Sediment + detritus		~98	~0.5	~0.2	~1.3	~0.4	~200	~2.5
Algae								
Phytoplankton	~92	~0.8	~2.5	~0.3	~4.4	~1	~40	~1
Macroalgae	~92	~2.5	~0.8	~0.3	~4.4	~0.9	~125	~1.1
Aquatic macrophytes	~94	~0.7	~1.5	~0.8	~3	~1	~65	~1
Oligochaete worms	~85	~2	~11	~1	~1	~3	~9	~0.35
Molluscs (soft parts)	80–85	1.5–3	~10	~2	~2	~4	~10	~0.25
Crustaceans								
Zooplankton	~80	~2.5	~13	~3.5	~1	~4.5	~7.6	~0.2
Crabs	~74	~8	~13	~3	~1.5	~4.5	~7.6	~0.2
Insect larvae	75–85	1–2	10–16	1–5	~2	3–6	~7.6	~0.2
Fish								
Lipid-rich (e.g. clupeids)	~72	~2.5	~14.5	~10	trace	~7.5	~7	~0.13
Lipid-poor (e.g. gadoids)	~77	~2.5	~17	~3.5	trace	~5.5	~6	~0.18

Carbohydrates

Carbohydrates were originally so named because it was believed that they were hydrates of carbon, that is, $C_x(H_2O)_y$ but it soon became clear that other compounds not showing the 2:1 ratio of hydrogen to oxygen had many of the characteristics and chemical properties assigned to the carbohydrates. For example, deoxyribose ($C_5H_{10}O_4$) has the chemical properties typical of the carbohydrates, as have a number of compounds that contain small proportions of nitrogen and sulphur in addition to carbon, hydrogen and oxygen. The carbohydrates are generally grouped into categories depending upon the size of the molecule – monosaccharides, oligosaccharides and polysaccharides (Table 1.2).

Monosaccharides are simple sugars that cannot be hydrolysed into smaller units, and the monosaccharides are usually classified according to the number of carbon atoms they possess, e.g. trioses ($C_3H_6O_3$) have three carbon atoms, tetroses have four carbon atoms, pentoses have five and hexoses have six. The monosaccharides are also classified according to their molecular structure. Thus, both glyceraldehyde and dihydroxyacetone are trioses, but they differ in molecular structure. Glyceraldehyde is an aldose, possessing an aldehyde group (—CHO), and dihydroxyacetone is a ketose, with a keto group (—CO) (Fig. 1.1). The monosaccharide molecule has either an active aldehyde or keto group and, consequently, these molecules act as reducing substances. They can be oxidized to produce various acids. In the case of glucose, glucuronic acid may be produced on oxidation (Fig. 1.2). The most important sugar acids are the uronic acids, especially glucuronic and galacturonic acids, which are components of large heteropolysaccharide molecules.

Table 1.2 Simplified classification of the carbohydrates

Monosaccharides	Oligosaccharides	Polysaccharides
Trioses	Disaccharides	– Homopolysaccharides (single type of monomeric unit)
$C_3H_6O_3$	e.g. Sucrose	
Glyceraldehyde	Lactose	
Dihydroxyacetone	Cellobiose	– Heteropolysaccharides (two or more types of monomeric unit)
Hexoses	Tetrasaccharides	
$C_6H_{12}O_6$	e.g. Stachyose	
Glucose		Storage – Starch
Fructose		– Glycogen
		– Laminarin
		Structural – Cellulose
		– Chitin
		– Pectins
		– Gums

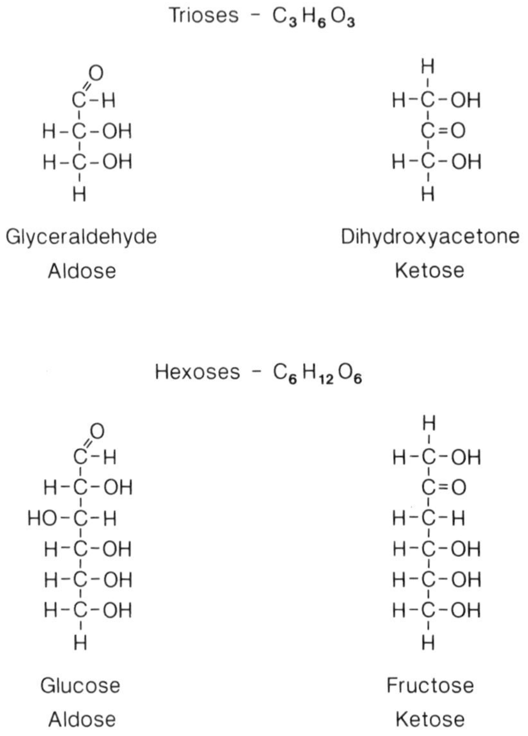

Fig. 1.1 Structure of triose and hexose monosaccharides indicating differences between an aldose and a ketose.

An important property of the monosaccharides is their reaction with phosphoric acid. A number of sugar phosphates occur naturally in both plants and animals (Fig. 1.2). The hexose phosphates play a very important role in cell metabolism, with two important compounds being glucose-6-phosphate and glucose-1-phosphate.

If the hydroxyl (—OH) group on carbon atom 2 of an aldohexose is replaced by an amino group (—NH_2), the resulting compound is an amino sugar (Fig. 1.2). Examples of amino sugars are glucosamine and galactosamine, both of which occur as components of glycoproteins.

Oligosaccharides are compound sugars that yield from two to six molecules of simple sugars on hydrolysis, and the term polysaccharide refers to a compound yielding a large number of monosaccharide molecules on hydrolysis. Polysaccharides are often also called glycans, and can be further classified as being either homopolysaccharides or heteropolysaccharides. Homopolysaccharides are condensation polymers of a single type of sugar, with, for example, hydrolysis of glucans yielding

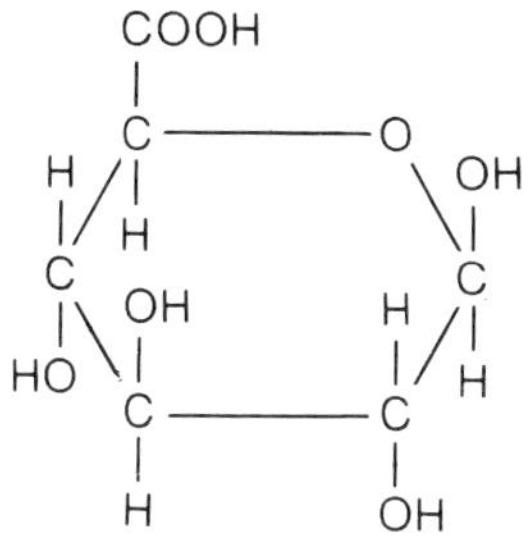

Glucuronic Acid

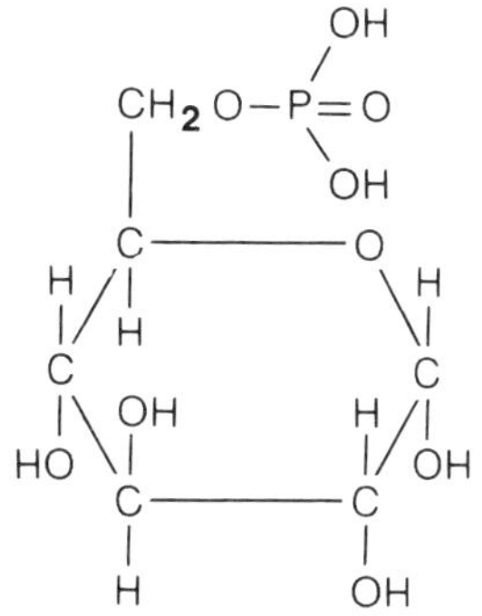

Glucose - 6 - Phosphate

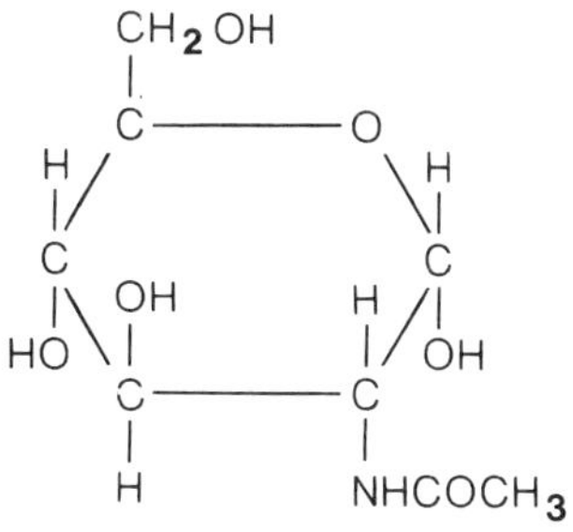

N-Acetyl glucosamine

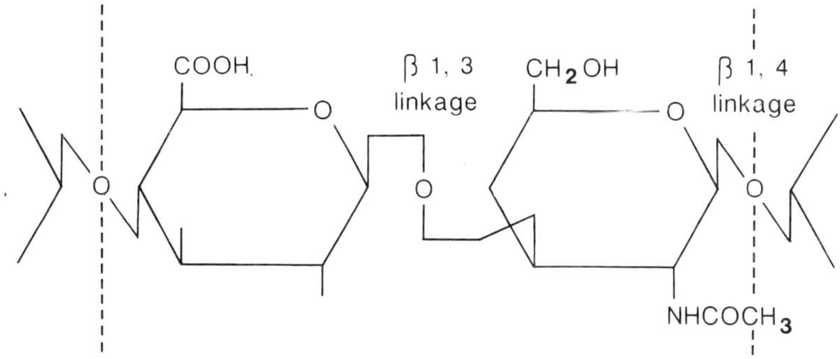

Glucuronic Acid N-Acetyl glucosamine

Hyaluronic Acid

Fig. 1.2 Some derivatives of glucose (a hexose). Note the different glycosidic linkages between the monomers in hyaluronic acid, a heteropolysaccharide consisting of glucuronic acid and N-acetyl glucosamine.

glucose. Heteropolysaccharides, on the other hand, are mixed polysaccharides which, on hydrolysis, yield mixtures of monosaccharides and derived products (Fig. 1.2).

In nature, the carbohydrates usually perform either a structural or an energy storage function, and they are usually present in biological materials in the form of long-chain polysaccharides. For example, cellulose is an important constituent of plant cell walls, and the exoskeletons of insects and crustaceans contain varying proportions of chitin as a strengthening agent. The most important storage product of plants is starch whereas in animals the storage carbohydrate is glycogen.

Both starch and cellulose consist of long chains of glucose molecules. The main difference between the two polysaccharides is the nature of the linkages between adjacent glucose molecules – alpha linkages in starch and beta linkages in cellulose (Fig. 1.3). This difference gives rise to differences in physical and chemical properties between cellulose and starch, and also affects the ease with which the two carbohydrates can be digested and absorbed by animals. Two forms of starch may be found in the storage granules of plants – amylose and amylopectin. Amylose is a

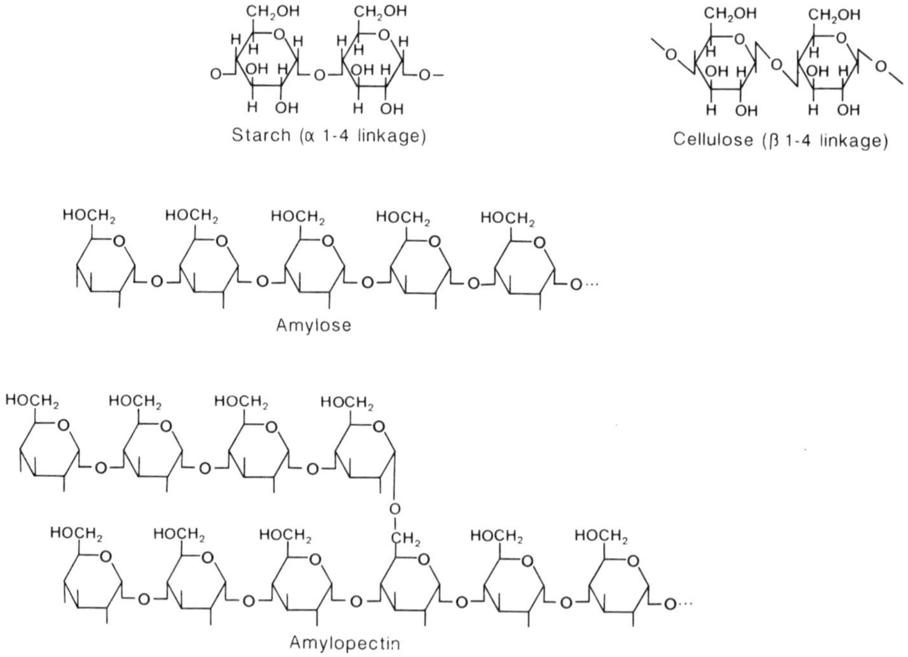

Fig. 1.3 Chemical structures of starch and cellulose, homopolysaccharides comprising glucose units. Note the different glycosidic linkages between the glucose monomers of starch and cellulose.

straight-chained molecule, in which the glucose units are glycosidically linked through carbon atoms at positions 1 and 4 (C 1,4 linkage), whereas amylopectin is a more complex molecule with side branches. In amylopectin there are both C 1,4 and C 1,6 glycosidic linkages. The relative proportions of the two molecules in the starch granules differ between plant species and this difference imparts different properties to the starches of the plants.

Carbohydrates are usually present in much lower proportions in animals than in plants, with the storage carbohydrate, glycogen, being found in the liver and muscle tissues. Glycogen is similar in structure to amylopectin, with the major difference between the two molecules being that glycogen has relatively large numbers of alpha 1,6 linkages, whereas amylopectin has only one of these linkages for every 30 or so alpha 1,4 linkages.

The fourth common polysaccharide, chitin, differs from cellulose, starch and glycogen in the nature of its monosaccharide subunits. Unlike the other polysaccharides, which have glucose as the basic unit, chitin is composed of long chains of an amino-sugar, N-acetyl glucosamine, which means that the chitin molecule contains atoms of nitrogen in addition to carbon, hydrogen and oxygen. Structurally, chitin is a close relative of cellulose; the hydroxyl (—OH) group at position 2 in the glucose residues of cellulose is replaced by an N-acetylamino ($—NHCOCH_3$) group in the N-acetyl glucosamine units that make up chitin.

Whilst cellulose, starch, glycogen and chitin are the most widespread of the polysaccharides, there are many other glycans that occur in nature. The algae, for example, contain a wide range of glycans as components of their cell walls, and may also have carbohydrates other than starch as the main storage product. For example, alginic acid may be an important component of the cell walls of algal species and can constitute up to 40% of the dry weight of the brown seaweeds (Phaeophyceac). Alginic acid is a linear copolymer of the two uronic acids (sugars with a carboxyl group on C-6) mannuronic acid and guluronic acid, linked via 1,4 glycosidic linkages. The relative proportions of the two acids making up the polymer vary between species, and this imparts different properties to the alginic acids extracted from the different brown algae. The red algae (Rhodophyceae), on the other hand, contain agar and carrageenans as structural elements in their cell walls. These polysaccharides represent a family of galactans, consisting of a spectrum of molecules having similar, but varying chemical structure. The molecules are made up of repeating sequences of 1,3 and 1,4 linked units of sulphated or methylated sugar residues. All of these algal polysaccharides are soluble in warm water and, on cooling, they may form firm gels. These colloidal and gelling properties are widely exploited in an industrial context, with algal polysaccharides being used as thickening or gelling agents in the food, pharmaceutical, cosmetic and textile industries.

The storage reserves of algae consist of a range of oils and carbohydrates, the exact composition of the reserves varying between the different phyla. The green algae (Chlorophyceae), for example, tend to store their energy reserves in the form of starch, predominantly amylose, whereas the red algae (Rhodophyceae) accumulate a peculiar form of solid carbohydrate, floridean starch, built up of glucose residues. The brown algae (Phaeophyceae) store their reserves as soluble carbohydrates, such as laminarin, and at least one alcohol, mannitol. Laminarin is a polymer of glucose subunits linked via beta 1,3 and beta 1,6 glycosidic linkages, and a polysaccharide having similar chemical structure, chrysolaminarin, is also found in many species of diatoms.

Lipids

The lipids are a rather heterogeneous class of water-insoluble organic compounds. Biochemical separation techniques yield fractions of polar and non-polar lipids that are distinguishable both on the basis of their abilities to dissolve in different solvents, and by the possession of differences in physical and chemical properties. The class of lipids most frequently encountered in nature is the acylglycerols. The acylglycerols represent the major storage lipids of both plants and animals, and they form the major part of the non-polar or neutral lipid group. The neutral lipids are formed by esterification of fatty acids with the alcohol, glycerol (Fig. 1.4). Whilst triacylglycerols make up the great bulk of neutral lipids found in nature, mono- and diacylglycerols are also found. Closely related to this group of compounds are the wax esters, which are formed by the esterification of fatty acids with long-chain alcohols. The wax esters are found as waterproofing agents on the outer surfaces of fruits and leaves or on the exoskeleton (cuticle) of insects. Wax esters are also major lipid storage reserves in some zooplanktonic crustaceans and hence may be an important energy source for a number of fish species.

Fatty acids differ in the numbers of carbon atoms in the molecule, and in the number and positioning of the double bonds between the carbon atoms. The fatty acids form distinct series, and a shorthand system has been devised for classifying these fatty acid series. For example, the shorthand formula 16:0 represents a fatty acid with 16 carbon atoms and no double bonds, a fatty acid lacking double bonds between the carbons being known as a *saturated* fatty acid. The shorthand formula 18:3(n−3) represents an 18 carbon fatty acid with three double bonds, the (n−3) denoting that the first double bond is found in the link between the third and fourth carbon atoms from the methyl end (Table 1.3). Fatty acids

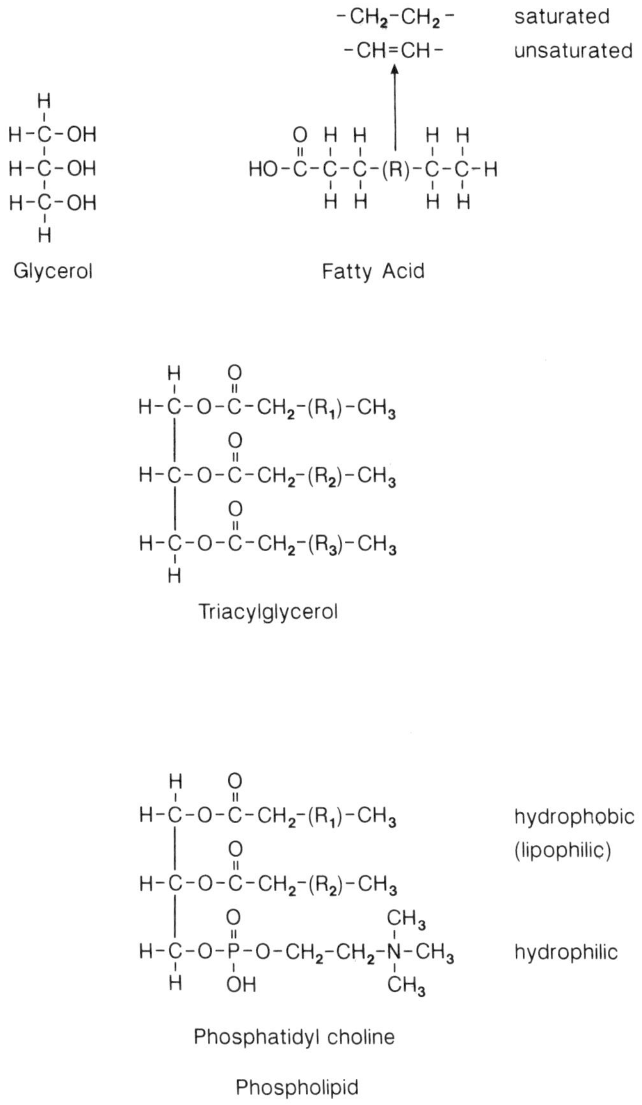

Fig. 1.4 Chemical structures of glycerol, fatty acids, triacylglycerols and phospholipids. (R) denotes the carbon chain of the fatty acid, which may or may not contain double bonds between carbon atoms. In the triacylglycerols and phospholipids the fatty acids (indicated by R_1, R_2 and R_3) esterified with the glycerol may be either the same or different acids.

with one or more double bonds are *unsaturated* fatty acids. Those having two to four double bonds in the molecule are often termed *polyunsaturated* fatty acids (PUFAs), and those with more than four double bonds may be described as being *highly unsaturated* fatty acids (HUFAs). These terms are used to indicate increasing degrees of unsaturation that can occur amongst the fatty acids. Fatty acids found in terrestrial plants and animals generally have relatively low degrees of unsaturation and have carbon chain lengths of 14–18, but longer-chain fatty acids, with up to 22 carbon atoms, are commonly encountered in aquatic organisms. The fatty acid components of the wax esters tend to be relatively short-chained, saturated fatty acids (14:0 or 16:0) and the alcohols are generally saturated (16:0) or have a single double bond in their carbon atom chains (20:1 or 22:1).

Both the chain length and the number of double bonds (degree of unsaturation) determine the physical and chemical properties of the fatty acids (Table 1.3) and the lipids into which they are incorporated. For example, triacylglycerols, such as pork fat and beef tallow, that contain saturated fatty acids are solid at room temperature, whereas the triacylglycerols derived from the seeds of a number of plant species, e.g. soya, sunflower, peanut and rapeseed, contain quite high levels of unsaturated fatty acids and are fluid oils at room temperature. Marine fish oils, which usually contain relatively high proportions of long-chain, polyunsaturated and highly unsaturated fatty acids, have low melting points and are fluid at room temperature.

Not only do terrestrial and aquatic lipids differ in the degrees of saturation of the fatty acids found in the triacylglycerols, but there are also differences in the unsaturated fatty acids typical of food chains in these environments (Table 1.4). The (n–6) series fatty acids are typically found in terrestrial and freshwater environments (e.g. plant seed oils are typically rich in 18:2(n–6)), whilst in marine ecosystems fatty acids of the (n–3) series tend to dominate. The food chain in fresh water is characterized by 18:2(n–6), 18:3(n–3) and 20:5(n–3) fatty acids, and freshwater fish may obtain relatively large amounts of 20:4(n–6) by consuming insects. The fatty acid profiles of marine phyto- and zooplankton tend to be dominated by 18:3(n–3), 20:5(n–3) and 22:6(n–3) fatty acids, and this pattern is generally reflected at higher levels in the food chain. Thus, when lipids are extracted from freshwater and marine fish species and the fatty acid profiles are compared, there will usually be found to be large differences (Table 1.5). These differences may, to a large extent, reflect differences in the fatty acid compositions of the prey present in the two environments, but abiotic factors, such as temperature and salinity, are also known to influence the fatty acid composition of the body lipids.

The vast majority of the lipid extracted from a fish or prey species will

Table 1.3 Examples of fatty acids, illustrating methods of classification and the effects of carbon chain length and degree of unsaturation on melting points

Shorthand formula	Structure	Melting point (°C)
Saturated fatty acids		
16:0	$CH_3(CH_2)_{14}COOH$	63
20:0	$CH_3(CH_2)_{18}COOH$	76.5
24:0	$CH_3(CH_2)_{22}COOH$	86
Unsaturated fatty acids		
18:1 (n–9)	$CH_3(CH_2)_7CH{=}CH(CH_2)_7COOH$	13.5
18:2 (n–6)	$CH_3(CH_2)_4CH{=}CHCH_2CH{=}CH(CH_2)_7COOH$	–5
18:3 (n–3)	$CH_3CH_2CH{=}CHCH_2CH{=}CHCH_2CH{=}CH(CH_2)_7COOH$	–11
20:4 (n–6)	$CH_3(CH_2)_4CH{=}CHCH_2CH{=}CHCH_2CH{=}CHCH_2CH{=}CH(CH_2)_3COOH$	–49.5

Table 1.4 Examples of fatty acid compositions of fish prey from terrestrial (earthworm – Oligochaete), freshwater (sample of mixed zooplankton) and marine (krill – mixed sample of euphausiids – and squid, *Todarodes sagittatus*) environments. Values are presented as percentages of the total fatty acids present. Compositions of two commercial dry pellet feeds for salmonids are shown for comparison. Note the relatively high proportions of the (n–6) series fatty acids in the terrestrial (earthworm) and freshwater (zooplankton) prey organisms

Fatty acid	Commercial feeds		FW zooplankton	Earthworm	Krill	Squid
18:2 (n–6)	3.9	4.9	3.8	8.5	2.1	–
20:4 (n–6)	0.2	0.2	7.8	12.2	1.5	2.7
20:5 (n–3)	8.9	7.8	17.6	7.8	15.6	14.1
22:6 (n–3)	6.3	5.8	11.7	–	14.7	14.6
Saturates	22.2	19.9	25.9	31.6	23.5	15.1
Monoenes	47.3	52.3	16.1	18.7	33.0	46.2
PUFA	30.5	25.9	58.0	49.6	43.5	38.6

Table 1.5 Examples of fatty acid compositions of fish species from marine (capelin, *Mallotus villosus*, and elver of American eel, *Anguilla rostrata*) and freshwater (American eel and perch, *Perca fluviatilis*) environments. Values are presented as percentages of the total fatty acids present. American elvers were sampled whilst entering freshwater from the sea. Note the relatively high levels of the (n–6) series fatty acids in the freshwater fish species, leading to these fish having much lower ratios of (n–3)/(n–6) than the marine species

Fatty acid	Perch FW	Eel FW	Elver	Capelin
18:2 (n–6)	3.5	8.9	1.3	1.4
20:4 (n–6)	11.7	4.4	2.9	trace
20:5 (n–3)	13.4	2.9	5.0	13.8
22:6 (n–3)	25.0	5.4	18.8	11.3
Saturates	24.9	23.3	29.6	28.7
Monoenes	14.4	40.0	30.3	38.8
PUFA	57.7	30.0	36.4	27.9
(n–3)	45.5	13.6	29.1	26.5
(n–6)	15.2	16.4	7.3	1.4
(n–3)/(n–6)	3.0	0.8	4.0	18.9

be made up of the non-polar, neutral, storage triacylglycerols, but the total lipids will also contain smaller quantities of other lipid classes, including the polar phospholipids that are important components of the cell membranes, where they form lipoprotein complexes. The majority of the phospholipids contain nitrogen and phosphorus in addition to carbon, hydrogen and oxygen. Phospholipids are esters of glycerol, in which two of the alcohol groups are esterified with fatty acids. The third alcohol group is esterified with phosphoric acid, which, in turn, is esterified by a nitrogenous base, the amino acid serine, the sugar alcohol inositol or by glycerol phosphate (Fig. 1.4). The nitrogenous bases may be choline or cholamine (amino ethyl alcohol), and the nature of the base or esterification compound provides the basis of the name of the specific families of phospholipids, e.g. phosphatidylcholine, phosphatidylethanolamine, phosphatidylserine and phosphatidylinositol. The phospholipids combine within the same molecule both the hydrophilic (water-loving) phosphate and nitrogen-containing groups, and the hydrophobic fatty acid residues. These molecules are therefore surface-active, and they play an important role as emulsifying agents in biological systems. Their surface-active nature also explains their function as part of various biological membranes. The major membrane phospholipids are phosphatidylcholine and phosphatidylethanolamine, whereas phosphatidylinositol appears to have a number of important roles in the transduction of hormonal signals through biomembranes, and in relation to gill function.

The sphingomyelins are not esters of glycerol, but are made up of fatty acids, phosphoric acids, choline and sphingosine. Because they contain phosphorus, and are closely associated with the phospholipids, both in occurrence and in biological function, the sphingomyelins are usually included within the general class phospholipids.

The fatty acid profiles of the phospholipids tend to be quite similar, irrespective of whether a fish comes from a marine or a freshwater environment, and the long-chained highly unsaturated fatty acids, particularly 20:5(n−3) and 22:6(n−3), tend to dominate in the membrane phospholipids. In phosphatidylinositol, on the other hand, the (n−6) series fatty acid 20:4(n−6) is present in large amounts. The fatty acid compositions of the phospholipids are, however, modified to some extent under different environmental conditions. For example, exposure to low temperature often leads to the increased incorporation of highly unsaturated fatty acids into the membrane phospholipids. Ambient salinity may also influence the fatty acid composition of fish lipids. For example, when euryhaline milkfish, *Chanos chanos*, were reared on the same diet in seawater and fresh water, fish from the two groups differed in fatty acid composition. The milkfish reared in seawater incorporated greater proportions of (n−3) PUFAs into the body lipids than did fish held in fresh water, with the differences being most pronounced in the lipids of the gills and kidney. Similarly, when salmonids undergo the parr–smolt transformation prior to migrating from the freshwater to the marine environment, they tend to show increased synthesis of PUFAs, and these fatty acids are preferentially incorporated into the membrane phospholipids. On the other hand, if the fish are prevented from migrating to the sea they undergo desmoltification and the fatty acid profile may revert to the 'freshwater' type. Thus, the phospholipids of the fish held in fresh water beyond the period of the parr–smolt transformation may contain higher proportions of saturated and monounsaturated fatty acids, and lower proportions of PUFAs and HUFAs, particularly those of the (n−3) series, than both smolt that are undergoing downstream migration and those fish which complete the migration to the sea. Whilst differences in phospholipid fatty acid compositions between the smolt and desmolt are apparent in several body tisues, the differences are most marked in the gut and gills, both of which are organs intimately involved in iono- and osmoregulation.

The polyunsaturated and highly unsaturated fatty acids (PUFAs and HUFAs) usually account for about half of the phospholipid fatty acids, whereas the triacylglycerols are usually dominated by saturates and monounsaturates. Polyunsaturated fatty acids may only constitute 30% or less of the fatty acids present in the storage lipids (Table 1.6).

Table 1.6 Fatty acid compositions of the polar and neutral lipid fractions extracted from the tissues of American eels and elvers, *Anguilla rostrata.* Values are presented as percentages of total fatty acids present. Note the higher percentages of the (n−6) series fatty acids in both the polar and neutral lipids of the eel (freshwater) compared with the elver (marine). There is a corresponding reversal with respect to the relative proportions of the (n−3) series fatty acids. Note that PUFAs constitute about 50% of the fatty acids present in the polar lipids of both the eel and the elver, but PUFAs make up less than 30% of the fatty acids in the neutral lipids

Fatty acid	American elvers		American eels	
	Neutral	Polar	Polar	Neutral
18:2 (n−6)	1.2	1.2	2.1	9.9
20:4 (n−6)	1.7	5.4	11.5	3.4
20:5 (n−3)	3.7	7.5	5.5	2.8
22:6 (n−3)	13.3	27.7	16.5	3.2
Saturates	34.2	24.8	23.4	22.7
Monoenes	31.9	18.7	25.3	43.0
PUFA	27.6	51.3	45.2	26.3

In the phospholipids the HUFAs are generally esterified to carbon 2 of the glycerol molecule, but di-HUFA phospholipids are also found. The HUFAs are cleaved from position 2 by the action of the enzyme phospholipase A_2 and the 20-carbon (20-C) HUFAs so released may act as precursors for eicosanoid (prostanoids, leukotrienes and lipoxins) synthesis. The synthesis of the prostanoids (prostaglandins, prostacyclins and thromboxanes) from the 20-C HUFAs (20:3(n–6), 20:4(n–6), 20:5(n–3)) is dependent upon the cyclooxygenase enzymatic pathway, whereas the metabolic pathway leading to the production of leukotrienes and lipoxins involves lipoxygenase as an enzymatic catalyst.

As a group the eicosanoids are biologically highly active, and this group of compounds is known to be involved in the regulation of a wide range of physiological processes. Various eicosanoids are known to have influences on the reproductive, respiratory and circulatory systems, or upon the organs involved in maintaining ion and water balance. Certain of the prostaglandins are, for example, known to be intimately involved in the regulation of the final stages of the reproductive cycle – ovulation and spawning – in a range of fish species. In addition, prostaglandins, leukotrienes and thromboxanes may have marked influences on the circulatory system via vasoconstrictor or vasodilatory effects. These compounds may, therefore, be involved in the regulation of the patterns of blood flow both between and within particular organs of the body.

Eicosanoid synthesis may be markedly stimulated following physical injury or cell damage caused by exposure to xenobiotics, such as heavy metals or organic pollutants. Thus, peroxidative damage to cell membranes caused by exposure to xenobiotics can lead to the release of HUFAs from the membrane phospholipids and thereby initiate a burst of enzymatic activity leading to the production of eicosanoids. An increase in the tissue levels of eicosanoids will usually result in the tissue showing an inflammatory response, this contributing to the initial line of defence against invasion by potential pathogens. Overstimulation of the inflammatory response can, however, be harmful and lead to further tissue damage. The efficacy of anti-inflammatory drugs, such as aspirin, is related to their inhibition of the cyclooxygenase enzymatic pathway. Steroidal anti-inflammatory agents, such as the naturally occurring corticosteroid cortisol, on the other hand, act by inhibiting phospholipase A_2 activity. Inhibition of phospholipase A_2 prevents, or markedly reduces, the release of the precursor fatty acids from the phospholipids, resulting in a fall in the rate of eicosanoid biosynthesis.

The terpenoids are a very large and important group of compounds in which the basic structural element is the isoprenoid unit (Fig. 1.5). Various condensation reactions involving the isoprenoid unit give rise to a range of biological compounds such as carotenoids, steroids, rubber and

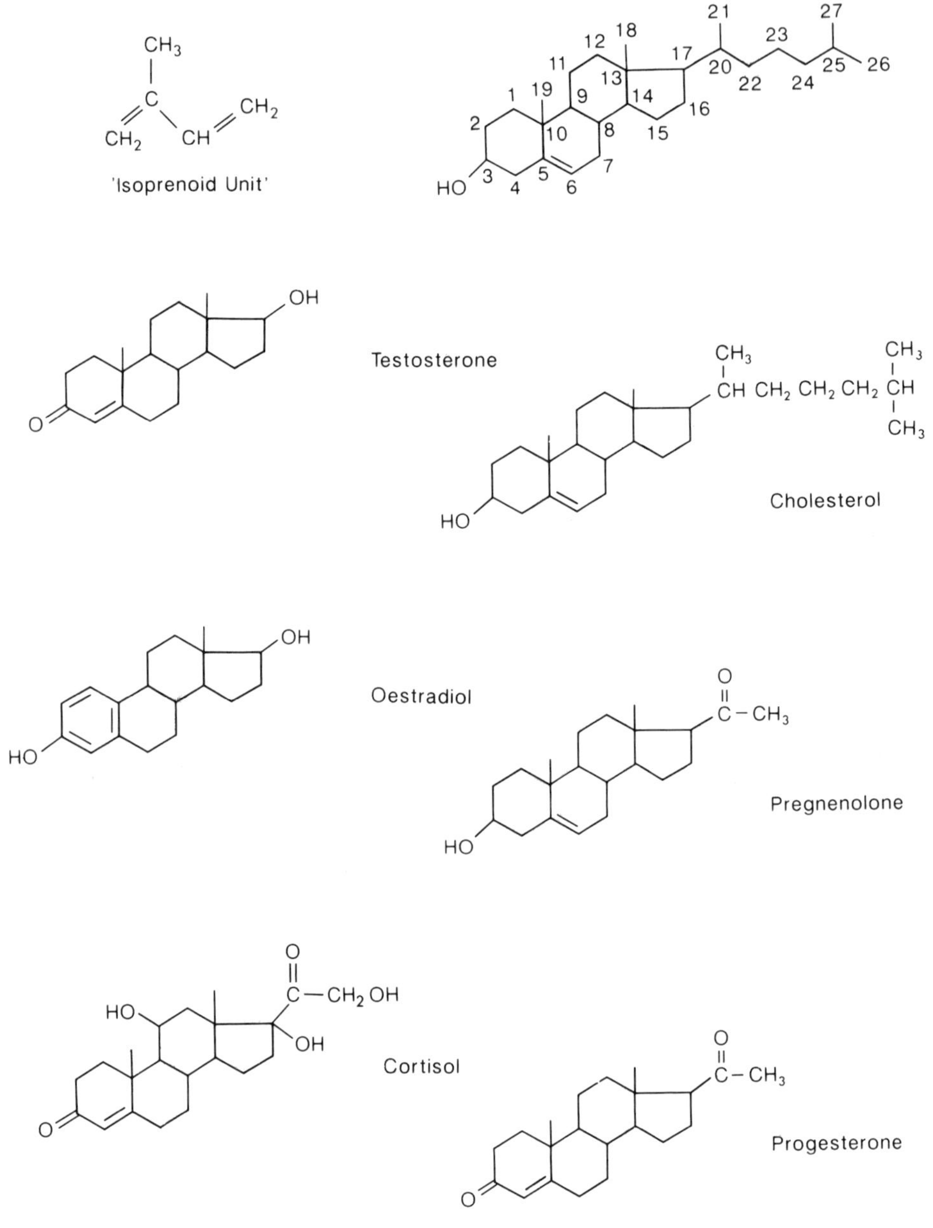

Fig. 1.5 Cholesterol is made up of isoprenoid units, the basic structural element of the terpenoid group of compounds. The chemical structures of some of the steroid hormones derived from cholesterol are also shown.

many simpler terpenes. The carotenoids are widely distributed in biological materials and may have a number of important metabolic functions, with beta-carotene, for example, being cleaved to form vitamin A (Fig. 2.8).

Another typical terpenoid product is cholesterol. Cholesterol, a 27-carbon (27-C) compound, is the basic precursor for all the steroid hormones. The first step in the production of steroids involves the conversion of cholesterol to pregnenolone, a 21-C compound, from which the steroid hormones can be derived. Thus, the 21-C pregnenolone, can, via a series of chemical steps, be converted into every type of steroid hormone (Fig. 1.5). The different steroid hormones can be broadly divided into:

1. progesterone, a 21-C compound, that both acts as a potent female sex hormone, and can be converted into other steroids;
2. the 21-C steroids of the interrenal, e.g. cortisol;
3. the male sex hormones (androgens), which are 19-C compounds, e.g. testosterone;
4. the female sex hormones (oestrogens), which are 18-C compounds, e.g. oestradiol.

Thus, cholesterol and pregnenolone are the starting points for the production of the steroids, but which hormone will be produced in specific endocrine cells depends upon the types of enzymes that are present within the cell. Different types of enzymes will modify the molecular structure of the steroid at different points, leading to the possibility of a range of end products from a single precursor. For example, the cells may contain hydroxylase enzymes which introduce a hydroxyl group (—OH) into the steroid at some specific carbon atom of the molecule – a 17-hydroxylase would introduce a hydroxyl group onto C atom 17, and an 11-hydroxylase leads to the introduction of a hydroxyl group onto C atom 11. Cortisol, for example, has hydroxyl groups on both carbons 11 and 17, whereas both testosterone and oestradiol have hydroxyl groups on carbon 17, but not carbon 11.

The corticosteroid hormones of the interrenal tissues play a part in both the regulation of salt and water balance, and in the metabolism of carbohydrates and proteins. The corticosteroids, and particularly cortisol, have been implicated in the 'stress responses' shown by fish following exposure to unfavourable conditions.

The gonadal steroids, androgens and oestrogens, have a variety of reproductive functions, including the induction of the production of vitellogenin in the liver, the development of secondary sexual characteristics and the initiation of a number of behavioural changes related to spawning activity.

The steroid hormones are removed from the blood as they circulate through the liver, and breakdown of the hormones takes place there.

Usually the steroids are coupled to either a sulphate ion or glucuronic acid, and are then excreted in the bile or urine. The metabolism of the steroids to sulphates or glucuronides leads to increased solubility of the compounds in water and, therefore, aids in increasing the ease with which they are excreted.

Proteins and amino acids

Proteins are large organic molecules that contain carbon, hydrogen, oxygen, nitrogen and often sulphur. The elementary composition of most proteins is very similar, approximate percentages being C = 50–55%, H = 6–8%, O = 20–23%, N = 15–18%, S = 0–4%. The fundamental structural unit of the protein molecule is the amino acid, of which there are 20 naturally occurring forms that are incorporated into proteins. These amino acids are joined by peptide bonds to give long repeating units making up the protein molecule (Fig. 1.6). All of the amino acids have the same basic design, with a carboxyl (—COOH) and an amino (—NH_2) group attached to the alpha carbon atom. The remainder of the amino acid molecule (denoted by R in the basic structural formula shown in Figs 1.6 and 1.7) can have different configurations, and the amino acids can be divided into a number of series depending upon the nature of

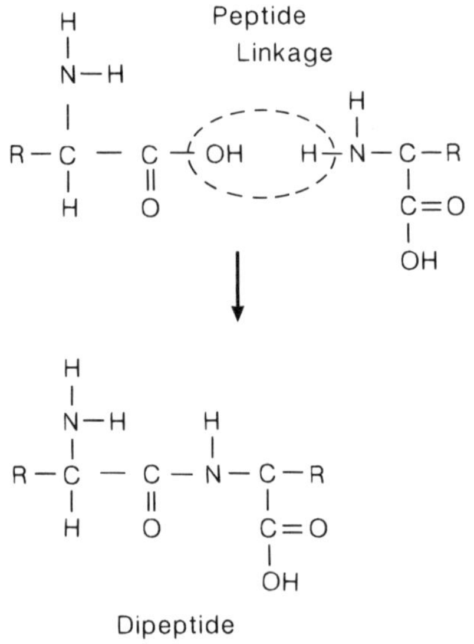

Fig. 1.6 Formation of a peptide linkage between two amino acids giving rise to a dipeptide. R denotes the side chain of the amino acid.

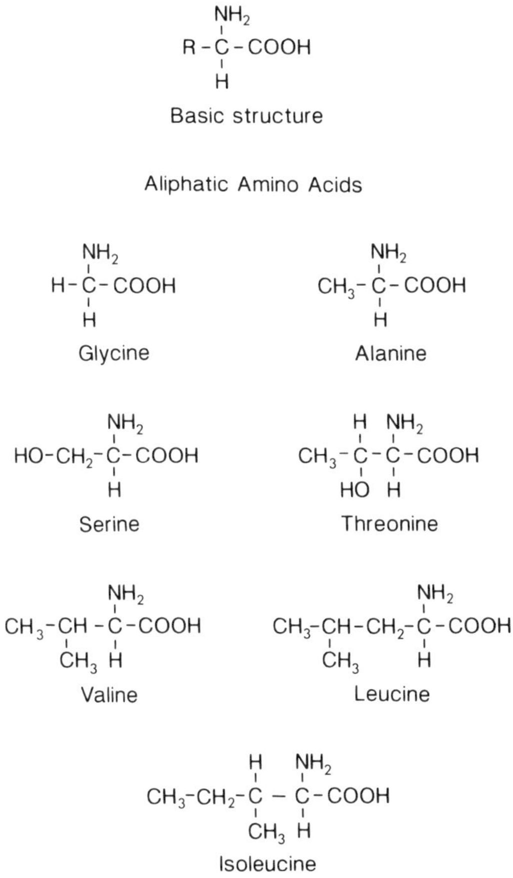

Fig. 1.7 Chemical structures of amino acids making up the various amino acid series. In the basic structural formula, R represents the side chain.

the R-group (Fig. 1.7):

1. aliphatic series (glycine, alanine, serine, threonine, valine, leucine, isoleucine);
2. aromatic series (phenylalanine, tyrosine);
3. sulphur-amino acid series (cysteine, cystine, methionine);
4. heterocyclic series (tryptophan, proline, hydroxyproline);
5. acidic series (aspartic acid, glutamic acid);
6. basic series (arginine, histidine, lysine).

These amino acids can be linked together in chains of different length, and also with differing orders of the amino acids in the chains, so it is easy to

Aromatic Amino Acids

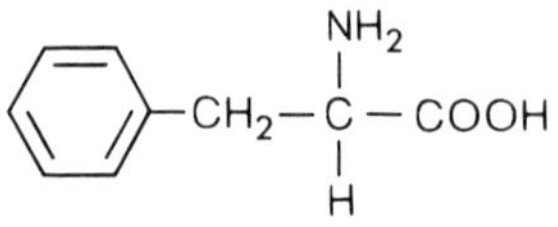

Phenylalanine

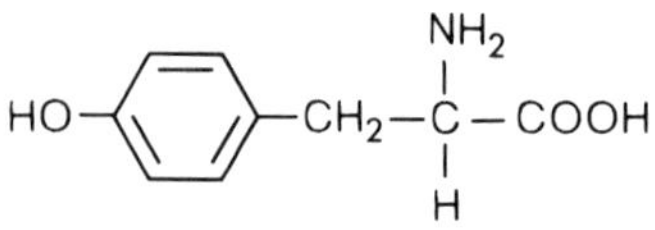

Tyrosine

Sulphur - Amino Acids

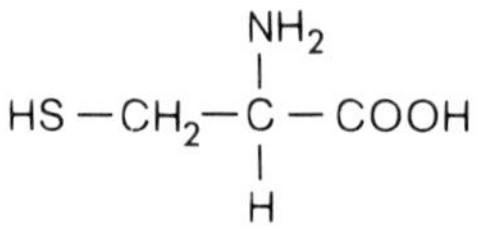

Cysteine

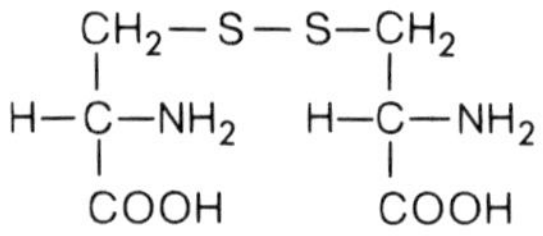

Cystine

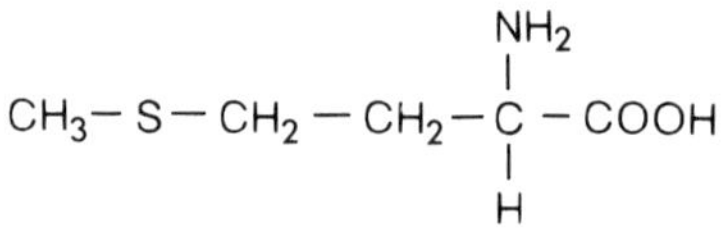

Methionine

Heterocyclic Amino Acids

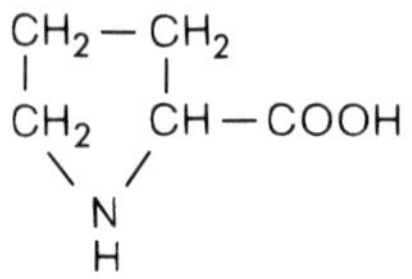

Proline

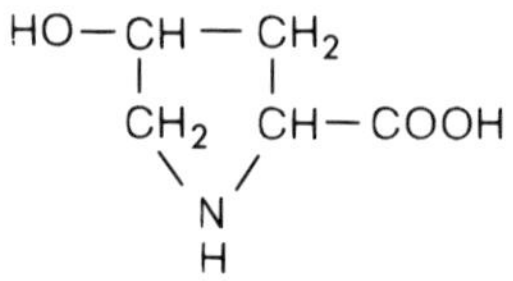

Hydroxyproline

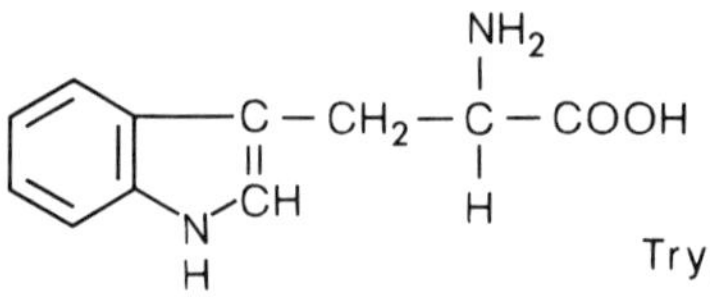

Tryptophan

Acidic Series Amino Acids

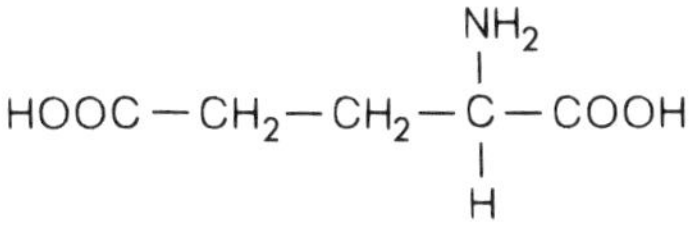

Glutamic Acid

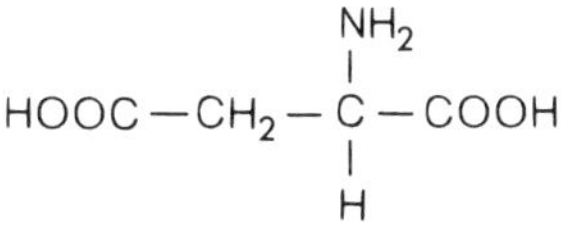

Aspartic Acid

Basic Series Amino Acids

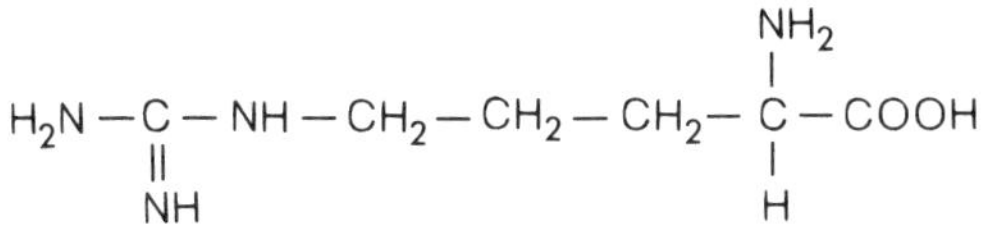

Arginine

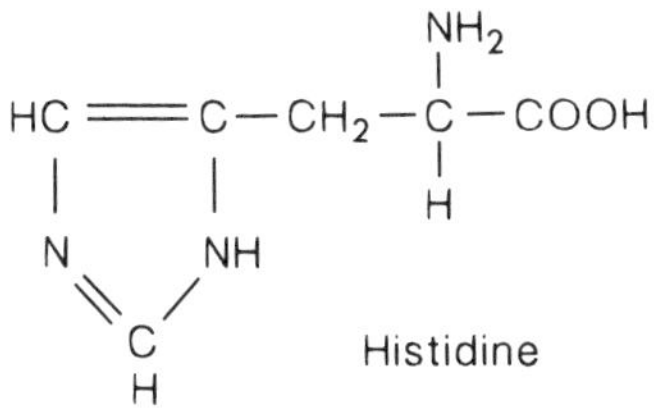

Histidine

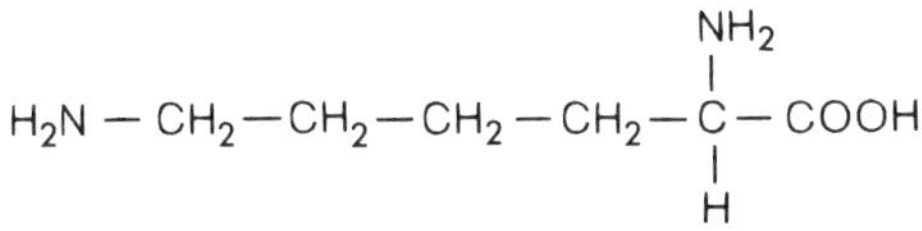

Lysine

see that with 20 amino acids available the possible number of combinations is almost limitless. Thus, the proteins are extremely heterogeneous, can have markedly different physical properties, and can play a range of functional roles within biological materials.

1.3 THE BASICS OF ENERGY METABOLISM

Compounds from the three main groups of nutrients (carbohydrates, lipids and proteins) are all capable of being used as metabolic fuels (respiratory substrates). When they are combusted in the presence of oxygen they will all yield carbon dioxide and water and there will be production of heat. For example, the combustion of glucose can be described by the equation:

$$C_6H_{12}O_6 + 6O_2 = 6CO_2 + 6H_2O \quad (1.1)$$

and the combustion of 1 gram-molecule of glucose (180 g) will produce approximately 2800 kJ (or 16 $kJ.g^{-1}$) as waste heat. The heat produced by the combustion of the different metabolic fuels depends upon their chemical composition, with the heat produced by the combustion of a 'representative' carbohydrate generally being assumed to be approximately 17 $kJ.g^{-1}$. In the combustion of a saturated fatty acid, such as palmitic acid (16:0):

$$CH_3(CH_2)_{14}COOH + 23O_2 = 16CO_2 + 16H_2O \quad (1.2)$$

there will be the production of about 10 000 kJ as heat, or 39 $kJ.g^{-1}$ (10 000/256) palmitic acid metabolized. The heat produced from the combustion of lipids from aquatic animals is generally slightly lower than that derived from terrestrial lipids. Combustion of a typical protein leads to a heat production of about 24 $kJ.g^{-1}$.

This combustion produces waste heat, which cannot be utilized by the fish, and what is required is a system by which this heat energy can be chemically bound and harnessed to perform work of various kinds. The way in which the energy is stored is in the form of high-energy phosphate bonds (*P), notably those attached to the molecule adenosine triphosphate (ATP). The formation of 1 gram-molecule of ATP from adenosine diphosphate and inorganic phosphate involves the capture of approximately 40 kJ.

The pathways for the production of ATP from ADP and inorganic phosphate are briefly summarized in Fig. 1.8, where it can be seen that the organic components derived from the different nutrients (carbohydrates, lipids, proteins) enter the cycle at different positions. These different points of entry will mean that there will be different ATP yields from the different substrates.

The reactions of the cycle can first be summarized by considering the fate of glucose as respiratory substrate. The earliest part of the metabolic processing of glucose is termed *glycolysis*, and this results in the production of two molecules of pyruvic acid from each molecule of glucose. At this stage, two molecules of ATP have been produced, and about 80 kJ of the 2800 kJ contained in the original 1 gram-molecule of glucose will have been stored. In the absence of oxygen (anaerobic conditions) the pyruvic acid is converted into lactate (Fig. 1.9).

Under aerobic conditions the pyruvic acid enters the citric acid cycle (also known as the tricarboxylic acid or Krebs cycle) where it is further metabolized (Fig. 1.10). The metabolic reactions of the citric acid cycle involve *oxidative phosphorylation*. The reactions of the aerobic metabolic pathway, the citric acid cycle and respiratory chain, lead to the produc-

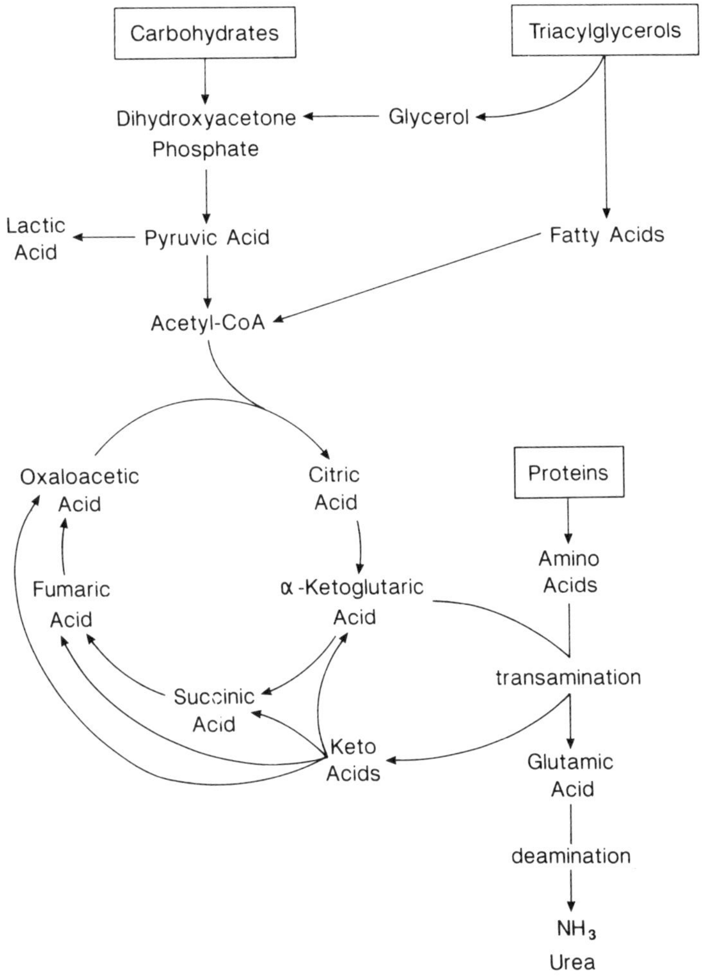

Fig. 1.8 Simplified scheme showing the points of entry of different classes of respiratory substrate into the citric acid (tricarboxylic acid or Krebs) cycle.

tion of an additional 34 ATP molecules. Thus, the aerobic metabolism of glucose leads to a total production of 36 ATP molecules – 2 ATP via glycolysis, plus 34 ATP produced aerobically.

This stresses the importance of the citric acid cycle, and the role of oxygen, in the energy storage processes. In the absence of oxygen, under conditions of anaerobic glycolysis, the ATP yield is only two molecules of ATP per molecule of glucose metabolized, and only about 3% of the energy available from the glucose is stored as high-energy phosphate

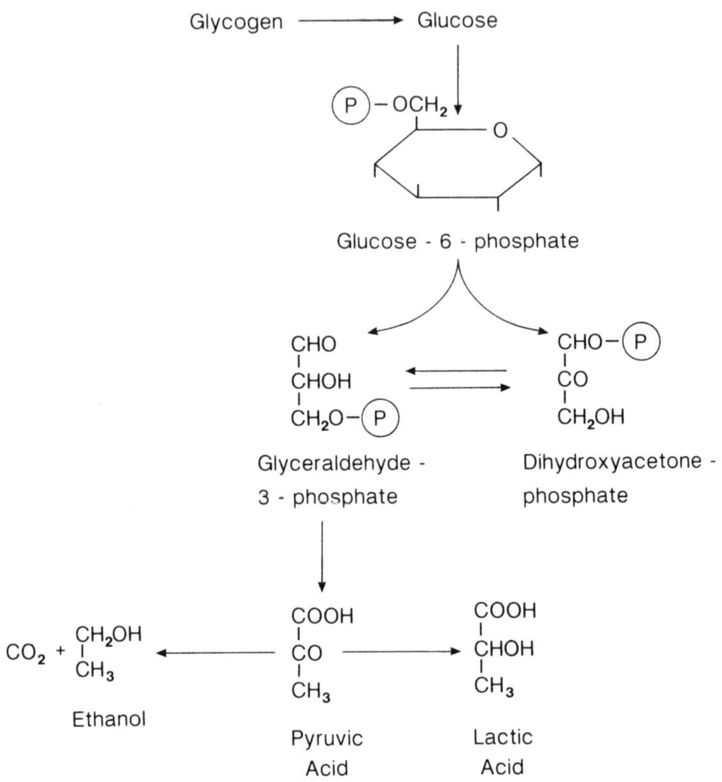

Fig. 1.9 Simplified scheme showing biochemical changes involved in the glycolytic pathway.

bonds. Under aerobic conditions, on the other hand, the yield is 36 ATP molecules and considerably more of the energy available from the glucose is harnessed in a form that can be used to produce useful work.

Both lipids and amino acids can enter biochemical pathways that lead to the production of glucose or glycogen (gluco- or glyconeogenesis), so they may enter the respiratory metabolic cycle via indirect routes. It is, however, more usual that they have an alternative route of entry.

Proteins are made up of 20 different amino acids, and there are 20 different multienzyme sequences, or pathways, for the oxidation of these amino acids. The catabolic pathways may be long and complex, with many intermediate steps, but they ultimately converge into a few terminal pathways leading to the citric acid cycle. The carbon skeletons of many of the amino acids are converted to acetyl CoA (acetyl coenzyme A) prior to entering the cycle. Conversion of ten of them occurs via pyruvate or acetoacetyl CoA as intermediates. A few amino acids are converted to

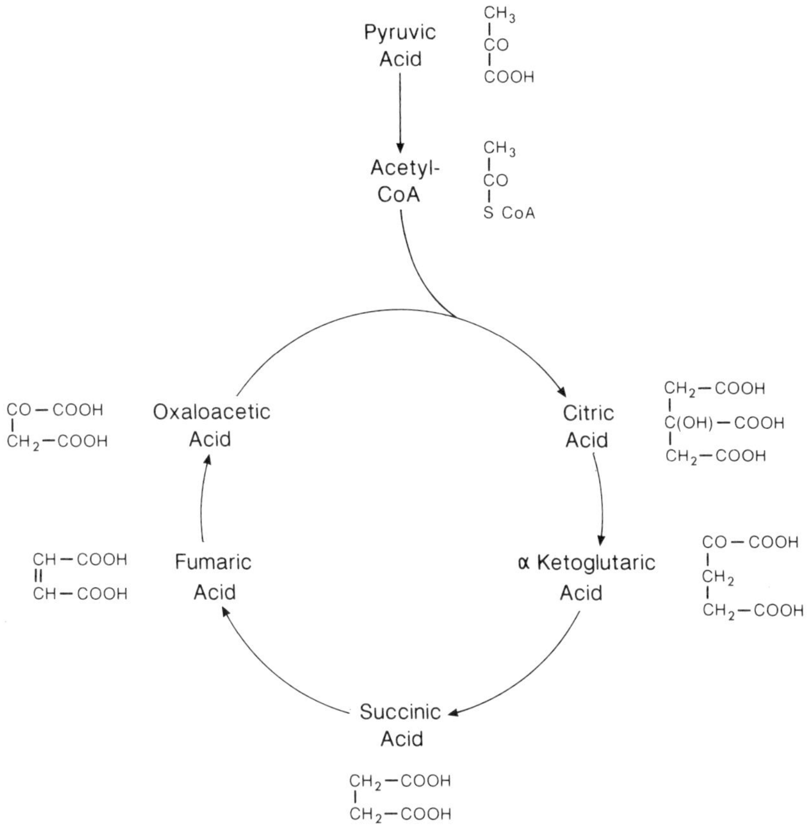

Fig. 1.10 Biochemical changes involved in the citric acid cycle.

alpha-ketoglutarate, some to succinyl CoA, and some to oxaloacetate for entry into the citric acid cycle. Two of the amino acids, phenylalanine and tyrosine, are so degraded that one portion of the carbon skeleton enters the citric acid cycle as acetyl CoA and the other as fumarate (Fig. 1.11).

When fatty acids are the fuel being combusted, the pathway of oxidation is termed the beta-oxidation spiral. This pathway results in the formation of acetyl CoA that can enter the citric acid cycle. Long-chain fatty acids are first activated by an energy-requiring step in which the fatty acid ester of CoA is formed enzymatically at the expense of ATP. There then occur a number of steps, which ultimately produce acetyl CoA and a fatty acid ester having two fewer carbon atoms than the original fatty acid. This ester can then become the substrate for a new round of reactions, ending with the production of another molecule of acetyl CoA

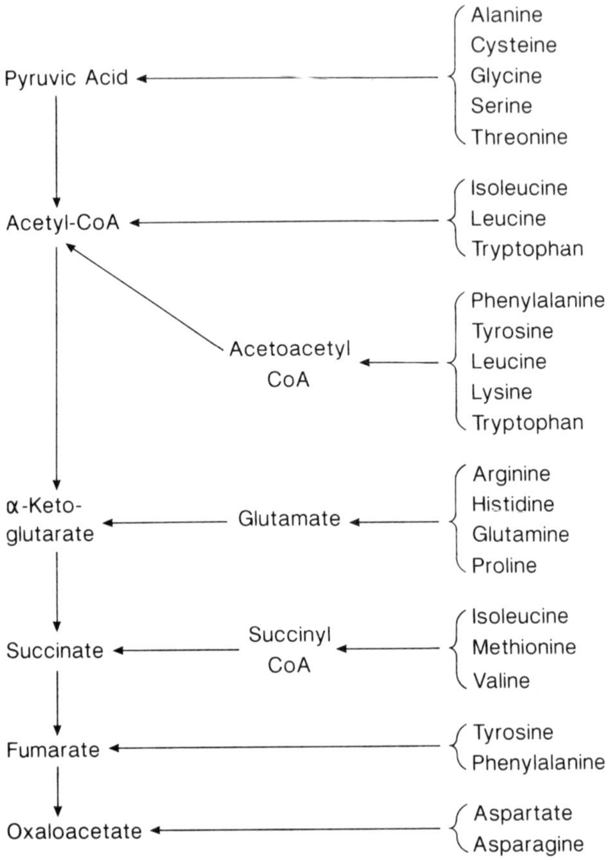

Fig. 1.11 Points of entry of different amino acids into the citric acid cycle.

(Fig. 1.12). Thus, at each passage through this spiral process, the fatty acid ester loses a two-carbon fragment as acetyl CoA, and two pairs of hydrogen atoms are also produced. For example, 16:0, palmitic acid, will undergo a total of seven such cycles, to yield altogether eight molecules of acetyl CoA and 14 pairs of hydrogen atoms. The palmitic acid must be primed, or activated, only once, right at the outset of the cycle, and this requires the expenditure of ATP. During the splitting off of the two-carbon acetyl CoA there is production of ATP. The acetyl CoA formed as a product of the fatty acid oxidation may then enter the citric acid cycle, so the complete oxidation of palmitic acid will result in the formation of 129 ATP, i.e. eight acetyl CoA are produced as the result of the beta-oxidation spiral, giving rise to $7 \times 5 = 35$ ATP, and the acetyl CoA entering the

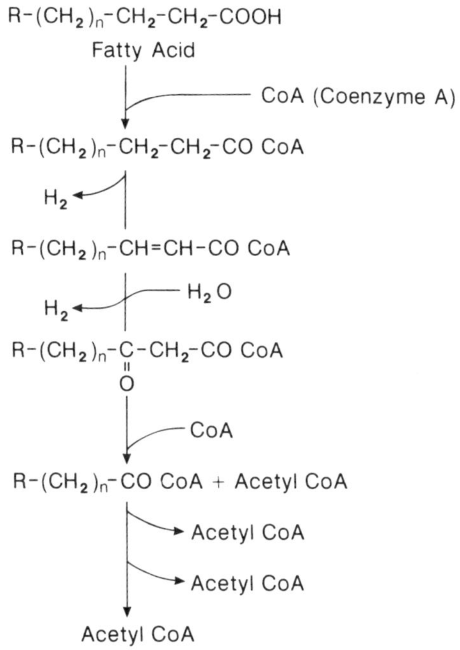

Fig. 1.12 The beta-oxidation spiral of fatty acids, resulting in the production of acetyl coenzyme A (Acetyl CoA).

citric acid cycle results in further ATP production ($8 \times 12 = 96$ ATP), but ATP is also required for the activation of the fatty acid oxidation, giving a net yield of 129 ATP.

For practical purposes, it can be considered that there are a relatively small number of metabolic pathways responsible for energy production, and that the different respiratory substrates ultimately enter the citric acid cycle by a limited number of routes. The most important point to be grasped is the central role of the oxidative pathways in energy metabolism, with the energetic yields of such pathways being considerably greater than those involved in the anaerobic metabolism of respiratory substrates.

Chapter two

Nutritional requirements

2.1 INTRODUCTION

Fish, and other animals, are required to use some of the organic materials obtained in the food as respiratory substrates to provide the energy to sustain processes such as the maintenance of bodily functions and activity. Once these energy requirements have been met the remaining dietary input can be utilized for growth. Carbohydrates, proteins and lipids can all be used as energy sources (metabolic fuels) by fish, but these organic compounds are not equally well suited for the promotion of growth. In order for growth to occur, there is a requirement for dietary protein, and the efficient functioning of the biochemical processes of the body also requires that the diet provide certain lipids and a range of other nutrients (vitamins and minerals).

2.2 PROTEIN AND AMINO ACID REQUIREMENTS

An adequate supply of dietary protein is essential if fish are to thrive and grow well. Results from laboratory studies conducted on several carnivorous fish species, such as salmonids, percids and marine flatfish, indicate that optimum growth is achieved when around half the food energy is supplied from protein. Thus, it is generally recognized that protein should provide 40–55% of the dietary energy for these species. Other species, such as cyprinids, tilapias and some ictalurids, are capable of maintaining excellent rates of growth when fed diets of lower protein content. For these species, diets formulated to contain 30–40% protein will be adequate to fulfil the requirements of the fish for a dietary protein supply.

Theoretically, fish do not have a dietary requirement for protein *per se*, but they do require a dietary supply of certain essential (indispensable) amino acids that contribute to protein structure. The fish are able to synthesize and interconvert some of the amino acids, but are incapable of *de novo* synthesis of others. Thus, from a nutritional point of view the amino acids can be divided into nonessential and essential (indispensable) groups. The essential amino acids are obviously those that the fish cannot synthesize, or cannot synthesize in sufficient quantity, to enable the maintenance of good rates of growth. Certain amino acids appear to be essential because the fish lack the biochemical machinery required to manufacture the chemical configurations of the carbon chain skeletons of these amino acids.

Which amino acids are essential and which are not can be determined experimentally either by conducting growth studies or by following the fates of labelled substrates that are injected into the fish. In growth studies, the fish are fed a purified diet containing all the amino acids, as the control, and other groups of fish are fed test diets similar to the control diet except that one amino acid has been excluded. Groups of fish that do not grow, or grow at rates markedly lower than those of the controls, are adjudged to have been fed a test diet deficient in an essential amino acid. In labelling studies, the test fish are injected with a radioactively labelled (usually ^{14}C) substrate, such as glucose or acetate, and it is assumed that the radioactive label will gradually be incorporated into organic compounds that the fish can synthesize. Thus, the fish are left for some period of time following injection to allow for the label to be incorporated into the organic compounds making up the body tissues. The fish are then killed and the amino acids in the body proteins analysed to determine whether or not they contain any of the radioactively labelled carbon (^{14}C). Amino acids that contain the label are considered to be nonessential, whilst those that do not contain any trace of label are considered essential.

Results of both types of study have revealed that ten of the amino acids are essential for fish species:

1. aliphatic series – threonine, valine, leucine, isoleucine;
2. aromatic series – phenylalanine;
3. sulphur-amino acid series – methionine;
4. heterocyclic series – tryptophan;
5. acidic series – no essential amino acids in this series;
6. basic series – arginine, histidine, lysine.

The types of study described above give qualitative information about which of the amino acids are essential, but they do not provide quantitative information, i.e. they give no indications about the levels of dietary amino acid inclusion necessary for the promotion of good rates of growth.

Data about quantitative amino acid requirements are usually obtained by conducting a series of growth studies in which diets are formulated to contain graded levels of the amino acid of interest. The growth responses of the fish are recorded and dose–response curves are constructed by equating growth with the amount of amino acid in the diet. The growth data may then be examined and analysed in a number of ways, with the following approaches being commonly adopted:

1. analysis based upon a 'broken line' response;
2. fitting of an asymptotic curve to the growth data;
3. analysis based upon saturation kinetics.

The broken line response relies on the assumption that there is a linear relationship between growth response and the dietary concentration of a limiting essential nutrient (Fig. 2.1), such as an essential amino acid. The dietary requirement for a particular essential amino acid may, therefore, be determined by calculating the break-point in the response curve obtained by plotting the growth of the fish against the dietary concentration of the amino acid. This break-point is seen as the point at which increased inclusion of the amino acid does not promote any further increase in the growth response, and a plateau is reached. The dietary requirement is determined as the point where there is intersection between the horizontal plateau line and the straight line representing the increase in growth increment resulting from increased dietary inclusion of the limiting amino acid.

Alternatively, the influences of increased nutrient inclusion on the growth response may be examined by fitting the data to an asymptotic

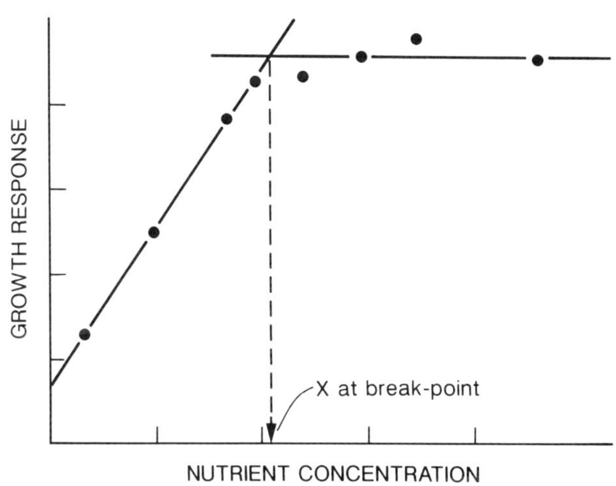

Fig. 2.1 Determination of the dietary requirement for an essential nutrient using analysis of the broken line response. The intersection of the two regression lines indicates the break-point.

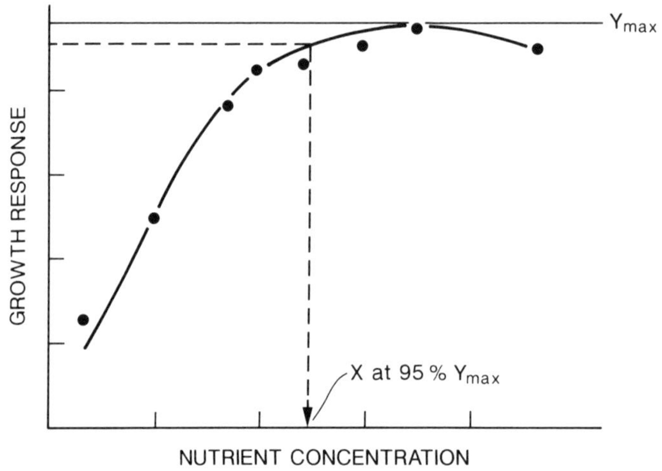

Fig. 2.2 Fitting of dose–response data to a polynomial equation to determine the dietary requirement of an essential nutrient. The dietary requirement is considered to be that nutrient concentration which promotes 95% of the maximum growth response.

curve, such as a second order polynomial (Fig. 2.2). The second order polynomial is represented by the equation:

$$Y = a + b_1X + b_2X^2 \tag{2.1}$$

where X is the nutrient concentration in the diet, Y is the growth response and a, b_1 and b_2 are constants. This curve is characterized by having a unique maximum of Y (Y_{max}), and the value of X that corresponds to Y_{max} is defined as the concentration of the dietary nutrient that produces maximum growth. The requirement for the nutrient is usually defined as that which gives a growth response corresponding to 95% Y_{max}.

There is an increasing trend towards the use of non-linear models in the analysis of the dose–response curves obtained in nutrient requirement studies (Fig. 2.3). These analyses often employ the concept of saturation kinetics, in which it is assumed that the growth response is dependent upon the result of the outcomes of a series of enzymatic steps, one of which is rate limiting. In an analysis of this type the growth data may be fitted to an equation of the form:

$$Y = (aK + Y_{max}X^b)/(K + X^b) \tag{2.2}$$

which has one dependent variable (Y the growth response), one independent variable (X the nutrient concentration) and four parameters (a, K, Y_{max} and b). The parameter a represents the intercept on the ordinate (Y when $X = 0$), Y_{max} is the maximum theoretical (asymptotic)

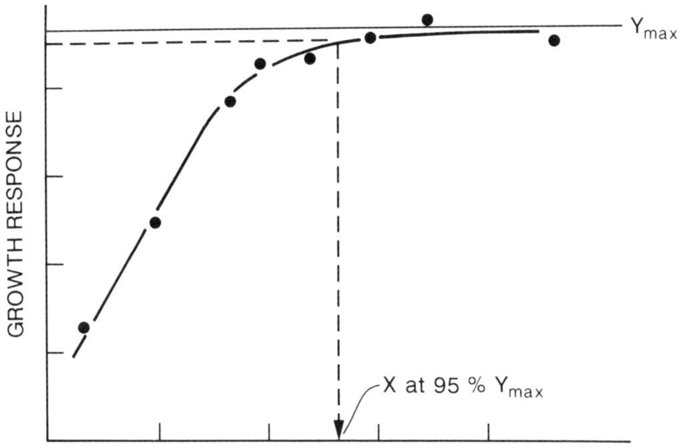

Fig. 2.3 Fitting of dose–response data to a four-parameter model that describes the data in terms of saturation kinetics.

response, b is the slope factor and K is the nutrition constant. Using the experimentally observed data pairs (*Y* for a given *X*) the four equation parameters may be determined with non-linear curve-fitting techniques.

In dietary requirement studies, where the test diets are formulated to contain relatively large quantities of purified crystalline amino acids, rates of growth recorded on diets formulated to contain all the essential amino acids at levels above the minimum requirement will often be found to be quite low. In other words, growth rates are lower when amino acid mixtures are fed than when the same amino acids are provided in the form of complete proteins. This may, in part, be due to the facts that when complete proteins are fed they are not completely digested to single amino acids before being absorbed, and that there are differences in the rates at which different amino acids are absorbed. Thus, the feeding of amino acid mixtures may not lead to the absorption of amino acids at the optimal rates, and in the ideal proportions, to promote the maximal possible growth.

It should be appreciated that experiments for the quantitative determination of amino acid requirements are extremely time- and resource-consuming, and the details of the essential amino acid requirements have been studied for only a very limited number of fish species. Species studied include some salmonids, cyprinids, channel catfish, *Ictalurus punctatus*, and some tilapias, all of which are of commercial importance in connection with fish farming. A summary of some of the findings relating to quantitative amino acid requirements is presented in Table 2.1.

Table 2.1 Essential amino acid requirements of various fish species. Values are presented on a percentage of dietary protein basis

Amino acid	Channel catfish *Ictalurus punctatus*	Carp *Cyprinus carpio*	Tilapia *Oreochromis niloticus*	Chinook salmon *Oncorhynchus tshawytscha*
Arginine	4.3	4.3	4.2	6.0
Histidine	1.5	1.8	1.7	1.8
Isoleucine	2.6	2.5	3.1	2.2
Leucine	3.5	4.1	3.4	3.9
Lysine	5.0	5.8	5.1	5.0
Methionine	2.3	1.8	2.7	4.0
Phenylalanine	5.0	5.7	5.5	5.1
Threonine	2.0	3.8	3.8	2.2
Tryptophan	0.5	0.8	1.0	0.5
Valine	3.0	3.5	2.8	3.2

There tend to be some differences in the estimated requirements for particular amino acids, both between species and between studies carried out by different workers on any given species. These differences may represent real differences in requirements between species, or between life history stages of the fish, but the discrepancies may also in part be attributable to methodological differences between studies.

It is also known that there are interactions between certain of the amino acids, so the estimated requirement for one amino acid may be influenced by the levels of inclusion of other amino acids in the test diets. For example, both phenylalanine and tyrosine are classified as aromatic amino acids, and adequate amounts of each are required if correct physiological functions are to be maintained. Fish can readily convert phenylalanine to tyrosine (i.e. tyrosine is not an essential amino acid), or, alternatively, dietary tyrosine can be used to meet the needs for this amino acid. Phenylalanine is an essential amino acid, and, from the above, it should be clear that the estimate of the quantitative requirement for phenylalanine will differ depending upon whether the test diet contains adequate, or low, levels of tyrosine. Similar interactions occur between the sulphur-amino acids. Methionine is an essential amino acid, but cystine can be synthesized from methionine, and if cystine is present in the diet it reduces the methionine requirement. In view of the interaction between methionine and cystine, fish are usually considered to have a total sulphur-amino acid requirement, rather than a specific methionine requirement, and the results obtained in experimental studies are often presented in these terms.

A number of workers have observed that fish may show improved

Table 2.2 Comparison of the essential amino acid requirements for growth with the compositions of egg yolk protein and tissues of rainbow trout, *Oncorhynchus mykiss*. Values are presented on a percentage of protein basis

Amino acid	Growth requirement	Yolk protein	Yolk-sac fry	Swim-up fry
Arginine	3.5	4.5	6.6	5.7
Histidine	1.6	2.2	2.9	2.5
Isoleucine	2.4	5.7	5.9	4.5
Leucine	4.4	9.1	9.9	7.9
Lysine	5.3	7.1	9.0	7.7
Methionine	1.8	1.2	3.3	2.9
Phenylalanine	3.1	3.8	5.6	4.5
Threonine	3.4	5.4	5.0	4.6
Tryptophan	0.5	–	–	–
Valine	3.1	8.2	7.7	5.9
Cystine	0.9	1.1	1.5	1.0

growth when their diets are supplemented with essential amino acids to simulate the composition of either the egg yolk or muscle proteins, and essential amino acid requirements have generally been shown to correlate reasonably well with the essential amino acid profile of whole-body protein (Table 2.2). Thus, body protein composition data may provide useful background information that can be used as a guideline about the essential amino acid requirement of a given fish species or specific life history stage.

2.3 LIPID AND FATTY ACID REQUIREMENTS

The lipids represent a concentrated energy reserve for the fish, but they also have other important roles, such as being the transport medium for lipid-soluble compounds (e.g. some vitamins), as structural elements in cell membranes, and as precursors for a number of important biologically active compounds (e.g. some hormones, pigments and growth factors).

It is the phospholipids that have functions in the determination of cell membrane structure and fluidity, and the phospholipids may also be precursors for other compounds. Consequently, research has been directed towards the examination of the structures and fatty acid profiles of the different phospholipids. The phospholipids are almost invariably found to have their fatty acids dominated by the long-chained highly unsaturated fatty acids (e.g. 20:5(n–3), 22:6(n–3), 20:4(n–6)), and it is these HUFAs that impart the particular structural and physiological properties to the phospholipids. Fish, and other animals, do not appear to be able to synthesize polyunsaturated fatty acids of the (n–3) and (n–6) series, so these fatty acids are essential nutrients that must be supplied in the diet. Terrestrial animals require quite large amounts of both (n–6) and (n–3) fatty acids, particularly the former, and these are generally supplied in the diet as 18:2(n–6) and 18:3(n–3). The animals are able to convert and chain elongate these fatty acids into the longer-chain HUFAs found in the phospholipids (Fig. 2.4). In the earliest work on fish nutrition it was assumed that the fatty acid requirements of fish were similar to those of the terrestrial mammals, and test diets were often formulated to include high levels of 18:2(n–6), and lower levels of the (n–3) series fatty acids.

Later studies showed that the majority of fish species have a greater requirement for the (n–3) series fatty acids than for the (n–6) series, and many of the fish species examined appear to be able to grow well on diets low in (n–6). For example, in a series of experiments carried out on rainbow trout, *Oncorhynchus mykiss*, it was found that the fish were capable of completing the full life cycle on diets containing (n–3) fatty acids, but could not grow when fed diets containing only (n–6) series fatty acids. Thus, the impression was gained that the (n–3), but not the (n–6), series fatty acids were essential for rainbow trout. However, when the fatty acid profiles of the tissue and egg phospholipids of these rainbow trout were examined, it was found that the phospholipids of those fish fed the supposedly (n–6)-free diet contained small amounts of the HUFA 20:4(n–6). Thus, the diet cannot have been truly (n–6)-free, but the minute amounts of (n–6) fatty acids present appeared to have been preferentially retained, chain-elongated and desaturated to 20:4(n–6) for incorporation into the phospholipids. The fatty acid 20:4(n–6), along with other HUFAs, may also act as a precursor for eicosanoid, e.g. prostaglandin, synthesis. This group of compounds is known to have a number of central physiological roles in the regulation of respiratory and reproductive functions. Thus, it appears that the rainbow trout may have a small, but nevertheless important, essential dietary requirement for (n–6) fatty acids, or more specifically for 20:4(n–6).

The essential requirement for fatty acids, appears, therefore, to be a requirement for the HUFAs of the (n–3) and (n–6) series. However, because some fish species possess the biochemical machinery required for the chain-elongation and desaturation of the fatty acids, their essential fatty acid requirements can be met by the inclusion of 18:3(n–3) and 18:2(n–6) in the diet. The exact levels at which these fatty acids must be supplied (i.e. the quantitative requirement) will be dependent upon the abilities of the fish to convert these fatty acids into the longer-chained HUFAs, but there are complex interactions in that high levels of some fatty acids can inhibit the chain-elongation and desaturation of others.

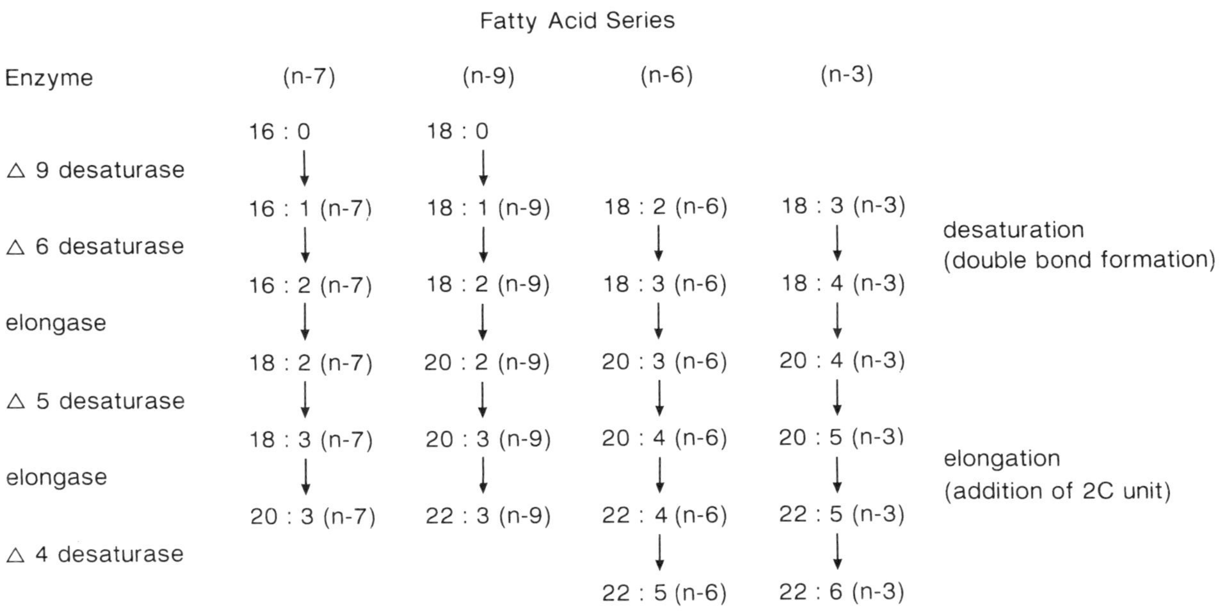

Fig. 2.4 Pathways for chain elongation and desaturation of fatty acids, resulting in biosynthesis of polyunsaturated fatty acids of different series. The enzyme Δ6 desaturase has a preference for unsaturated fatty acids as substrate. In the absence of 18:3(n−3) and 18:2(n−6), 18:1(n−9) is desaturated and elongated and long-chain fatty acids of the (n−9) series accumulate in the tissues (this effect can be used as an indication of essential fatty acid deficiency). Formation of the long-chained HUFAs is thought to involve a Δ4 desaturase enzyme, but the existence of this enzyme remains in doubt. An alternative pathway, involving chain elongation to 24-C fatty acids, followed by desaturation, and then reconversion to 22-C fatty acids via the beta-oxidation spiral has been proposed, e.g. 22:5(n−3) → 24:5(n−3) → 24:6(n−3) → 22:6(n−3).

The critical enzyme in the desaturation and chain-elongation pathway of the fatty acids appears to be Δ6 desaturase, which acts to desaturate 18-C fatty acid substrates. The enzyme appears to have greatest affinity for substrates with the greatest number of double bonds. Thus, 18:3(n−3) with three double bonds is desaturated at the highest rate, followed by 18:2(n−6) and then 18:1(n−9). In the presence of either of the two dietary fatty acids 18:3(n−3) and 18:2(n−6), little desaturation of 18:1(n−9) will occur. In other words, presence of the 18-C PUFAs effectively inhibits the desaturation of 18:1(n−9). In the absence of an adequate dietary supply of the 18-C fatty acids of the (n−3) and (n−6) series, 18:1(n−9) will be desaturated and the longer-chain, more unsaturated members of the (n−9) series will be incorporated into the tissue lipids. Thus, the accumulation of (n−9) fatty acids, particularly 20:3(n−9), in the tissue lipids can be taken as an indication that inadequate levels of dietary (n−3) and (n−6) fatty acids have been included in the diet, and that there is an essential fatty acid deficiency.

Because the 18:3(n−3) and 18:2(n−6) fatty acids are converted to longer-chain HUFAs before being incorporated into the phospholipids, it might be expected that the HUFAs would be more effective in meeting the essential fatty acid requirements of the fish than the shorter-chain members of the same series. This 'enhanced' essential fatty acid ability of the HUFAs has been confirmed in experimental studies carried out on a number of fish species. For example, in studies with rainbow trout, the fatty acids 20:5(n−3) and 22:6(n−3) were found to give positive growth effects at dietary inclusion levels as low as 0.25–0.5%, whereas when 18:3(n−3) was used as the major source of (n−3) fatty acids, inclusion levels of 1–2% were required. Many marine species of fish appear either to have considerably reduced, or to lack, ability to chain-elongate and desaturate the short-chain (n−3) and (n−6) fatty acids. Thus, these species require the fatty acids to be provided in the diet as the long-chain HUFAs, such as 20:5(n−3) and 22:6(n−3).

The possibility that there may be chain-elongation, desaturation and interconversion of fatty acids, along with the fact that the interconversions of some fatty acids can be inhibited by the presence of others, makes the quantitative determination of the fatty acid requirements of fish a somewhat complex exercise. Consequently, the expression of essential fatty acid requirements in terms of percentage dietary composition may be misleading, and it is becoming more usual to give requirements as a proportion of total dietary lipid. For example, the requirement of sea bream, *Pagrus major*, for (n−3) HUFAs was found to increase from about 1.5% to almost 4% of the diet as dietary lipid content was increased from 10% to 20%. When expressed in terms of lipid content, however, the requirement for (n−3) HUFAs ranged between 15% and 20% of the total lipid, irrespective of the level of inclusion of lipid in the diet. Similarly, whilst the inclusion of 1–2% 18:3(n–3) in the diets of rainbow trout is usually sufficient to promote good growth, these levels may be inadequate if the diet also contains large amounts of other fatty acids, especially those of the (n–6) series. Consequently, the recommendation for essential fatty acids has been amended to 20% of dietary lipid as 18:3(n–3) or 10% of dietary lipid as the longer-chained HUFAs, 20:5(n–3) or 22:6(n–3). This modification has been made in order to ensure that a dietary balance is maintained between the (n–3) and (n–6) series fatty acids, when lipid-rich diets are fed.

Problems may arise in the determination of fatty acid requirements because of the methodological restrictions imposed by the carrying out of such studies. The quantitative determination of essential fatty acid requirements necessitates the feeding of test diets in which the fatty acid content is strictly defined. Thus, the fatty acids have often been provided in the purified form of methyl-esters. This form of dietary presentation differs from that normally encountered by the fish, because the vast majority of the fatty acids in the ingested lipids will usually be present in triacylglycerols or phospholipids. A significant proportion of the essential fatty acids ingested in the food is likely to be present in the phospholipids, and it is open to question whether or not methyl-esters of fatty acids will be digested and absorbed as efficiently or in the same manner as the dietary phospholipids.

2.4 VITAMIN REQUIREMENTS

Vitamins are organic chemical compounds essential for the maintenance of health and growth of the fish, but they are required in relatively small amounts. Eleven water-soluble and four lipid-soluble vitamins are known to be required by fish species (Table 2.3). Eight of the water-soluble vitamins, the B complex, have coenzyme functions and are required in small quantities, whereas the other water-soluble vitamins – ascorbic acid (vitamin C), myo-inositol and choline – are required in larger amounts (about 100 mg per kg diet or more) and these compounds are sometimes referred to as the macrovitamins (Fig. 2.5). The lipid-soluble vitamins are vitamins A, D, E and K.

Water-soluble vitamins

The water-soluble vitamins consist of the B-complex vitamins, vitamin C and a few compounds of more doubtful status. Small amounts of these vitamins may be stored in the liver, but tissue storage is limited and

Table 2.3 Vitamin requirements, or levels of recommended dietary inclusion, for a range of fish species. Values are presented as either mg vitamin per kg dry diet or as IU (international units) vitamin per kg dry diet

Vitamin	Channel catfish, *Ictalurus punctatus*	Carp, *Cyprinus carpio*	Chinook salmon, *Oncorhynchus tshawytscha*	Dietary recommendation (salmonids)
Water soluble				
Thiamin	1	R*	~15	15
Riboflavin	9	~15	~25	30
Pyridoxine	3	~6	~20	20
Pantothenic acid	15	~50	~50	50
Niacin	14	28	~200	200
Folic acid	R	–	~10	10
Cyanocobalamin (B_{12})	R	–	~0.02	0.02
Myo-inositol	NR†	440	~400	400
Choline	400	~4000	~800	3000
Biotin	R	1	~1.5	1.5
Ascorbic acid (vitamin C)	60	R	~150	200
Lipid soluble				
Vitamin A	~2000 IU	10 000 IU	~2500 IU	2500 IU
Vitamin D	~500 IU	–	–	2400 IU
Vitamin E	50	100	30	50
Vitamin K	R	–	–	10

*R, vitamin required, but quantitative requirements uncertain.
†NR, vitamin probably not required to be included in the diet.

excesses are eliminated via the kidneys. Generally, a regular dietary intake of the water-soluble vitamins is needed unless metabolic needs can be met by intestinal microbial synthesis.

Ascorbic acid (vitamin C)

A wide variety of animal species are capable of synthesizing ascorbic acid from glucuronic acid, but many fish species appear either to lack or to have only limited ability to synthesize ascorbic acid. These species must, therefore, obtain their supplies of ascorbic acid in the diet. Ascorbic acid is a strong reducing agent, and acts synergistically with vitamin E and selenium to prevent oxidative damage (peroxidation) of the phospholipids present in cell membranes. The vitamin is also required for the biosynthesis of collagen and cartilage, tissue synthesis and repair. It may be involved in iron metabolism, and many of the cellular hydroxylase

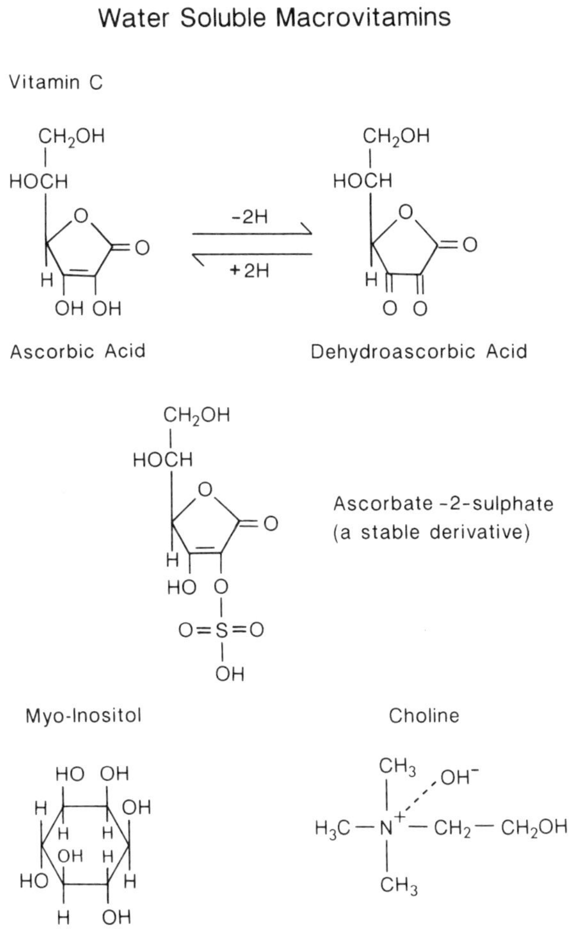

Fig. 2.5 Chemical structures of the water-soluble macrovitamins.

enzymes appear to require the presence of ascorbic acid in order to display maximal activity. Because ascorbic acid is involved in the synthesis of collagen and cartilage, a deficiency of this vitamin can lead to the development of skeletal deformities, reduced efficiency of wound healing, and non-specific symptoms such as anorexia and lethargy.

Myo-inositol

The water-soluble compound myo-inositol is often classified as a growth factor, rather than a vitamin. It does not seem to be an essential dietary constituent for a range of species, but is required by others. Myo-inositol seems to be synthesized by the intestinal microflora, and synthesis may also occur in the liver of some fish species. Myo-inositol is found as a constituent of phospholipids, and is present in the red blood cells. A deficiency of myo-inositol can lead to the fish becoming anaemic, and anorexic.

Choline

Choline is sometimes considered to be a B-complex vitamin, although the status of choline as a vitamin is somewhat in doubt. The choline molecule has three methyl ($—CH_3$) groups, and choline has an important function as a methyl donor in a variety of biochemical reactions. Choline reacts with acetyl CoA to form the neurotransmitter substance acetylcholine, and it is also a constituent of certain phospholipids and related compounds. As a component of phospholipid, choline is involved in cell membrane structure and in the transport of lipids.

The water-soluble B-complex vitamins (Fig. 2.6) all possess specific coenzyme functions in cellular metabolism.

Thiamin

This vitamin is required for the normal metabolic function of all animal species. Thiamin is absorbed from the intestine and is converted to thiamin pyrophosphate in the liver. The phosphorylated form is required for the decarboxylation of pyruvic acid and ketoglutarate in the intermediary metabolism of carbohydrates and lipids. Thiamin is also involved in the direct cellular metabolism of glucose.

Riboflavin

This vitamin is synthesized by plants and micro-organisms, but animals do not synthesize and store large quantities of riboflavin, so it is required as a dietary constituent. Riboflavin is absorbed from the intestine, phosphorylated in the liver and then transported to the various tissues of the body. The primary functions of riboflavin involve energy metabolism as a component of two flavoprotein coenzymes, flavin mononucleotide (FMN) and flavin adenine dinucleotide (FAD). FMN and FAD function in the electron transport chain as well as in the oxidase and reductase enzyme systems required for the catabolism of fatty acids, amino acids and pyruvic acid. Thus, this water-soluble vitamin is needed for the maintenance of a wide range of metabolic functions.

Pyridoxine

Pyridoxine, pyridoxal and pyridoxamine are related compounds having vitamin activity in animal species. Pyridoxine is absorbed from the intestine, phosphorylated in the liver and is then oxidized to the physiologically active form pyridoxal phosphate. Many of the enzymatic reactions in which amino acids are decarboxylated by aminotransferase enzymes require pyridoxine derivatives as a coenzyme. Pyridoxine is also involved in other reactions in which proteins, lipids and carbohydrates are metabolized, as well as having a coenzyme function in the synthesis of certain neuroendocrine substances.

Pantothenic acid

This vitamin is widespread in nature, and is synthesized by plants and micro-organisms. Pantothenic acid is absorbed from the intestine and is found in the tissues as a component of coenzyme A (CoA) (Fig. 2.7) and acyl carrier protein (ACP). The vitamin acts metabolically as a component of CoA to transfer acyl groups in enzymatic reactions involving the metabolism and release of energy from respiratory substrates – carbohydrates, lipids and proteins. CoA is required in the synthesis and oxidation of fatty acids, oxidation of pyruvate, conversion of choline to acetylcholine and for the conversion of oxaloacetate to citrate. CoA is also needed for the formation of two-carbon units from amino acids and carbohydrates that can be metabolized by way of the citric acid cycle. As a component of CoA, pantothenic acid is also required for the synthesis of steroids. ACP serves a role in fatty acid synthesis.

Biotin

Biotin is synthesized by intestinal bacteria, and this may be the major source of the vitamin in the majority of fish species. The vitamin is a component of several coenzymes, and plays an important role in both carboxylation and decarboxylation reactions. It is involved in some of the metabolic steps of fatty acid synthesis, in the oxidation of carbohydrate, and in the harnessing of the energy released from the three substrates – carbohydrates, lipids and proteins. Biotin is also intimately involved in purine synthesis.

Niacin

A number of species are capable of synthesizing this vitamin from the amino acid tryptophan, but the general importance of this conversion in

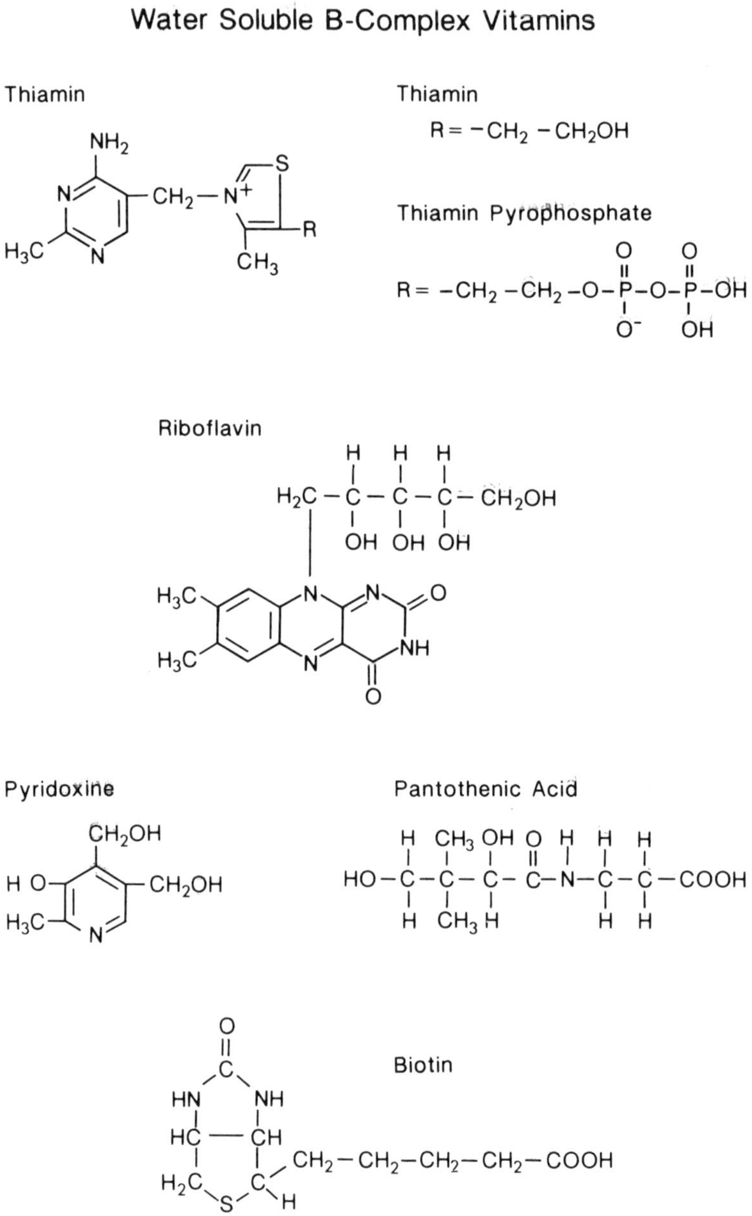

Fig. 2.6 Chemical structures of the water-soluble B-complex vitamins.

Niacin

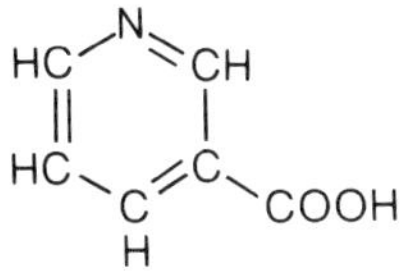

Nicotinic Acid

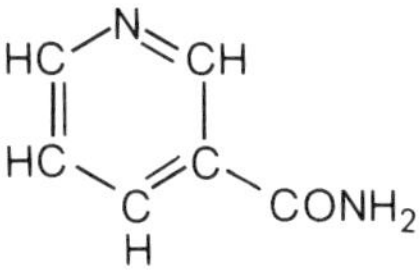

Niacinamide

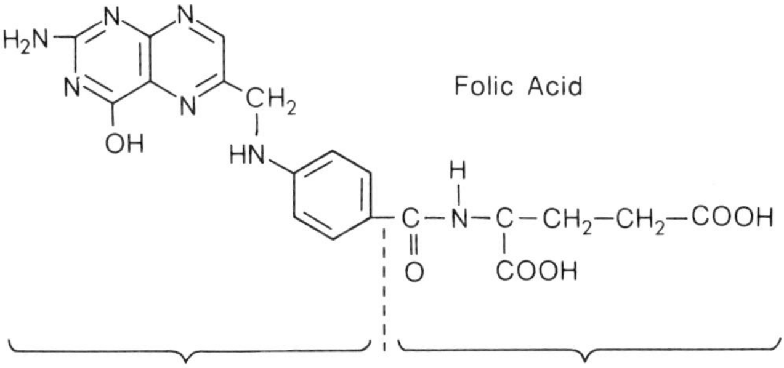

Folic Acid

Pteroic Acid

Glutamic Acid

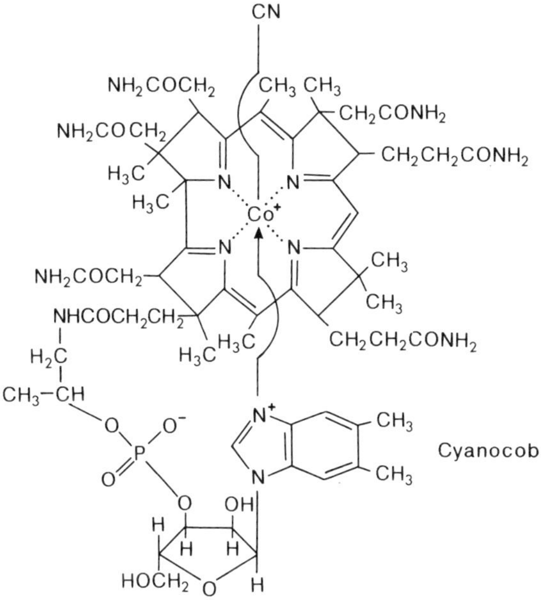

Cyanocobalamin

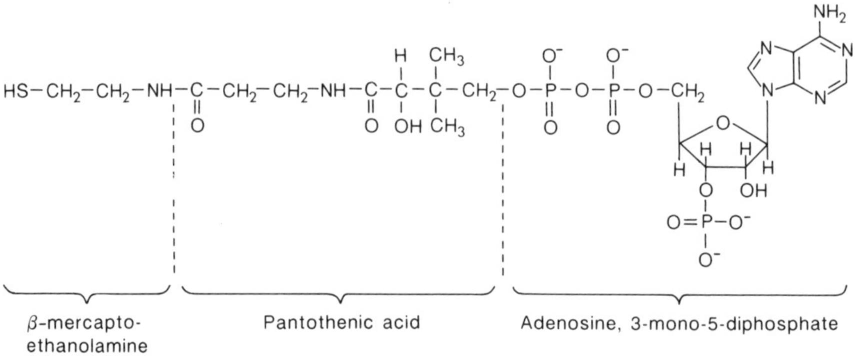

Fig. 2.7 Structure of coenzyme A, showing incorporation of the B-complex vitamin, pantothenic acid.

fish species is not known. Niacin is absorbed from the intestine, small quantities are stored in the liver, and excesses are excreted via the kidneys. The vitamin is an essential component of the two compounds – nicotinamide adenine dinucleotide (NAD) and nicotinamide adenine dinucleotide phosphate (NADP) – that are involved as hydrogen donors and acceptors in the harnessing of the energy released during respiratory metabolism. Thus, NAD and NADP are essential in certain energy-yielding and synthetic pathways, including fatty acid catabolism and synthesis, oxidative deamination, glucose and glycerol catabolism and the citric acid cycle.

Folic acid

Folic acid and related compounds are synthesized by the intestinal microflora, and conversion to the biologically active form, tetrahydrofolic acid, occurs in the liver (and in the bone marrow of some species). The active form of the vitamin is a coenzyme that acts as an intermediate carrier of single-carbon units. Because folic acid is also required for the transfer of methyl groups (—CH_3) between organic molecules, it is involved in the interconversion of amino acids. It is, for example, required for histidine degradation and for the synthesis of purines, pyrimidines, glycine, serine and creatine. The vitamin is essential for the synthesis of the nucleic acids (DNA and RNA). Folic acid is necessary for the normal formation of red blood cells, and a deficiency of folic acid leads both to the fish becoming anaemic and to abnormalities in the red blood cells. The role of folic acid in the prevention of anaemia appears to be related to purine synthesis.

Cyanocobalamin

The synthesis of this vitamin seems to be restricted to micro-organisms, and microbial synthesis of cyanocobalamin and related compounds occurs in the intestine of many fish species. Absorption of the vitamin from the gut seems to depend upon the presence of a special carrier, and following absorption the vitamin is transported to the liver where it can be stored. Cyanocobalamin is involved as a coenzyme in many metabolic functions. The vitamin is needed for the metabolism of lipids and propionic acid, and is involved in the metabolism of methionine. It is essential for normal growth, maturation of red blood cells, DNA synthesis and for the maintenance of normal functioning of nervous tissue. Cyanocobalamin is also involved in the recycling of tetrahydrofolic acid, so in many respects, the roles of folic acid and cyanocobalamin in fish metabolism are complementary. These two vitamins may, therefore, be able to substitute for each other in some physiological functions.

Lipid-soluble vitamins

The lipid-soluble vitamins are required to be supplied in the diet in comparatively low quantities. In feedstuffs, these vitamins are found in association with the lipids and the vitamins are absorbed from the gut along with the dietary lipids. The lipid-soluble vitamins may be relatively easily stored and accumulated in the body lipids. As a consequence, dietary overfortification with the lipid-soluble vitamins can give rise to toxicity symptoms (hypervitaminosis) owing to high levels of tissue accumulation.

Vitamin A (Fig. 2.8)

This vitamin is only found in animal tissues, although carotenoids, which are pre-cursors of the vitamin, are widely distributed in nature amongst plants and micro-organisms. Several different forms of vitamin A (retinol) occur in animals, with the compound transretinol being considered to have high biological vitamin A activity. Other forms in animals, and the carotenoids of plants, have varying degrees of vitamin A activity. Vitamin A and beta-carotene (from plants) are absorbed from the intestine along with the dietary lipids. Beta-carotene is converted into vitamin A in the intestinal mucosa before being transported to the liver, which is the primary storage site for the vitamin. Vitamin A is essential for normal

Vitamin A

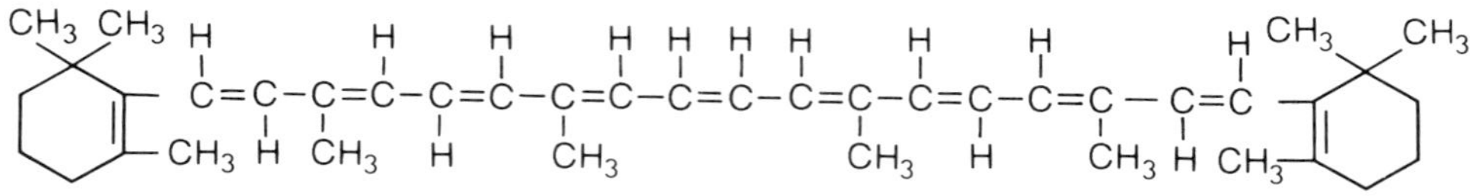

Beta-carotene

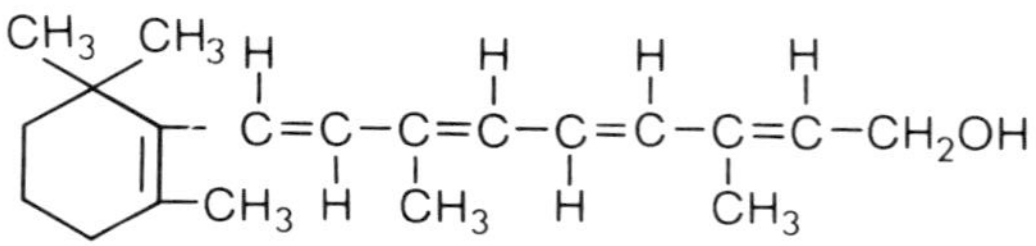

$Retinol_1$

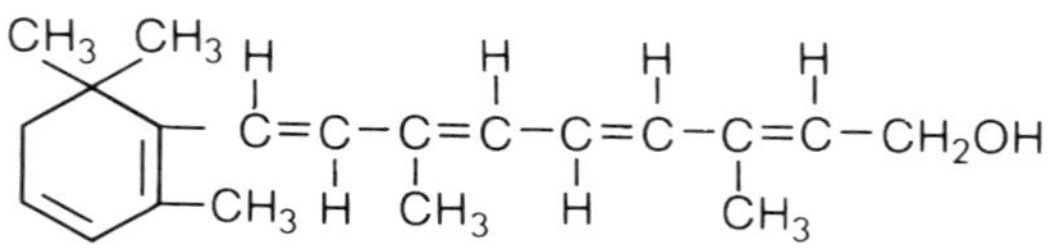

$Retinol_2$

Fig. 2.8 Chemical structures of beta-carotene and retinol.

structure and functions of the eye because it is required for the formation of rhodopsin (visual pigment). The vitamin has a role to play in gill function through involvements in the metabolism of mucopolysaccharides. Vitamin A is also necessary for the maintenance of epithelial structure and function.

Vitamin D (Fig. 2.9)

Several lipid-soluble sterol derivatives are known to have vitamin D activity. Of these compounds, ergocalciferol and cholecalciferol are the most important from a nutritional viewpoint. Ultraviolet radiation of the provitamins ergosterol (found in plants) and 7-dehydrocholesterol (found in animal tissues) yield ergocalciferol and cholecalciferol, respectively. Animals that are regularly exposed to sunlight will produce vitamin D, but a dietary source may also be required. Dietary vitamin D is absorbed from the intestine along with the lipids, and is transported to the liver for storage. Some vitamin D may be excreted in the bile. In the liver vitamin D is converted to the compound 25-hydroxycalciferol, but conversion to the biologically active form 1,25-dihydroxycalciferol takes place in the kidneys of the majority of animal species. Vitamin D is intimately involved in the mobilization, transport, absorption and utilization of calcium and phosphorus. Vitamin D deficiency results in disturbances to calcium metabolism, and can lead to tetanic contractions of the muscle tissue.

Vitamin E (Fig. 2.10)

One family of organic compounds, the tocopherols, is collectively known as vitamin E, but different tocopherols show different levels of vitamin activity. Vitamin E acts as a lipid-soluble antioxidant, and as such, serves to protect the phospholipids in the membranes of cell walls and subcellular organelles from oxidative destruction. The requirement for vitamin E depends upon the levels of highly unsaturated fatty acids in the phospholipids, the higher the HUFA content the greater the requirement for vitamin E to protect these fatty acids from oxidative degradation. The vitamin E will work in concert with other antioxidants, such as ascorbic acid and selenium. A deficiency of vitamin E manifests itself as increased fragilty of red blood cells and lesions in the muscles, which may eventually lead to the fish showing signs of muscular dystrophy.

Vitamin K (Fig. 2.11)

Certain quinone derivatives display vitamin K activity. This group of compounds appears to be involved in the synthesis of plasma proteins

Vitamin D

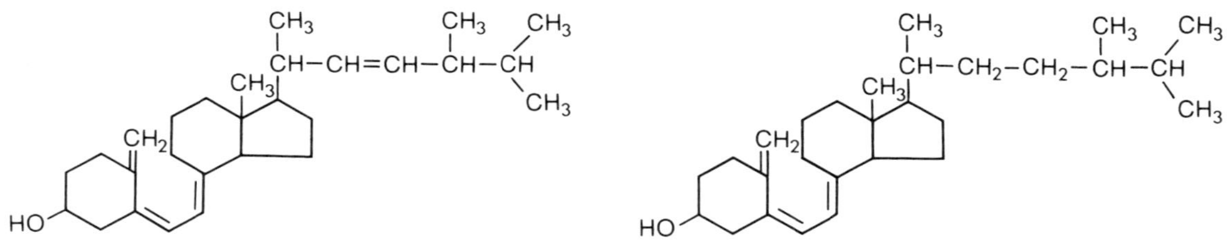

Ergocalciferol

Cholecalciferol

Fig. 2.9 Chemical structures of vitamin D – ergocalciferol and cholecalciferol.

Vitamin E

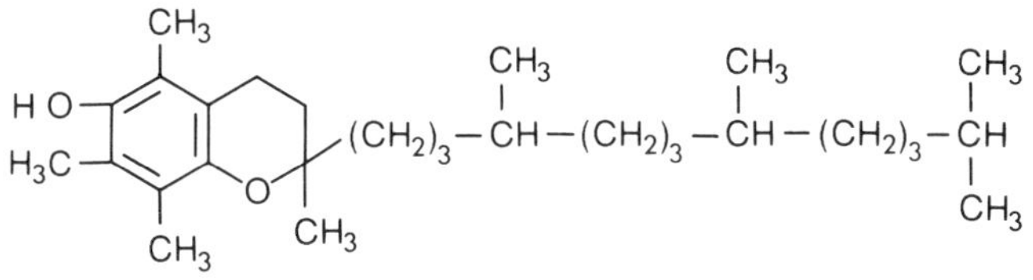

Alpha -Tocopherol

Fig. 2.10 Stucture of alpha-tocopherol (vitamin E).

Vitamin K

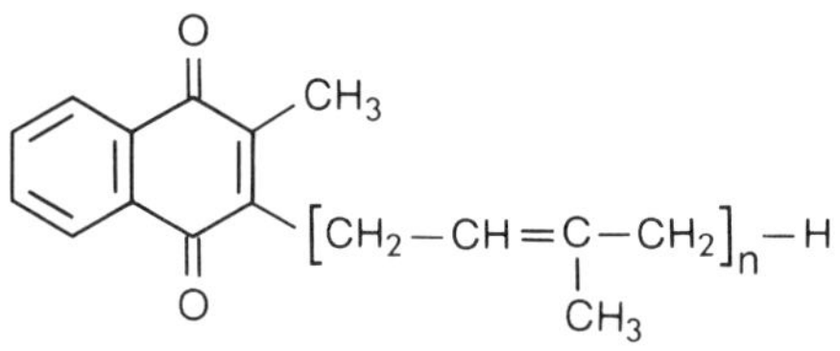

n may equal 6, 7 or 9 'isoprenoid units'

Fig. 2.11 Chemical structure of vitamin K.

required for blood clotting. Vitamin K is, therefore, essential in order to ensure adequate clotting and coagulation of the blood. The vitamin also seems to be implicated in a number of the oxidative phosphorylation reactions of cellular metabolism, and may have cellular transport functions.

2.5 MINERAL REQUIREMENTS

Not all of the inorganic elements found in the fish body are required to be included in the diet. Fish are surrounded by water that contains mineral salts in solution, and fish are capable of absorbing some dissolved minerals from the water. The majority of the mineral requirements are, however, probably met through the diet rather than by direct absorption from the water; this is almost certainly the case for freshwater species. The quantitative mineral requirements of fish have, however, been very poorly defined (Table 2.4).

Table 2.4 Mineral requirements, or recommended levels of dietary inclusion, for a range of fish species. Values are presented as mg mineral per kg dry diet

Mineral	Channel catfish, *Ictalurus punctatus*	Carp *Cyprinus carpio*	Tilapia, *Oreochromis niloticus*	Chinook salmon, *Oncorhynchus tshawytscha*	Dietary recommendations (salmonids)
Calcium	4500	–	–	–	3000
Phosphorus	4500	~7000	9000	R*	6000
Magnesium	400	~500	~750	–	500
Iron	30	150	–	–	30–60
Copper	5	3	~4	–	~5
Manganese	~2.5	13	12	R	~20
Cobalt	–	0.1	–	–	5–10
Zinc	20	~20	10	R	30–100
Iodine	–	–	–	R	~0.3
Selenium	–	–	–	R	~0.3

*R, mineral required, but quantitative requirements uncertain.

Calcium and *phosphorus* are important constituents of skeletal materials, as well as having a number of metabolic functions. Both calcium and phosphorus are required in quite large amounts (approximately 5 g per kg dry diet). The requirement for dietary supply of *magnesium* also appears to be relatively large, but only small traces (inclusion in mg proportions per kg dry diet) of other minerals appear to be required. These other minerals have, nevertheless, important physiological and biochemical roles, and many are required in order for a diversity of enzyme systems to function effectively. *Copper, manganese, cobalt, zinc,* and *selenium* are all trace elements with important metabolic functions, *iron* is a component of the respiratory pigment haemoglobin, and *iodine* is required for the production of thyroid hormones.

When exposed to water of low ionic strength, fish of several species (e.g. salmonids, tilapias, eels, cyprinids and ictalurids) may display proliferation and/or enlargement of the 'chloride cells' on the gill and opercular epithelia. Because the chloride cells of freshwater fish are thought to be involved in the absorption of ions, an increase in the surface area of this cell type exposed to the external environment would be expected to result in an increase in the absorptive capacity for various ions. Proliferation and enlargement of chloride cells have been shown to occur following exposure of the fish to waters that are deficient in either monovalent (e.g. Na^+, K^+) or divalent (e.g. Ca^{2+}, Mg^{2+}) cations. In some species, the ionic absorptive capacity of the gut may also be increased when the fish are held in waters of low ionic strength.

Despite the fact that the fish may adapt to ion-deficient waters by displaying an increased capacity for ionic absorption, this may be insufficient to enable the fish to remain in ionic balance. Thus, the fish may suffer a net loss of ions unless they can be replaced through the diet, and there may, therefore, be expected to be interactions between the levels of water-borne minerals and the dietary requirement.

It appears, for example, that the requirement of channel catfish for potassium may be met by either water-borne or dietary sources. When the fish were held in water containing 4 $mg.l^{-1}$ potassium no dietary requirement for potassium could be demonstrated, but when reared in potassium-deficient water, channel catfish may be dependent upon the diet having a potassium content of about 2.5 $g.kg^{-1}$ if they are to continue to grow well.

Experiments conducted on carp, *Cyprinus carpio,* fed on low-magnesium diets have shown that the fish respond to the dietary deficiency by increasing the surface area of the chloride cells exposed to the external environment. This presumably leads to an increase in ionic absorptive capacity, but the fish may nevertheless be unable to meet their requirement via this route, and they can suffer a reduction in growth compared with fish that are provided with magnesium in the diet.

The possible interaction between levels of water-borne minerals and the dietary requirement is further exemplified by the requirements of juvenile rainbow trout weighing about 1 g, for magnesium. When the fish were held in water containing high levels (1000 $mg.l^{-1}$) of dissolved magnesium, approaching the concentrations found in seawater, the fish became anorexic, body loadings of magnesium increased and considerable mortality was observed within the space of a few days. This suggests that magnesium was entering the fish from the water and that the high levels of water-borne magnesium resulted in toxicity. On the other hand, when fish were held in water with a relatively low magnesium concentration (1.5 $mg.l^{-1}$) anorexia and poor growth was observed in groups of fish fed on diets containing less than about 400 mg per kg diet. In this case magnesium deficiency seems to have been the problem for fish fed diets low in magnesium whilst being held in fresh water with a low magnesium content. When the fish were held in water containing higher concentrations of magnesium (20–150 $mg.l^{-1}$), a dietary inclusion of 100 $mg.kg^{-1}$ was more than adequate to meet the requirement for good growth, and dietary inclusion levels of 700 mg per kg diet gave rise to increased body loading of magnesium. Thus, it is clear that the fish are able to meet some, if not all, of their requirement for magnesium from water-borne sources, and the level to be included in the diet will depend upon the magnesium content of the water in which the fish are reared.

In many studies of mineral requirements, the possible contribution of the water-borne mineral source has not been adequately investigated, leading to marked disparity in the published values of the dietary requirements for minerals in fish. For example, the range of dietary magnesium requirement of the rainbow trout may be 50–1300 mg per kg diet, being about 50 mg per kg diet when the water contains high levels of dissolved magnesium, and possibly up to 1300 mg per kg diet when the fish are held in magnesium-free water.

Chapter three

Dietary ingredients and feed formulation

3.1 INTRODUCTION

Commercial feed manufacturers aim to develop feed formulations that are as cheap as possible in relation to the levels of fish growth achieved. Consequently, a good deal of research has been conducted in order to determine to what extent expensive, often protein-rich, ingredients can be replaced by cheaper components that supply metabolic substrates in the form of carbohydrates and lipids. The final goal will be to minimize the levels of inclusion of the most expensive ingredients without this having any negative consequences for the health and growth performance of the fish. In addition, the manufacturing process must ensure that the feeds produced are stable, do not disintegrate rapidly in water, and do not permit rapid leaching of water-soluble components.

The types of formulated feeds used in commercial fish farming have characteristics that are radically different from those of the natural prey organisms consumed by fish. Commercial feeds differ from natural prey with respect both to moisture and energy content, and to nutritional composition. Many of the natural prey organisms consumed by carnivorous fish species have a relatively high water content (70–90%) and the energy content is generally within the range 3–8 kJ per g wet weight, whereas the majority of commercially produced dry feeds contain 10% or less moisture and have an energy content of 20–25 $kJ.g^{-1}$. These commercially produced dry feeds can be broadly divided into compressed and extruded pellet types. Both of these dry feed types require that binding agents be included in the formulation. Because it is generally carbohydrates that are used as binders, the commercially produced dry feeds tend to contain higher levels of carbohydrates than do natural prey organisms (Table 3.1).

Table 3.1 Chemical compositions (% dry weight) of natural prey organisms from freshwater (mixed samples of zooplankton and insect larvae) and marine (prawn, *Pandalus borealis*, and herring, *Clupea harengus*) environments, and those of a range of commercially produced dry pellet feeds for salmonids

Component	Freshwater zooplankton	Insect larvae	Prawn	Herring	Pellet
Protein	48	56	55	47	40–55
Lipid	21	14	17	39	15–23
Carbohydrate	4	9	–	–	10–20
Ash	12	10	14	8	7–12
PE:TE*	0.56	0.66	0.58	0.40	0.43–0.58

*Protein energy: total energy, the proportion of the dietary energy provided by protein.

3.2 BINDING AGENTS AND FEED PRODUCTION

The carbohydrates included in commercially produced dry feeds act as both cheap sources of respiratory substrate and as binding agents. The carbohydrates present are mainly long-chain polysaccharides of plant origin, predominantly starch and cellulose. Fish do not produce digestive enzymes capable of the hydrolysis of cellulose, and any cellulase activity isolated from the alimentary tract of the fish will be derived from the gut microflora. Thus, cellulose is usually considered to be almost indigestible for fish. On the other hand, fish do produce and secrete starch-digesting enzymes – amylases – but there are considerable interspecific differences between fish in the ability to digest and utilize starches. Different forms of treatment will also affect the degree to which a fish can digest and absorb these long-chain polysaccharides. For example, only about 25–50% of raw starch is digested and absorbed by salmonids. Steam heating of the starch granules causes gelatinization and digestibility increases to about 50–75%. In nutritional studies, digestibility is defined as the amount of a nutrient ingested minus that lost in the faeces, with digestibility usually being expressed in percentage terms. Complete gelatinization of starch, which may occur during the heat treatment of the extrusion process, will further increase digestibility.

Alginates, which are components of the cell walls of kelp and other algae, possess good binding properties and so are often employed as binding agents in fish feeds. Alginates, and other binders such as carboxymethylcellulose, gums, lignins and bentonites, are frequently used as binding agents in wet and moist feeds. In dry feeds, gelatinized starch is usually the binding agent of choice.

Bentonites are aluminosilicate minerals, or clays, of varying composition that have been quite widely used as binders in the manufacture of animal feeds. In some studies with fish, the inclusion of sodium bentonite in the diet, at levels of 5–10% on a weight basis, has been reported to lead to improvements in both growth rates and feed conversion, but results from other studies have failed to confirm this. These differences between studies may be related to the high cation exchange and adsorption capacities of bentonites, which imparts the ability to bind to any aflatoxins that may be present in the feed. Aflatoxins are highly toxic, carcinogenic feed contaminants produced by *Aspergillus* spp. fungi, and such fungi may develop on feed ingredients such as corn, cottonseed and other plant-derived meals. The toxins produced will not be uniformly distributed in a feed batch, and whilst some grains, kernels, pellets or feed particles may be heavily contaminated, others will be completely free of toxin. Only very low levels of contamination may be required in order for there to be deleterious effects on growth performance, and clinical signs, such as hepatomas, may be induced in rainbow trout fed on diets containing aflatoxin concentrations of 1 $ng.g^{-1}$ or less.

The inclusion of bentonites and related aluminosilicate minerals in diets has been reported to reduce the deleterious effects that aflatoxins have on the growth performance of pigs and poultry, so the positive effects reported following the inclusion of bentonites in fish feeds may be related to the ability of the mineral to bind the fungal toxins. However, the fact that bentonites have a high adsorptive capacity does not necessarily imply that the inclusion of these aluminosilicates in feed formulations will be universally beneficial. Net improvement may be minimal if vitamins, trace minerals and other essential nutrients are also adsorbed and bound to the bentonites, making them unavailable to the fish.

Steam pelleting, utilizing compression, produces a dense food pellet that sinks quite rapidly. The steam pelleting process involves the use of moisture, heat and pressure to cause agglomeration of the feed ingredients into larger homogeneous particles. Correct treatment conditions and ingredient composition are important if firmly bound particles are to be produced. Feed ingredients should be finely ground prior to steam pelleting, and the inclusion of a certain amount of starch from grains is vital in order to ensure good pellet binding. On the other hand, the inclusion of fibrous material and lipids is detrimental to binding, and the amount of lipid included in the pellet mix should not exceed 7%. Consequently, additional lipid must be sprayed onto the feed after pelleting, and in this way, pellet lipid levels of 16–20% can be achieved.

Steam is added to the finely ground dry feed ingredients to increase the moisture content of the mixture to 15–20%, and to raise the temperature to 70–85 °C. The increase in temperature causes partial gelatinization of the starch, which, in turn, aids in the binding together of the feed ingredients. The mixture is then passed through a pellet die, and the moisture content is reduced by forcing air over the warm pellets immediately after they leave the pelleting machine. Compressed, steam-pelleted feeds must be carefully processed to achieve satisfactory water stability and prevent disintegration.

The production of extruded pellets requires more elaborate equipment and the use of more energy for processing than does steam pelleting, so extruded pellets are more expensive to produce than are compressed steam pellets. In the extrusion process the feed ingredients are moistened, precooked, extruded and dried. The pellets produced are of relatively low density. Depending upon production conditions, extrusion processes can result in either floating or slowly sinking pellets.

Usually, the finely ground feed ingredients are pretreated with either steam or water and the resulting mixture, containing 20–25% moisture, may or may not be precooked before being transferred to the extruder barrel. The temperature of the compacted mixture rapidly increases to 105–150 °C under pressure and this causes almost complete gelatinization of the starch. As the mixture is forced out of the extrusion barrel, and the pellet is subjected to reduced pressure, part of the moisture in the hot mixture vaporizes and this causes expansion of the feed pellets. These pellets contain more moisture than compressed pellets and external heating is required in order to reduce the moisture content to a final level of about 10%. Thus, after extrusion, the pellets are usually passed through a drying tunnel.

One advantage of extruded pellets is that they can soak up oil sprayed onto the surface, and lipid contents of 20% and over can be achieved. In addition, the complete gelatinization of the starch produces a firmly bound pellet with good water stability, and there is a lower loss of dust and fines than is the case with steam-pelleted feeds.

Many fish farmers use dry feeds, but some farmers prefer to rely on wet- or moist-pellet feeds manufactured on the farm. Wet- and moist-pellet feeds are made by combining wet and dry meal mixes in different proportions, ranging from 90:10 to 50:50 (wet:dry). The wet ingredients will usually consist of ground fresh or frozen fish (capelin, argentine and small gadoids), squid and shrimp/prawn waste, fish-processing waste and/or acid-preserved hydrolysed fish waste (silage). The dry mixes contain fishmeal, cooked starch, vitamin and mineral premixes and alginate binder. When the proportion of wet material is high (80–90%), the dry mix must contain an effective binder, and alginate (0.5–2.0%) is often the binding agent of choice.

Many commercial feed manufacturers now provide complete dry mixes, containing binder, vitamins and minerals, which when added to wet ingredients in fixed proportions (10%, 25% or 40%) will produce a pellet having reasonable water stability. Thus the farmer is spared the effort of manufacturing the dry mix, but the onus is on the farmer to select and blend wet materials that ensure that the fish are provided with a nutritionally adequate diet. These wet- and moist-pellet feeds are made at the fish-rearing site and should be used within a day or two of manufacture.

3.3 PROTEIN SOURCES

Commercial feed manufacturers endeavour to formulate feeds that contain protein sources to fully meet the essential amino acid requirements of the fish. Feeds containing high levels of good-quality fishmeal will usually fulfil the needs, but most feeds also contain varying proportions of other protein sources of single-cell, animal or plant origin (Table 3.2).

Fishmeals and oils are produced from industrial fish species (Table 3.3), with yields and compositions of the finished products being influenced both by the species and by the seasonal variations in compositions of the fish used as raw materials. Several types of fishmeal are produced, and it is increasingly common that 'special quality meals' (e.g. NorSeaMink, LT-meal) are used in the manufacture of feeds for carnivorous species, such as salmonids. Special quality meals can be produced from a variety of fish

Table 3.2 Digestibility of the protein and energy contents of a number of ingredients commonly used as protein sources in the manufacture of dry feeds for a range of fish species. Digestibility trials were conducted using rainbow trout, *Oncorhynchus mykiss*, as the test species

Foodstuff	Digestibility (%)	
	Protein	Energy
Fishmeals		
Anchovy	85	91
Herring	87	92
Animal by-product meals		
Bloodmeal	69	70
Meat meal	75	72
Poultry by-products	70	66
Plant products		
Wheat middlings	68	53
Alfalfa meal	61	45
Soybean meal	75	75
Torula yeast	82	73

Table 3.3 Chemical compositions of a variety of 'industrial' fish species commonly used in the manufacture of fish meals and oils

	Percentage composition			
Species	Water	Protein	Lipid	Ash
'Winter' capelin, *Mallotus villosus*	76.6	13.5	8.4	1.9
'Summer' capelin, *M. villosus*	72.2	12.4	14.6	1.7
Brisling, *Sprattus sprattus*	68.5	16.1	13.0	2.7
Sandeel, *Ammodytes marinus*	72.0	16.6	9.5	2.2
Mackerel, *Scomber scombrus*	62.7	17.4	20.6	1.9
Blue whiting, *Micromesistius poutassou*	76.3	16.7	3.5	3.5
Norway pout, *Trisopterus esmarkii*	73.0	16.3	8.5	2.7
Polar cod, *Boreogadus saida*	76.9	12.9	6.4	2.6

species, but there are standards required to be met with regard to transport and preservation of the raw materials, and with respect to production procedures. Raw materials destined for the production of special quality meals should be transported under cold storage, without the addition of chemical preservatives, and should be delivered as fresh as possible. Some autolysis and deterioration, measured as an increase in volatile nitrogen (mostly present as trimethylamine), is permitted. Levels of volatile nitrogen should not exceed 90 mgN and 50 mgN per 100 g in raw material used in the production of NorSeaMink and LT-meal respectively (an upper limit of 30 mgN per 100 g is normally imposed on fish for human consumption).

Fishmeals are produced by 'cooking' the raw materials at a temperature of 70–95 °C, followed by pressing, separation and drying. During the 1970s, the LT- (low temperature) process was developed. This allows cooking both at lower temperatures (50–60 °C) and for a shorter duration than in the traditional process. The LT-process, nevertheless, gives improved separation of the oil and water fractions, and the milder treatment results in the production of meals of improved quality when compared with those produced using traditional methods. Fishmeals of 'special quality' will usually contain 5–10% moisture, up to 10% lipid and at least 70% protein. Antioxidant is added to the finished product to improve storage properties.

Whilst the drying processes which result in the production of fishmeal are convenient for preserving industrial fish, other methods of preservation of fish and fish waste are also available. Commercial fish farmers who manufacture their own wet or moist feeds on site may store wet ingredients in frozen block form. A freeze-storage facility is quite expensive to run, and the length of time the ingredients can be held under such conditions without undergoing deterioration is limited to a few weeks or, at most, months. An alternative method of preserving fish and fish waste is by means of treatment with acid. When fish is acid-preserved prior to hydrolysis, the product is known as fish silage.

Production of fish silage involves the thorough mixing of the fish raw material with the acid preservative to create conditions unsuitable for the growth of spoilage bacteria and other micro-organisms. Either mineral (e.g. sulphuric) or organic (e.g. propionic, formic) acids can be used as preservative agents so long as the pH of the ensiled mixture remains low (a pH of approx. 4 is usual), and antifungal agents are added. Use of mixtures of propionic and formic acids allows the preservation of fish waste at a pH of about 4.5, and using these acids the pH of the silage remains stable for prolonged periods of time. The silage produced using these organic acids has a fluid consistency that enables easy pumping and tank transport of the ensiled mass.

As an alternative to the direct use of inorganic or organic acids in the production of silage, the fermentative capabilities of lactic acid bacteria may be exploited. When bacterial fermentation is used in the production of silage a carbohydrate source, such as molasses or dextrose, is added as a substrate and then the mixture is seeded with a culture of lactic acid bacteria. It is obvious that the success of this method is dependent upon the addition of sufficient carbohydrate substrate and bacterial culture to enable fermentation to proceed rapidly. Rapid fermentation is required to ensure that the pH of the ensiled mixture falls quickly to levels that inhibit the establishment and proliferation of spoilage micro-organisms. Addition of about 5% by weight of carbohydrate and 1% by weight of bacterial culture to the raw ingredients to be ensiled appears to lead to rapid fermentation accompanied by a decline in pH, and this leads to a hindrance of the development of spoilage bacteria that could result in putrefaction. There may be a number of advantages in producing microbially fermented silages, compared with the production of silage by direct addition of acids. Firstly, cultures of lactic acid bacteria are relatively easy to maintain and the fermentation process can be controlled to ensure that the batches of silage produced have similar physical and nutritional properties. Another advantage of microbially fermented silages is that there is no need for the producer to handle concentrated acids, because it is the fermentation of the carbohydrate substrate to lactic acid *in situ* which results in the acidification of the ensiled mixture. Finally, use of bacterial fermentation allows silage to be produced without the addition of large quantities of extraneous liquid to the raw ingredients.

During acid preservation, the proteolytic enzymes present in the fish or fish waste hydrolyse the proteins, resulting in a liquid product. The ensiled mass will tend to form three fractions: an aqueous fraction containing the hydrolysed proteins, a lipid fraction and a bottom sediment. This product can be used as it is, the different fractions may be separated, or the silage may be concentrated by partial drying to remove some of the water content.

Fish silage is usually mixed with other ingredients, such as fishmeal, vitamins, minerals and binder, before being fed to the fish in the form of a moist pellet. Whilst ensiling provides a cheap and effective method for the preservation of industrial fish and the waste from fish-processing plants, there may be some limitations to the use of fish silage in the production of feeds. Firstly, because the silage is liquid, a number of dry ingredients must be added to form a moist pellet having reasonable properties of water stability. Secondly, the process of acid preservation can lead to the destruction of some essential nutrients, notably the amino acid methionine. Finally, the acids added to the ensiled mixture may affect palatability of the feed, leading to reduced acceptance by the fish.

In areas where there is a ready supply of silage produced from fish viscera and offcuts, it may be deemed desirable to dry the product prior to inclusion in animal feeds. In such cases the wet silage can be mixed with a range of other protein sources such as fishmeal, animal by-products meal, soybean or canola meal, or various other oilseed meals before being dried.

Oilseed meals (derived from peanut, sunflower and cottonseed) are deficient in the amino acids lysine and methionine, so are not particularly suitable as the major protein source in a fish feed. Soybean and canola meals have better essential amino acid profiles than other oilseed meals, and soybean meal is frequently included as a major protein source in dry feeds for species such as channel catfish, grass carp, *Ctenopharyngodon idella*, and milkfish. Whilst the essential amino acid profiles of soybean and canola meals may be satisfactory, other factors may limit the levels at which these meals can be included in feeds for some carnivorous species, such as salmonids. High levels of soybean meal appear to affect the palatability of salmonid feeds, leading to reduced food consumption and poorer growth.

Soybeans contain several antinutritional factors, including protease inhibitors, lectins, goitrogens, antivitamins, phytates, saponins, various oligosaccharides and antigenic proteins (allergens). Some, such as the trypsin inhibitors, may be destroyed or inactivated by the heating and drying procedures involved in the production of soybean meal. These treatment procedures may, however, in themselves, influence the digestibility of the final product (Table 3.4), and considerable care is required in the choice of processing methods. Some antinutritional factors, such as phytic acid and the oligosaccharides raffinose and stachyose, are, however, less affected by the normal processing procedures. Phytic acid can both reduce the availability of minerals and reduce the protein digestibility

Table 3.4 Effects of different forms of processing and heat treatment on the digestibility of soybean meal. Digestibility trials were conducted using rainbow trout, *Oncorhynchus mykiss*, as the test species

Treatment	Digestibility coefficient (%)	
	Protein	Energy
Heat treatment		
127 °C 10 min	40	51
175 °C 10 min	70	72
204 °C 10 min	78	77
234 °C 10 min	72	70
Steam processing		
0.35 kg cm^{-2} 5 min	37	42
0.70 kg cm^{-2} 10 min	80	75
1.05 kg cm^{-2} 10 min	75	69

of the diets, and the presence of high levels of oligosaccharides may also have some negative effects upon digestibility. Thus, the potential for the use of conventional soybean meal in diets for some carnivorous species of fish may be somewhat limited, and soybean protein concentrate might be expected to be a better alternative. The use of protein concentrates surmounts many of the problems associated with the presence of the antinutritional factors, and provides the fish with a source of protein having an essential amino acid profile that, with the exception of methionine, would seem to fulfil the known requirements for growth. Protein concentrates may, however, still contain proteinaceous allergens, such as glycinin, that can give rise to mucosal inflammation and thereby result in major disturbances in the digestive processes.

Thus, although soybean may be a valuable source of protein, its full nutritional potential may only be attained following processing to remove, denature or inactivate the various antinutritional factors present. Consequently, much effort has been expended in devising processing techniques for improving the nutritional value of soybeans, and modern processing techniques may employ a range of chemical, enzymatic and physical treatments. Conventional soybean meal is prepared by hexane extraction and steam treatment, although extractions using other solvents, such as ethanol, may also be incorporated into the manufacturing process. Treatments of soybeans with alkali and various enzyme preparations have also been tested in an attempt to ameliorate the effects of the antinutritional factors, and recent findings suggest that once the trypsin inhibitors and allergens are removed the nutritional value of soybean meal is very similar to that of high-quality fishmeal.

Meals derived from grains and grain by-products are generally rich in starch, and some grains are included in commercial feeds to improve binding properties. The proteins present in most grains are, however, poorly balanced and are deficient in lysine. It is now common practice for grains to be partially pre-cooked or extruded before they are added to the remaining ingredients of a commercial fish feed. This pretreatment serves to increase the gelatinization of the starch, leading to improved binding properties, and also increasing the biological availability of the carbohydrate. Pretreatment may not, however, result only in positive effects, because starch and non-reducing sugars, such as sucrose, present in the grains may be hydrolysed to form reducing sugars. Reducing sugars (glucose, fructose, lactose, maltose) may react with a free amino group on an amino acid and this chemical reaction can have important nutritional consequences. The essential amino acid lysine is particularly prone to this type of reaction, known as non-enzymatic browning (Maillard reaction), so that considerable care must be taken in order to keep lysine losses to a minimum during the course of pretreatment and extrusion processes.

Meals derived from various animal by-products may often have a relatively high ash content, and digestibility is often lower than that of fishmeal (Table 3.2). Thus, availability of the amino acids may be poorer than a chemical analysis would suggest. Animal by-product meals are often rich sources of the essential amino acids lysine and histidine, but they may be deficient in the sulphur-amino acids.

The published values for the amino acid compositions of ingredients and feeds are based upon chemical analyses of the proteins. The results of these analyses may suggest that the essential amino acid requirement is met, but a chemical analysis will not reveal to what extent the different nutrients and amino acids are available to the fish. There is relatively little information about the availabilities of specific amino acids constituting the various protein sources, but it is known that protein digestibilities vary widely from feedstuff to feedstuff (Tables 3.2 and 3.4). When estimating availability of nutrients and amino acids to fish, the type of processing and storage of feed ingredients must also be taken into consideration. For example, heat treatment (95 °C), freezing and drying can result in reduced availability of sulphur-amino acids, and storage of protein mixtures at 37 °C for 40 days has been reported to lead to an 80% reduction in lysine availability.

3.4 LIPID SOURCES

Some lipids must be added to the diet in order to meet the requirement of the fish for certain essential fatty acids, but lipids are also cheap sources of respiratory substrates that can be used in energy metabolism. Many different lipids are used in formulated feeds for fish species, including lipids of both plant and animal origin, and those from both terrestrial and aquatic environments. Terrestrial lipids of animal origin, such as pork fat and beef tallow, will tend to contain high levels of saturated fatty acids, and are poor sources of essential fatty acids of the (n−3) and (n−6) series. Lipids derived from the seeds of a number of plant species, such as soya and sunflower, contain quite high levels of unsaturated fatty acids of the (n−6) series, but the best sources of the essential (n−3) series fatty acids are lipids of marine origin. All of these lipids contain fatty acids with carbon chain lengths of 14 or more, and all can be used as substrates in respiratory metabolism. There is also evidence that short-chain acids, 6-C to 12-C, can be used effectively as respiratory substrates by some fish species, such as tilapias and ayu, *Plecoglossus altivelis*, so that dietary inclusion of small amounts of these short-chain acids may have beneficial effects upon growth and food conversion.

The interest in lipid nutrition of fish has not only been concerned with the investigation of lipids as energy sources and in ensuring that the diet provides sufficient essential fatty acids for the promotion of good rates of growth of fish under commercial farming conditions. The possible advantages of fish lipids in human nutrition have attracted increasing attention, as has the realization that there are potential marketing benefits in producing farmed fish with fatty acid profiles similar to those of their wild-caught counterparts. There is an increasing amount of evidence indicating that there are links between nutrition and certain human diseases, e.g. the association between high-fat diets, rich in saturates, and the incidence of heart disease. The facts that fish flesh has a high protein content, and has lipids that are rich in monounsaturated and polyunsaturated fatty acids, have been used in nutritional campaigns advocating the increased consumption of fish products at the expense of animal flesh that has a higher content of saturated fatty acids. The long-chain HUFAs, particularly 20:5(n−3) and 22:6(n−3), are deemed to be especially important for human health, with respect to aiding both in the prevention of heart disease and in the reduction of inflammatory disorders. Fish are the only sources of these HUFAs that are regularly consumed by humans.

The long-chain HUFAs of both the (n−3) and (n−6) fatty acid series, 20:5(n−3) and 20:4(n−6), can serve as precursors of eicosanoids, which are a group of highly active biological compounds. The number of biologically active eicosanoids is now known to exceed 20, and includes prostaglandins, thromboxanes, prostacyclins and leukotrienes. The (n−3) eicosanoids, derived from 20:5(n−3), have homologues derived from 20:4(n−6) with which they compete for receptor sites in the various tissues. There is also interaction between the (n−3) and (n−6) fatty acids during biosynthesis of the eicosanoids, with, for example, the presence of (n−3) fatty acids leading to a competetive attenuation of the rate of (n−6) eicosanoid formation.

Upon stimulation of a tissue the formation of the (n−6) eicosanoids from 20:4(n−6) often occurs as a rapid burst, and when rates of synthesis are very rapid, the excess active eicosanoids can result in pathophysiology. The presence of (n−3) fatty acids leads to reduced rates of (n−6) eicosanoid synthesis, and the (n−3) eicosanoids are both less rapidly synthesized and appear to give a less vigorous response than the (n−6) eicosanoids when bound to specific receptors. Thus, 20:5(n−3) appears to be a partial agonist of 20:4(n−6) both at the level of eicosanoid biosynthesis and at the level of the intensity of receptor activation induced by the eicosanoids formed.

Excessive physiological responses to eicosanoid signalling are associated with human health disorders and pathologies, including thrombosis (e.g. heart attacks and ischemic strokes) and immune–inflammatory disease (e.g. arthritis and asthma). For example, thromboses appear to be linked to an explosive biosynthesis of the (n−6) eicosanoid, thromboxane, and any substance that can reduce the rate of biosynthesis of (n−6) thromboxane would be expected to have beneficial effects. The ingestion of relatively high levels of dietary (n−3) HUFAs has been shown to lead to reduced thromboxane formation, and would therefore be expected to lead to long-term health benefits.

The participation of (n−6) eicosanoids in mediating a range of major human disease processes means that the risks of disease would be expected to become higher as 20:4(n−6) makes up increasingly greater proportions of the HUFA. Diets rich in the (n−6), but poor in the (n−3), series fatty acids will result in 20:4(n−6) being the major HUFA, but a change in the diet to include (n−3) fatty acids would decrease the proportion of (n−6) eicosanoid precursors in the HUFAs. Consequently, a dietary input of (n−3) fatty acids is viewed as being of considerable importance with respect to the maintenance of good health in humans.

The lipids present in the edible portion (muscle or fillet) of fish will be a mixture of phospholipids and storage triacylglycerols, the relative proportions of each depending upon the species. For example, fish such as gadoids, cyprinids and tilapias store comparatively little of their energy reserves in the form of triacylglycerols in the muscle, so most of their muscle lipids will be structural phospholipids. On the other hand, species such as salmonids, clupeids and scombrids store considerable amounts of lipid in the muscle tissue, so that the majority of the muscle lipid in these species will be storage triacylglycerols. The fatty acid profiles of fish triacylglycerols tend to reflect the composition of the diet, and because wild fish consume the natural prey occurring in marine and freshwater environments, the fish fatty acids will be similar to those of their prey. Prey organisms in aquatic environments tend to have relatively high contents of (n−3) fatty acids, so these fatty acids will be well represented in the muscle lipids of the fish. The 20 and 22-carbon fatty acids may account for 15–20% of the total fatty acids in the muscle lipids of wild fish. The fatty acid profiles of the fillets of farmed fish will be governed by the composition of the diet provided, and the dietary lipids will be of particular importance in determining the fillet fatty acid composition in those species, such as salmonids, that store considerable quantities of triacylglycerols in the muscle tissue. Farmed fish that have been fed diets in which plant oils and animal fats have been used as the main lipid sources will have triacylglycerols high in 18:2(n−6), monounsaturated and saturated fatty acids, which is obviously unfavourable from a human health and marketing point of view. On the other hand, if the farmed fish are fed diets formulated to include marine fish oils, and lipids from aquatic organisms (e.g. squid oil) as the major lipid sources, the fillet of the fish would be expected to contain relatively high levels of long-chain HUFAs.

3.5 VITAMIN DESTRUCTION AND DIETARY ADDITIONS

Vitamin requirements and nutrition have attracted considerable attention. This is because the consequences of vitamin deficiency can be severe despite the fact that vitamins are required in small doses and make up only a very minor proportion of the diet on a weight basis. Typical non-specific deficiency symptoms include loss of appetite, growth depression and increased susceptibility to disease. More specific symptoms indicative of particular vitamin deficiencies encompass various metabolic disorders, damage to eyes, gills, muscle and connective tissues, increased fragility of red blood cells and anaemia, which eventually lead to increased mortality if the deficiencies are not remedied.

Problems arise for the manufacturers of fish feeds because most of the vitamins are extremely labile organic compounds that are easily oxidized and destroyed by relatively mild treatments, e.g. exposure to heat, light, or humidity (Table 3.5). Some of the ingredients used in feeds fed to farmed fish (e.g. by-catch fish, offal from the fish industry) may contain enzymes that lead to destruction of some of the vitamins. For example, the tissues of a number of fish species contain the enzyme thiaminase, and long-term storage of feeds manufactured using fish as a major ingredient could lead to the feed becoming deficient in thiamin. The thiaminase activity can, however, be destroyed by heating, so either cooking of the fish, or the mixing of the final dietary mixture immediately prior to feeding, would solve the problem.

Table 3.5 Factors affecting the stability of some vitamins. +, Stability affected, with there being some destruction of the vitamin; –, stability unaffected or only slightly affected

Vitamin	Heat	Light	Oxygen	Humidity	Acid	Alkali
Water soluble						
Thiamin	+	–	+	+	–	+
Riboflavin	–	+	–		–	+
Pyridoxine	+	+				
Pantothenic acid	+	–	–	+	+	+
Niacin	–	–	–	+	–	–
Folic acid	+	+	+		+	–
Cyanocobalamin	–	+	+		–	–
Biotin	+	–	–		+	+
Ascorbic acid (vitamin C)	+	+	+		–	+
Lipid soluble						
Vitamin A	+	+	+	+	+	–
Vitamin D	+	+	+		–	+
Vitamin E	+	+	+		–	+
Vitamin K	–	+	–		+	+

The instability of the vitamins means that there can be considerable losses of activity during the processing and storage of fish feeds. For example, the manufacture of feed pellets by steam pelleting and extrusion procedures can lead to the destruction of certain heat-sensitive vitamins. Ascorbic acid (vitamin C) is particularly heat-sensitive, and 25% or more may be destroyed in the steam pelleting process. Manufacture of pellets by the extrusion process can lead to an ascorbic acid loss of 50% or more, and destruction of vitamin A and the B-complex vitamins thiamin, riboflavin, niacin and pyridoxine can amount to 30–40%. There may be considerable oxidation of ascorbic acid during processing, but some of the oxidation products retain vitamin C activity. Ascorbic acid is readily oxidized to dehydroascorbic acid during processing and storage. This compound may, however, be absorbed and reconverted to ascorbic acid in the tissues of the fish. Thus, dehydroascorbic acid retains vitamin activity and can be used as a vitamin C source by fish species.

In addition to destruction of vitamins during processing, further losses may occur during the course of long-term storage, so there is a real danger that manufactured feed pellets could become deficient in one or more of the vitamins. The possible risks of this occurring are reduced by the overfortification of the diets with vitamins, and the vitamins may also be applied to the exterior of the pellets after they have undergone the destructive procedures of steam pelleting or extrusion.

Problems with destruction of the vitamins during processing and storage, leading to considerable uncertainties about the exact levels of vitamin activities present in feed mixtures, have led to attention being directed towards the improvement of stability of the vitamins. The problem has been investigated by two different approaches. One approach involves the testing of relatively stable vitamin derivatives in the hope that a stable compound having similar properties to the unstable vitamin will be found, and relatively stable derivatives of most of the vitamins are now regularly used in the manufacture of most feeds. The alternative approach involves the examination of the efficacy of a number of coatings in the protection of the vitamin against destruction.

Both approaches have been used in an attempt to reduce the losses of the highly unstable, water-soluble vitamin ascorbic acid that occur during processing and storage. A range of coatings, including ethyl cellulose, lipids and synthetic polymers, have been tested, with varying degrees of success. There is evidence to suggest that polymer coating may serve to protect and improve the stability of the ascorbic acid without there being any loss of biological activity. Some 20% or so of the coated ascorbic acid may be lost during processing, but 50% of the biological activity may remain even after 3–4 months storage at room temperature.

Considerable attention has been directed towards the examination of the biological activities of stable ascorbic acid derivatives, particularly ascorbate-2-sulphate and ascorbate-2-monophosphate. Although ascorbate-2-sulphate is much more resistant to destruction than ascorbic acid, the biological activity of this derivative appears to be low, possibly owing to a limited absorption of this vitamin-derivative from the gastrointestinal tract of the fish. On the other hand, the ascorbate-2-monophosphate derivative has been shown to have a biological activity similar to ascorbic acid, and it is also relatively resistant to destruction during processing and storage. Other phosphate derivatives of ascorbic acid also appear both to be biologically available and to show vitamin properties, so these compounds would seem to be viable substitutes for ascorbic acid in the manufacture of fish feeds.

The efficacy of the phosphate derivatives of ascorbic acid as sources of vitamin C activity may be related to the presence of phosphatase enzymes in the tissues of the alimentary tract and other body organs. Because the alkaline phosphatase activity of the brush border membrane of the intestine is high and the enzyme shows low substrate specificity, there may be considerable hydrolysis of the phosphate derivatives to ascorbic acid before entry into the blood stream. On the other hand, acid phosphatase is a ubiquitous enzyme in blood and in the cell lysosomes, so any of the phosphate derivatives that are absorbed intact from the gut might be rapidly hydrolysed to ascorbic acid once they enter the blood.

3.6 MISCELLANEOUS FEED COMPONENTS

In addition to the feed ingredients that either contain essential nutrients, or act as sources of respiratory substrate, commercial fish feeds may contain other additives of various types. These additives may include antioxidants, pigments, antifungal agents, flavourings and attractants, and various drugs or antibiotics.

Antioxidants

Oxidation of fatty acids, particularly the long-chained polyunsaturates, present in the lipids can cause reduction of the nutritional value of the feed, and antioxidants are added to prevent lipid oxidation. Oxidation of unsaturated fatty acids results in the feed going rancid, with the development of off-flavours and odours. The oxidation products may be toxic to the fish, especially if the feed is low or deficient in vitamin E and selenium. Vitamin E can act as a natural antioxidant, but the synthetic antioxidants BHA (butylated hydroxyanisole), BHT (butylated hydroxytoluene) and ethoxyquin are usually included in formulated feeds at levels of about 0.015%.

Pigments

In certain countries the colour red has auspicious connotations, and red-coloured products are particularly highly prized. For some species of fish, such as the sea bream, tilapias and salmonids, a red body or flesh colour is viewed as being highly attractive and such coloration is often taken as being indicative of a 'quality product'. These types of fish and fishery products may be eaten on special feast days and are also very desirable as gifts. Consequently, fish and fishery products that have a bright red or pink hue may fetch higher market prices than equivalent products that have lower levels of pigmentation and are paler in colour.

The pink or red colour of the flesh of salmonids is a characteristic feature of this group of fishes. This coloration is due to absorption and deposition of carotenoid pigments. Carotenoids are the most numerous and widespread pigments to occur in nature, but only plants and protists are able to synthesize carotenoids. Salmonids, like other animals, do not synthesize carotenoids *de novo,* nor are salmonids capable of extensive oxygenation of carotenoids, and oxygenated carotenoids are often deposited with only limited modification. In wild salmonids the carotenoids are present as a mixture of astaxanthin and its esters, these carotenoids being obtained, primarily, by the consumption of zooplankton and crustaceans. These carotenoid sources are not generally available to farmed fish. Thus, in order to get the colour the consumer associates with the flesh of the salmonid group of fishes, a pigment source must be provided in the diet. The flesh of farmed salmonids will usually be found to contain some astaxanthin and its esters, but the pigment picture may be more complex for farmed than for wild salmonids because a wide range of feed ingredients, containing a variety of pigments, may be used in formulating feeds. In addition, alternative forms of carotenoids, such as canthaxanthin, have been added directly to formulated feeds as a pigment source for farmed salmonids.

Carotenoid supplementation of diets is achieved by adding either pigment-containing natural products, extracts of natural products or chemically synthesized pigments. The pigment sources most often used in wet and moist feeds are zooplankton, krill and wastes from shrimp/prawn/crab processing. Synthetic pigments, predominantly canthaxanthin, have hitherto usually been added to commercially produced dry feeds, but there is increasing use of astaxanthin in these feeds. Certain algae and yeasts, such as *Phaffia*, contain astaxanthin pigments and some species have been investigated as being potential carotenoid sources for farmed salmonids. Selection experiments carried out with *Phaffia* yeasts have led to the development of strains which are particularly rich in pigments. The exact pigment concentrations and bioavailability of the yeast carotenoids are, however, markedly influenced by factors such as the growth conditions, harvesting and processing of the *Phaffia*. Nevertheless, the pigments of plant and micro-organism origin may have considerable potential as carotenoid sources for fish.

In order to meet the consumer preference for red coloration, salmonid flesh should contain at least 4 mg pigment per kg. Because the carotenoids may fade in colour during processing and storage, compensation must be made for this by producing farmed salmon with flesh carotenoid concentrations elevated above the minimum acceptable level. Pigments supplied in the diet are, however, relatively poorly retained (2–10%) in the flesh of salmonids, and pigments are included at levels of 40–100 mg per kg dry feed. Since the cost of carotenoid pigments is high, dietary inclusion at these levels leads to pigments representing a considerable proportion of the overall cost of the feed.

Antimicrobial agents

Unless stored frozen, diets containing 15% moisture or more are subject to attack by bacteria and moulds. Microbial growth can be rapid at room temperature, and a number of moulds produce substances that are toxic to fish. If moist feeds are stored frozen, or are fed almost immediately after manufacture, there is little danger of microbial deterioration. If longer-term storage is anticipated, antimicrobial agents must be added to the feed to prevent spoilage. Many compounds are used in the feed industry to inhibit fungal and microbial growth, and these include benzoates, propionates and formic acid.

Flavourings and attractants

Even though a dietary formulation may appear to be nutritionally complete and well balanced, the diet will not produce the desired growth of the fish if palatability is poor. Fish appear to be very sensitive to certain tastes in their food, and can lose appetite for several days if diet is changed abruptly. Squid and prawn/shrimp have been reported to stimulate the appetite of a range of marine species and salmonids, and inclusion of high levels of fishmeal is considered to improve feed acceptance in many fish species.

Attempts to isolate the active ingredients which stimulate appetite have not yet been completely successful. Research on feeding attractants has been concentrated on the investigation of the attractant effects of mixtures of amino acids and other organic substances (betaine-glycine, inosine, urea and trimethylamine) that are found in the body fluids and tissues of natural prey species. In some cases, certain amino acids, such as alanine and arginine, together with betaine-glycine have proved to be effective stimulants, but in other cases the addition of mixtures of this type seems to have had little effect on food acceptance. Certain derivatives of inosine are commonly used as flavour enhancers in the preparation of foods for human consumption, and there are some reports that the addition of inosine monophosphate, nucleotides and nucleosides to feed mixtures may promote increased appetite of the fish under some circumstances. Tertiary sulphonium compounds, such as 2-carboxyethyl dimethyl sulphonium bromide (DMPT), have also been reported to have an appetite- and growth-stimulating effect in some species, but the modes of action of these feed additives remain unknown. In some cases, the addition of feeding stimulants to diets has resulted in there being an increase in appetite by the fish, but in others an improved growth response has been recorded despite little or no stimulation of appetite. In the latter instance the improvement in growth, following the addition of the stimulants to the diet, must have been brought about either through improvements in digestive and absorptive efficiency or by induction of changes in metabolic enzyme systems.

Drugs and antibiotics

Drugs and antibiotics have been exploited as feed additives in the livestock industry owing to their growth-promoting properties, but compounds of this type are not routinely included in fish feeds. Medicated feeds are, however, used to treat and mitigate disease.

Drugs and antibiotics will usually be obtained on prescription, and medicated feeds will be given to the fish under the supervision of a veterinarian. The majority of disease-causing bacteria are sensitive to a variety of antibiotics, sulpha-drugs and nitrofurans, but because many of these drugs are used in human medicine and resistant strains of bacteria may develop, it is vital that such compounds not be used indiscriminately. Drugs may also be added to feeds in order to treat fish that are heavily infected with intestinal parasites, such as tapeworms.

Exogenous enzyme preparations

In recent years it has become increasingly common to add crude enzyme mixtures to animal feeds formulated for livestock such as poultry and pigs. The enzymes are added to the feeds either to improve nutrient availability or to destroy antinutritional factors that may be present in some of the feed ingredients. The crude extracts added to the feed may contain a range of different enzymes, including β-glucanases, pentosanases, amylases and phytases. The majority of these enzymes are carbohydrases that can degrade the polysaccharides present in plant cell walls, and thereby improve the quality of diets formulated to contain cereals and other feed ingredients of plant origin.

The main cell wall polysaccharides present in barley and oats are β-glucans, whereas pentosans occur in the cell walls of wheat and rye. When diets containing these cereals are fed to livestock, there may be a danger that the presence of the refractory polysaccharides leads to reduced nutrient availability. There may also be digestive disturbances which give rise to symptoms associated with malabsorption. The inclusion of enzyme preparations in the feed may remedy matters, particularly in cases where diets have been formulated to contain barley and rye.

Efforts have been made to improve the digestibility of a range of dietary components by the addition of exogenous enzymes to the diet in order to augment endogenous enzyme activity. Crude preparations of proteases, amylases and oligosaccharidases have been added to livestock feeds, and some feeding trials have been carried out with fish to test the effects of adding exogenous proteases and amylases to the diet. The results of the feeding trials with fish have shown that enzyme addition generally leads to little or no improvement in either weight gain or efficiency of feed utilization in the fish fed the supplemented diets, so the advantages to be gained by adding exogenous proteases and amylases to fish feeds may be minimal.

Failure to digest and absorb the short-chain carbohydrates, such as raffinose and stachyose, present in the diet may have serious consequences, because the presence of such oligosaccharides in the digesta may lead to marked increases in the rates of growth of the microbial populations present in the hindmost parts of the gut. Raffinose and stachyose are α-galactosides that are present at quite high levels in a range of plants such as lupins and legumes (peas and beans). These carbohydrates are present in soya, but much of the oligosaccharide content may be removed by ethanol extraction during the preparation of soybean meal. Nevertheless, the addition of α-galactosidase enzyme to the diet may be expected to be beneficial when feeds have been formulated to include high proportions of legumes or soybean meal. Enzymes of this type are, for example, marketed for inclusion in pet foods intended for consumption by cats and dogs, both of which are carnivorous species by nature.

The phytate present in feed ingredients of plant origin, such as canola meal and rice bran, may have negative effects upon nutrient availability via binding with calcium, zinc and protein. Since phytate is also potentially a major source of phosphorus, the supplementation of diets with phytase may not only lead to an increase in the availability of calcium, zinc and dietary protein, but may also result in the release of phosphorus in a utilizable form.

Mineral mixtures

The majority of raw ingredients used in the manufacture of feeds used on fish farms will contain sufficient minerals without there being any need for additional supplementation. Nevertheless, mineral deficiency or imbalance symptoms can occur. This will usually be the result of problems with mineral bioavailability, rather than being due to low levels of mineral inclusion in the diet. Changes in mineral bioavailability may arise as a result of the binding of minerals by a variety of food components, inhibition of uptake from the gut owing to competition between different minerals for transport sites, or the presentation of the mineral in a form that cannot be utilized by the fish. Thus, a mineral premix is often added to commercially produced feeds, leading to overfortification of the diet with minerals.

Chapter four

Nutrient interactions and fish health

4.1 INTRODUCTION

Incorrect formulation of commercial feeds, or the destruction of nutrients during processing and storage, can result in dietary deficiencies that lead to the fish displaying disease symptoms. Under some circumstances, however, deficiency symptoms may appear even when the feed appears to contain adequate levels of a given nutrient. In such cases it may be suspected that interactions between nutrients are affecting bioavailability. Interactions between different nutrients may, however, affect bioavailability and metabolism in both a positive and a negative fashion.

4.2 INTERACTIONS BETWEEN NUTRIENTS

Factors leading to reduced nutrient bioavailability can result in disturbances to physiological processes and in extreme cases can lead to the development of deficiency symptoms. Bioavailability may, for example, be reduced because the binding of nutrients hinders digestive and absorptive processes or because competition between closely related nutrients for intestinal transport sites leads to reduced uptake. Interactions can, however, also result in there being stimulatory effects on absorption from the gut, such as in the case of the absorption of the lipid-soluble vitamins along with the fatty acids and monoacylglycerols.

Amino acid interactions

There are known to be four major transport systems for amino acids in the intestine, these being classified as:

1. a system transporting most neutral amino acids;
2. a system specific for the transport of basic amino acids and cystine;
3. a system transporting acidic amino acids;
4. a system transporting imino acids (proline and hydroxyproline), glycine and related molecules (e.g. betaine).

From this it follows that several amino acids may be transported by the same system, leading to the possibility of competition between amino acids for binding to transport sites. For example, methionine, valine and threonine have transport systems in common, and absorption of methionine can occur to the detriment of transport of valine and threonine. Some amino acids can be transported by more than one system and, in addition, there are known to be independent transport systems for some di- and tri-peptides, with absorption of the di- and tri-peptides being more effective than that of the free amino acids.

Many observations over a long period have provided evidence that amino acid mixtures simulating proteins may be nutritionally inferior to the proteins themselves, and it has been suggested that the superiority of a protein to an equivalent amino acid mixture could be related to temporal factors in the absorption of the amino acids. It has long been recognized that simultaneous presentation of amino acids to the tissues is essential for effective protein synthesis. There is also evidence indicating that during the absorption of amino acid mixtures, presentation of the amino acids to the tissues is less simultaneous than when the products of protein or protein partial hydrolysate digestion are absorbed. This is probably due to the fact that when amino acid mixtures are presented, there may be competition between amino acids for intestinal transport sites and the different free amino acids are absorbed from mixtures at very different rates. Thus, when free amino acids are absorbed from mixtures, some reach early and high peak concentrations in the blood, whilst rates of absorption and increases in blood concentrations of other amino acids tend to lag behind. This could lead to an imbalance in the supply of amino acids to the tissues and, in turn, to inefficiencies in protein synthesis.

On the other hand, there are a multiplicity of transport systems available for the absorption of the products (e.g. free amino acids, di- and tri-peptides, larger peptides) of whole protein digestion. Thus, both the temporal displacement occurring during the absorption of free amino acids from mixtures, and any disturbance to tissue uptake, might be largely or completely obviated when amino acids are presented in the form of whole proteins, rather than free amino acid mixtures. Thus, a 'deranged' pattern of absorption of amino acids into the body seems to be, at least partially, responsible for the nutritional inferiority of amino acid mixtures when compared with the same amino acids presented in peptide or whole-protein form.

It is, therefore, open to question whether studies based upon the feeding of mixtures of free, crystalline amino acids to the fish can provide a true reflection of the dietary requirements for essential amino acids. Furthermore, it should not be reasonable to expect that additions of free amino acids will always prove to be completely effective in compensating for deficiencies when 'poor quality' proteins are used in the formulation of fish feeds.

Vitamin–vitamin interactions

The vitamin folic acid (or folacin) is relatively abundant in natural feedstuffs and may also be synthesized by the intestinal microflora. Folic acid is converted into its active form tetrahydrofolic acid in the liver, with the active form of the vitamin being required for the transfer of methyl groups ($—CH_3$) between organic molecules. It plays a role in the synthesis of purines, pyrimidines, glycine, serine and creatine. Purines are required for the synthesis of nucleic acids (DNA and RNA), and are also necessary for the normal formation of red blood cells. Thus, a deficiency of folic acid may lead both to the fish becoming anaemic and to abnormalities in the red blood cells.

In many respects the roles of the vitamin cyanocobalamin seem to complement those of folic acid. Cyanocobalamin is, for example, involved in the synthesis of labile methyl groups, and is also required for the synthesis of nucleic acids. This vitamin is needed for the development and maturation of red blood cells, and also plays a role in the recycling of tetrahydrofolic acid. Thus, the two vitamins folic acid and cyanocobalamin may closely interact, and they may be able to substitute for each other in some metabolic functions.

Vitamin–mineral interactions

Vitamin D is a general term used to describe several lipid-soluble sterol derivatives. Of these, ergocalciferol (found in plants) and cholecalciferol (found in animals) are the most important from a nutritional standpoint. Dietary vitamin D is absorbed from the intestine along with the lipids and is transported to the liver where it is converted to 25-hydroxycalciferol. Conversion to the active form, 1,25-dihydroxycalciferol, probably occurs in the kidney tissue. The primary function of vitamin D appears to be related to the absorption of calcium from the intestine, with the vitamin being required for the synthesis of a calcium-binding protein. Some vitamin D metabolites appear to be actively involved in the endocrine regulation of calcium uptake and metabolism, with dietary deficiencies of this vitamin resulting in serious disturbances of calcium metabolism. Vitamin D may also be involved in the metabolism of inorganic phosphate, and be required for the effective functioning of alkaline phosphatase. Thus, vitamin D appears to be intimately linked to calcium and phosphate metabolism, and a deficiency of this vitamin may lead to disturbances in mineral metabolism.

Several other vitamin–mineral interactions exist, most notably those involving the influences of ascorbic acid upon mineral bioavailability. Ascorbic acid has been reported to affect both the toxicity and tissue retention of water-borne copper, but apparently has few effects upon the uptake and metabolism of dietary copper. Ascorbic acid also has a central role to play in the metabolism of iron, with fish fed diets deficient in vitamin C becoming anaemic even though tissue iron stores, and dietary iron levels, would appear to be adequate.

Mineral–mineral interactions

The bioavailability of some minerals can be adversely affected by diet composition, with increased levels of dietary calcium and phytate seeming to have detrimental effects on availability of divalent ions. For example, it seems that a calcium–zinc–phytate complex forms in the intestinal tract of the fish and this reduces bioavailability of the zinc. High levels of phosphorus in the diet may also have negative effects, possibly owing to the formation of an insoluble precipitate between zinc, calcium and phosphorus that would reduce or prevent absorption from the intestine.

Thus, problems with mineral bioavailability may arise when diets are formulated with protein sources, such as soybean meal, meat-and-bone meal and some fishmeals, that contain either phytic acid or quite high levels of calcium phosphate. For example, it has been suggested that the tricalcium phosphate, $Ca_3(PO_4)_2$, present in some fishmeals may inhibit the availability of dietary zinc to salmonids and channel catfish, leading to poor growth, cataracts and skeletal deformities. Whilst the effects of high levels of dietary tricalcium phosphate do not seem to be as severe in carp as in salmonids and channel catfish, mineral bioavailability is reduced. The inclusion of high levels of tricalcium phosphate (7% by weight) in the diet has been shown to lead to a slight reduction in the absorption of magnesium, and pronounced reductions in the absorption of manganese (from 70–80% to 15–20%) and zinc (from 85–90% to 45–55%). In addition to reducing the availability of other minerals, any dietary calcium and phosphate provided in the form of the tricalcium phosphate salt will be almost unavailable to the fish.

Interactions involving several nutrient classes

Because there are a number of factors that can influence oxidative destruction of dietary ingredients, and the oxidative damage to fish tissues, there may be complex interactions between the requirements for the 'natural antioxidants' and dietary composition. Lipid peroxidation is a chemical process that results in the oxidative deterioration of polyunsaturated fatty acids, and thereby will lead to destructive changes in both the dietary lipids and the lipids of the cell membranes of the fish tissues. Fish fed diets rich in polyunsaturates, and depositing large quantities of these fatty acids in their tissues, may be expected to be most at risk of suffering damage owing to this form of oxidative deterioration. Iron is one of the metals primarily involved in the induction of lipid oxidation, and ferrous iron is a more potent catalyst of lipid peroxidation than ferric iron. Thus, a combination of high dietary levels of polyunsaturates coupled with high concentrations of ferrous iron might be expected to have deleterious effects on the stability of the dietary lipids and upon the growth, health and survival of the fish.

In biological systems, lipid peroxidation can be prevented, or at least hindered, by several complementary antioxidant systems (Fig. 4.1). The lipid peroxidation chain reaction is initiated by the removal of a hydrogen atom from a polyunsaturated fatty acid, giving rise to a lipid radical. In the presence of oxygen there will be a rapid reaction leading to the formation of a hydroperoxide radical. Hydroperoxide radicals, in turn, remove hydrogen atoms from other fatty acids, leading to a continuation of the chain reaction. Protection of the lipids against peroxidation can be provided by two broad classes of antioxidants – the preventive and the chain-breaking antioxidants. Preventive antioxidants reduce the rate at which lipid radicals are formed, i.e. they hinder the initiation of the lipid peroxidation chain reaction. The most important preventive antioxidants are enzymes, such as catalase and glutathione peroxidase, that exert their effects via reduction of the hydroperoxide radicals. Glutathione peroxidase activity is dependent upon adequate supplies of selenium, a micronutrient usually obtained in the diet. The chain-breaking antioxidants owe their antioxidant activity to their ability to bind with and trap hydroperoxide radicals. Ascorbic acid (vitamin C) is an important water-soluble chain-breaking antioxidant, and the most important lipid-soluble antioxidants are the tocopherols (vitamin E and related compounds). The antioxidant properties of ascorbic acid and vitamin E appear to be synergistic, with the ascorbic acid seeming to play a major role in the regeneration of vitamin E rather than acting directly to trap the hydroperoxide radicals. Thus, in this instance there is evidence of a series of interactions between selenium, vitamin E and vitamin C. The requirement for these contributors to

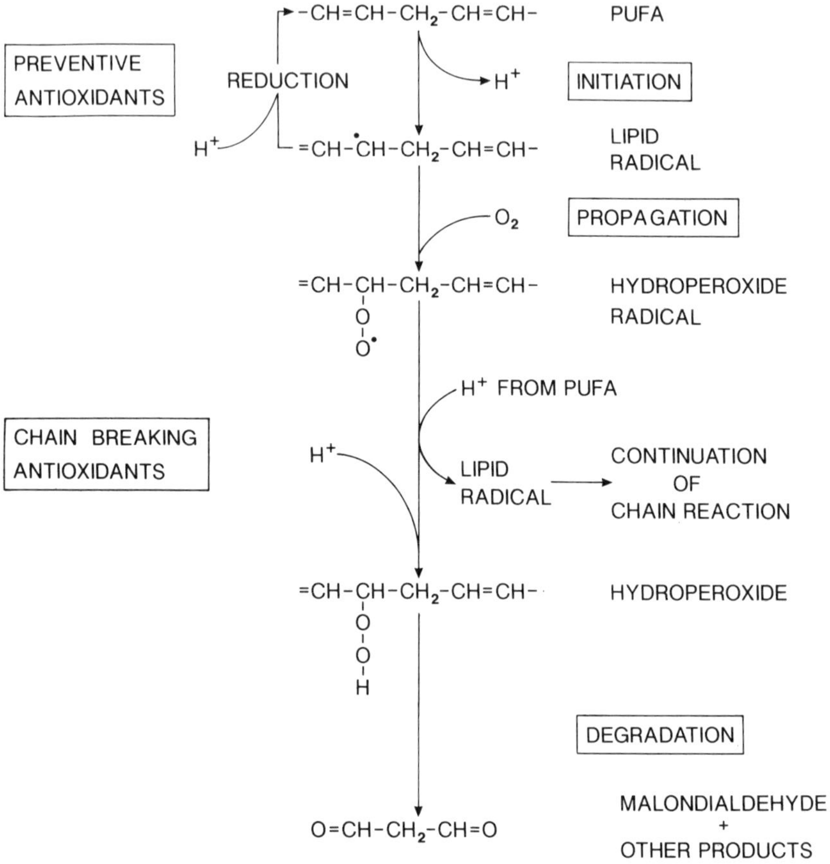

Fig. 4.1 Destruction of polyunsaturated fatty acids (PUFAs) by the lipid peroxidation pathway. Note the points of intervention of the two types of natural antioxidants – the preventive antioxidants and the chain-breaking antioxidants.

natural antioxidant activity will, in turn, depend upon the levels of polyunsaturates at risk from peroxidation and the level of exposure of these fatty acids to initiators of the peroxidation reaction, e.g. ferrous iron, heavy metals and various organic pollutants.

4.3 NUTRITION AND DISEASE

The earliest studies of the nutritional influences upon disease in fish were related to the description of deficiency symptoms, especially those associated with vitamin and mineral deficiencies. Vitamin and mineral nutrition attracted considerable attention because, despite the fact that these micronutrients are required in small doses, the consequences of deficiency can be severe. Typical non-specific deficiency symptoms include loss of appetite and growth depression. More specific symptoms indicative of particular micronutrient deficiencies encompass various metabolic disorders, cataracts, damage to gills, muscle and connective tissue, increased fragility of red blood cells and anaemia. These will lead to increased mortality if the deficiencies are not remedied.

There is an increasing body of evidence demonstrating that very mild forms of deficiency, which may not be reflected in reduced growth or the development of specific pathological symptoms, can have detrimental effects upon the fish if they are exposed to poor environmental conditions or other potential stressors. In other words, whilst diets for farmed fish have primarily been formulated in order to promote good rates of growth, there is an increasing awareness that diet composition can affect the general health of the fish and influence susceptibility to disease.

Under commercial farming conditions the fish may be subjected to a series of procedures, such as netting, grading and transport, that are potential stressors. Many other factors, related to both the abiotic and biotic environments, e.g. pollutants, poor water quality and high stocking densities, may elicit stress responses in fish, and if there is chronic exposure to the stressor the health and survival of the fish may be compromised. Exposure of fish to environmental stressors induces a characteristic series of responses mediated via the hypothalamic–pituitary–interrenal (HPI) axis (Fig. 4.2). Adverse stimuli (stressors) result in the almost immediate release of two types of endocrine factor – catecholamines and the corticosteroid, cortisol. The catecholamines initiate a series of short-term physiological changes, whereas the effects of cortisol tend to be more long-lasting. Should the exposure to the stressor be of limited duration, recovery will generally occur within the space of a few hours, but chronic exposure to stressors may have serious long-term effects on the fish. From a nutritional point of view it is known that exposure of fish to stressors, such as pollutants and poor water quality, leads to a rapid depletion of tissue reserves of ascorbic acid (vitamin C), such that the dietary requirements are elevated in fish held under poor environmental conditions.

Vitamins

Vitamin C has long been implicated in disease resistance in homeothermic animals, and during the 1970s it was observed that channel catfish fed diets deficient in ascorbic acid (vitamin C) were more susceptible to infection by *Aeromonas hydrophilia* than were fish fed on vitamin-supplemented

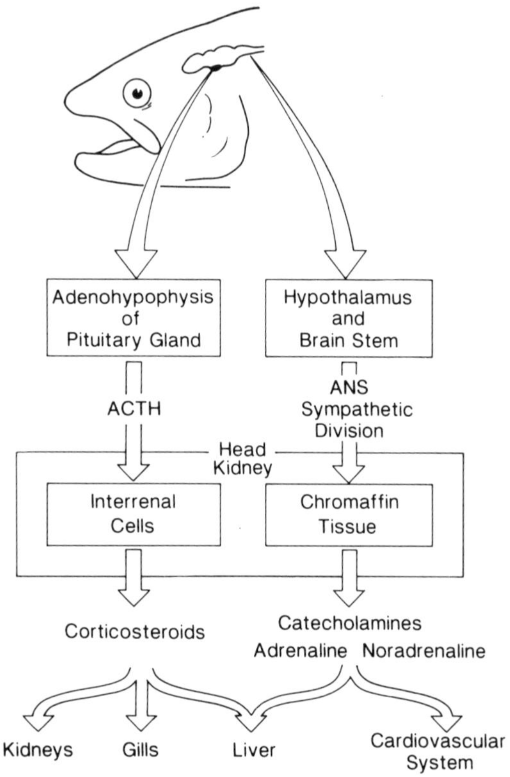

Fig. 4.2 The hypothalamic–pituitary–interrenal (HPI) axis in fish, illustrating nervous and endocrine links between the central sites of integration and the periphery. ACTH, adrenocorticotropic hormone; ANS, autonomic nervous system.

diets. Since that time there has accumulated a considerable body of evidence that the feeding of megadoses, 5–20 times the growth requirement, of vitamin C to channel catfish and salmonids can contribute to increased resistance against a range of bacterial diseases, such as vibrios, diseases caused by *Edwardsiella* and *Aeromonas* species, and infection with *Renibacterium salmoninarum*, which gives rise to bacterial kidney disease (BKD). Megadoses of ascorbic acid may also reduce the mortality rates of salmonids exposed to protozoan parasites, such as 'white spot', *Ichthyophthirius multifiliis*, and vitamin C may also contribute to the resistance shown to viral disease.

Other vitamins, such as the tocopherols (vitamin E), retinol (vitamin A) and pyridoxine, have also been implicated in disease resistance in fish, but the possible contributions of these vitamins have been much less studied than have the effects of ascorbic acid. Vitamin A supplementation of the diet has been shown to decrease the susceptibility of animals to a variety of infections. Available evidence indicates that at least some of this protection may be due to the enhancement of the activities of phagocytic cells, since both phagocytosis and intracellular killing are increased in animals fed diets supplemented with levels of vitamin A. The protective effects of vitamin E may also be due to effects on the phagocytic cells, because when fed at moderate doses, vitamin E has been shown to enhance the proliferation and bactericidal activity of phagocytes.

There is some evidence that the pyridoxine requirement of salmonids is dependent upon dietary protein level, because the feeding of high-protein diets results in the more rapid development of deficiency symptoms than does the feeding of low-protein diets to the fish. In pyridoxine-deficient fish, both the biosynthesis of nucleic acids and rates of protein synthesis are depressed, so it would be expected that pyridoxine deficiency would have profound effects upon disease resistance. Available evidence suggests that the supplementation of high-protein diets with additional pyridoxine may indeed increase the resistance of some salmonids to bacterial infections, but the mechanisms by which resistance is enhanced remain to be elucidated.

Minerals

Numerous minerals, including iron, selenium, copper and zinc are known to have important influences upon immune responses and disease resistance in homeotherms, but the effects of mineral nutrition on disease resistance in fish have been comparatively little studied. There is some evidence that iron may have immuno-regulatory properties, because antibody production is severely impaired in iron-deficient animals. On the other hand, overloading of the body tissues with iron may lead to increased susceptibility to disease, and the feeding of diets containing high levels of iron (200 or more mg iron per kg feed) has been shown to adversely affect the health of salmonids.

Macronutrients

To what extent the macronutrients influence disease resistance in fish is almost unknown, because very little research concerning the effects of protein and carbohydrate on disease resistance mechanisms has been conducted. There is, however, some evidence that different sources of macronutrients may affect the susceptibility of fish to disease. For example, it has been observed that outbreaks of bacterial kidney disease (BKD) led to high mortality rates in groups of fish fed diets containing corn gluten meal, whereas substituting cottonseed meal for the corn gluten meal seemed to confer protection. The incidence of BKD was not influenced by the source of dietary protein (cottonseed vs. corn gluten meal), but the non-specific stress of the chronic disease seemed to be increased in the fish that were fed corn gluten, and thus higher mortalities were observed when compared with fish that were fed diets containing cottonseed meal. That there was increased stress in the fish fed on diets containing corn gluten meal appeared to be confirmed by the finding of depleted tissue levels of ascorbic acid and depressed protein metabolism in these fish.

Stimulation of the immune system

Chronic exposure to environmental stressors increases the susceptibility of the fish to disease via immuno-suppressive effects of repeated or prolonged elevations of plasma cortisol levels. Recent work has been aimed at counteracting these negative effects through the enhancement of the immune response by the manipulation of dietary components. Vitamins, especially vitamins C and E, and fatty acids are considered to be the key dietary components having influences upon the immune system. These nutrients have been implicated in both cell-mediated and humoral immunity. Thus, whilst there is evidence to suggest that high dietary doses of vitamins C and E can lead to improved protection of the fish against infectious diseases under some conditions, much more work is still required in order to elucidate the potential nutritional links between the immune system and disease resistance.

An alternative approach to enhancing the disease resistance and overall health of the fish may involve the oral delivery of immune potentiators. Whilst these are not nutrients *per se*, some form of dietary manipulation may be required in order to prevent destruction and to ensure adequate absorption of these protective factors from the gut. Liposomes are bipolar layers that can be made to surround a drug or immunogen, and the size and fragility of these globular particles can be regulated by temperature and lipid bilayer characteristics. Thus, the inclusion of liposomes in diets may represent a novel form for drug presentation and oral vaccine delivery. Encapsulation of drugs and vaccines into small, acid-resistant beads may protect the contents from destruction in the stomach, and ensure delivery of the drug or vaccine to the intestine. Whilst the encapsulation of immunogens with acid-resistant coats may allow more intact material to reach the intestine, it may be necessary to include detergents, saponins and other soap-like substances in order to enhance absorption and uptake. Various bacterial peptides and yeast-wall glucans are considered to be promising candidates as immune potentiators, and trials to test the efficacy of these agents are being conducted in several research laboratories.

Chapter five

Fish nutrition – concluding summary

Fish are ectothermic animals and their energetic requirements for metabolism are much lower than those of endothermic vertebrates (e.g. birds, mammals) of similar size. The metabolic requirements can be met by the use of carbohydrates, lipids or proteins as respiratory substrates. There is no dietary requirement for carbohydrate, with compounds of this nutrient group being used solely as a metabolic energy source. A number of fish species appear to have a relatively limited ability to use carbohydrates as a respiratory substrate, and respiratory metabolism in fish appears to be, predominantly, based upon lipids and proteins.

Both lipids and proteins may be used as respiratory substrates, but there is a dietary requirement for certain fatty acids and amino acids if the fish are to survive and grow. The nutrient requirements have been investigated for no more than about 30–40 fish species, the majority of which are exploited as aquaculture species, e.g. salmonids, cyprinids, tilapias, ictalurids. Even for these species, however, nutritional information is far from complete.

The same amino acids have been found to be essential dietary constituents for all species examined, but whilst a qualitative requirement has been demonstrated, the quantitative requirements remain almost unknown.

There appear to be both qualitative and quantitative differences between fish species in their requirements for fatty acids. A number of species have a greater requirement for the fatty acids of the (n–3) series than for those of the (n–6) series. Other species, e.g. tilapias, may be able to utilize (n–6) fatty acids effectively, and they appear to have a reduced requirement for dietary (n–3) fatty acids. In addition, fish species differ in their abilities to transform and chain-elongate fatty acids, and these differences determine whether or not a particular fatty acid will display 'essential fatty acid' properties or not. For example, salmonids have the ability to desaturate and chain-elongate the shorter-chained 18-C fatty acids of the (n−6) and (n−3) series. Thus, these fatty acids can fulfil the function of essential fatty acids for salmonids. On the other hand, a number of marine species of fish are unable to transform the 18-C fatty acids to long-chained HUFAs, and the long-chained, highly unsaturated fatty acids must be obtained in the diet.

There is an essential requirement for certain vitamins and minerals. Whilst the absolute requirements for these micronutrients are small, a deficiency in one or more of these nutritional factors usually results in serious disruptions to metabolism and growth. Reliable quantitative information about micronutrient requirement is available for no more than a handful of fish species.

PART ONE: FURTHER READING

Books

Cowey, C.B., Mackie, A.M. and Bell, J.G. (eds)(1985) *Nutrition and Feeding in Fish*, Academic Press, London.

Halver, J.E. (ed.)(1989) *Fish Nutrition*, Academic Press, London.

Hara, T.J. (ed.)(1992) *Fish Chemoreception*, Chapman & Hall, London.

Wilson, R.P. (ed.)(1991) *Handbook of Nutrient Requirements of Finfish*, CRC Press, Boca Raton, FL.

Review articles and papers

Anderson, D.P. (1992) Immunostimulants, adjuvants and vaccine carriers in fish: Applications to aquaculture. *A. Rev. Fish Diseases*, **2**, 281–307.

Barton, B.A. and Iwama, G.K. (1991) Physiological changes in fish from stress in aquaculture with emphasis on the response and effects of corticosteroids. *A. Rev. Fish Diseases*, **1**, 3–26.

Blazer, V.S. (1992) Nutrition and disease resistance in fish. *A. Rev. Fish Diseases*, **2**, 309–23.

Camire, M.E., Camire, A. and Krumhar, K. (1990) Chemical and nutritional changes in foods during extrusion. *CRC Rev. Food Sci. Nutr.*, **29**, 35–57.

Campbell, G.L. and Bedford, M.R. (1992) Enzyme applications for monogastric feeds: A review. *Can. J. Anim. Sci.*, **72**, 449–66.

El-Sayed, A-F.M. and Teshima, S-I. (1991) Tilapia nutrition in aquaculture. *CRC Rev. Aquat. Sci.*, **5**, 247–65.

Evans, L.V. (1986) Seaweed bioproducts. *Sci. Progr.*, **70**, 287–303.

Greene, D.H.S. and Selivonchick, D.P. (1987) Lipid metabolism in fish. *Progr. Lipid Res.*, **26**, 53–85.

Henderson, R.J. and Tocher, D.R. (1987) The lipid composition and biochemistry of freshwater fish. *Progr. Lipid Res.*, **26**, 281–347.

Hilton, J.W. (1989) The interaction of vitamins, minerals and diet composition in the diet of fish. *Aquaculture*, **79**, 223–44.
Lands, W.E.M. (1992) Biochemistry and physiology of n-3 fatty acids. *FASEB J.*, **6**, 2530–6.
Pickering, A.D. (1993) Endocrine-induced pathology in stressed salmonid fish. *Fish. Res.*, **17**, 35–50.
Sandnes, K. (1991) Vitamin C in fish nutrition – A review. *FiskDir. Skr. Ser. Ernæring*, **4**, 3–32.
Sardesai, V.M. (1992) Nutritional role of polyunsaturated fatty acids. *J. Nutr. Biochem.*, **3**, 154–66.
Sardesai, V.M. (1992) Biochemical and nutritional aspects of eicosanoids. *J. Nutr. Biochem.*, **3**, 562–79.
Torrissen, O.J., Hardy, R.W. and Shearer, K.D. (1989) Pigmentation of salmonids – Carotenoid deposition and metabolism. *CRC Rev. Aquat. Sci.*, **1**, 209–25.

Part Two

Physiological Energetics: Feeding, Metabolism and Growth

Chapter six

Introduction to energetics

6.1 INTRODUCTION

The foundation for the study of fish bioenergetics and growth was laid during the late 1940s and early 1950s. The past four decades have seen a continued increase in research efforts devoted to the study of fish growth and bioenergetics. Interest has been generated both from fisheries management, and, not least, by the recent upsurge in the aquaculture industry.

The basic principle of bioenergetics is relatively simple to grasp, and can be stated as follows – all energy acquired through the ingestion of food is ultimately lost as wastes in faeces or by excretion, used in metabolic processes or deposited as new body tissue (growth or energy gain). Thus, bioenergetics is concerned with the study of rates of energy intake and transformations within the organism, and provides the physiological framework for the study of relationships between feeding rates and growth rates of fish subjected to different environmental conditions.

6.2 THE ENERGY BUDGET EQUATION

The study of bioenergetics involves the examination of energy gains, losses and transfers within the whole organism, leading to the partitioning of ingested energy into the major physiological components of the energy budget equation. In its simplest form the equation can be represented as:

$$E(\mathrm{In}) = E(\mathrm{Out}) + E(\mathrm{P}) \qquad (6.1)$$

where $E(\mathrm{In})$ is the energy ingested as food, $E(\mathrm{Out})$ represents energy losses and $E(\mathrm{P})$ is energy retained as production or growth. The equation is usually expanded to the more general form:

$$R = F + U + M + P \tag{6.2}$$

Of the energy gained as food (R) some is lost as faeces (F), and some is expended in providing energy for a range of bodily functions (metabolism) (M). Metabolism results in the production of carbon dioxide, water, waste heat, and in some cases intermediate partially oxidized products, which also represent a source of energy loss (U). If body mass is to be held constant the energy intake from external sources (food) (R) must equal the requirement for maintenance and activity, and when energy intake is in excess of these needs, growth or energy storage (P) can occur. If energy input from food is insufficient to cover these energetic requirements, previously acquired body stores are metabolized and the fish will lose weight. Metabolism (M) can be subdivided in order to account for the energy losses representing different physiological processes. Thus, M may be divided into the metabolic costs of maintaining basic bodily functions, metabolism related to activity, and the metabolic costs associated with digestion, absorption and processing of food. It is also usual to differentiate between energy retained (P) as somatic (body) growth (P_S) and energy which is channelled into the production of gametes (reproductive growth) (P_R).

Experimental studies are aimed at quantifying the different components of the energy budget equation in order to present as complete a picture as possible of the physiological transformations and pathways of energy partitioning occurring within the fish. In studies of fish energetics it is usual to express these energy transformations in terms of physiological rates, with the flow of materials being calculated in energy units (kJ). The sources of ingested energy are the lipids, proteins and carbohydrates present in the food, and each of these substrates yields a different amount of energy when metabolized. Complete oxidation of these substrates yields approximately 24 $kJ.g^{-1}$, 39 $kJ.g^{-1}$ and 17 $kJ.g^{-1}$ for protein, lipid and carbohydrate, respectively, although the energy yield from marine lipids (fish oils etc.) may be slightly lower than that from lipids of terrestrial animals.

The energy (heat) losses of fish, and other aquatic organisms, are difficult to measure using direct methods. In bioenergetics studies with fish it is, therefore, more usual to use determinations of oxygen consumption as an indirect measure of energy metabolism. Metabolic rates in terms of energy units can be estimated from rates of oxygen consumption using oxycalorific conversion factors. The conversion factor to be used will depend upon the type of respiratory substrate (lipid, protein, carbohydrate) being metabolized. Conversion factors are determined using information about the amounts of oxygen consumed and the amounts of carbon dioxide and heat released when each substrate is combusted in a bomb calorimeter. For example, the combustion of one mole of glucose gives the following thermochemical equation:

$$C_6H_{12}O_6 + 6O_2 = 6CO_2 + 6H_2O \tag{6.3}$$

with the release of 2833 kJ as heat. Thus, the combustion of one mole of glucose requires six moles of oxygen (192 g), so the amount of heat released is 2833/192 or 14.76 kJ per g oxygen consumed. In this reaction the ratio of carbon dioxide produced to oxygen consumed, a ratio called the respiratory quotient (RQ), is 1.00.

When fatty acids and lipids are oxidized the ratio of the amount of carbon dioxide produced to oxygen consumed differs from that for carbohydrate oxidation:

$$CH_3(CH_2)_{14}COOH + 23O_2 = 16CO_2 + 16H_2O \tag{6.4}$$

and the respiratory quotient for lipid oxidation is 0.7 (i.e. 16/23). The oxycalorific coefficient employed when lipids are metabolized is approximately 13.7 kJ per g oxygen consumed.

The situation for proteins is more complicated. When proteins are metabolized, there will be production of nitrogenous and other compounds, in addition to carbon dioxide and water. This is illustrated in the following example:

$$C_{43}H_{70}O_{14}N_{12}S_{0.3} + 44.5O_2 = 43CO_2 + 12NH_3 + 17H_2O + 0.3H_2S \tag{6.5}$$

The oxycalorific coefficient appropriate for use when proteins are metabolized will depend upon the type of nitrogenous excretory products (ammonia, urea, uric acid) resulting from protein catabolism (Table 6.1). The majority of fish species excrete ammonia as the predominant nitrogenous excretory product, and under such circumstances, the oxycalorific coefficient for protein will be approximately 13.36 kJ per g oxygen consumed. When protein is used as substrate, and the metabolic products are ammonia, carbon dioxide and water, the respiratory quotient will be approximately 0.95.

Thus, the respiratory quotient (RQ) gives information about the type of substrate being metabolized, but unfortunately an animal will rarely be respiring a single substrate to the exclusion of all others. Usually the RQ will be found to be between 0.7 and 1.0; an RQ close to 0.7 indicates that lipid predominates as the substrate, whereas an RQ near 1.0 suggests that carbohydrate is the major respiratory substrate. If determinations of oxygen consumption, carbon dioxide production and nitrogenous excretion are made, it is possible to calculate the amounts of protein, lipid and carbohydrate used as respiratory substrate, and thereby to provide an estimate of the composite oxycalorific coefficient. Fish rely, predominantly,

Table 6.1 Respiratory quotients (RQ) and oxycalorific coefficients for different respiratory substrates

Respiratory substrate	End products	RQ	Oxycalorific coefficient (J per mg O_2)
Carbohydrate	CO_2, H_2O	1.0	14.76
Lipids	CO_2, H_2O	0.7	13.72
Protein	CO_2, H_2O, ammonia	0.95	13.36
Protein	CO_2, H_2O, urea	0.84	13.60
Protein	CO_2, H_2O, uric acid	0.72	13.60

upon lipids and proteins as respiratory substrates, and the appropriate oxycalorific coefficient for use in estimating energy metabolism in fish species has been suggested to be close to 13.59 kJ per g oxygen consumed (which converts to 19.4 kJ per l oxygen consumed). The use of an oxycalorific coefficient to estimate energy expenditure requires that metabolism occurs aerobically, and if there is a contribution from anaerobic metabolism the use of oxycalorific coefficients will not give a realistic estimate of the overall energy expenditure. When metabolism is occurring both aerobically and anaerobically, the use of an oxycalorific coefficient will lead to underestimation of the level of energy expenditure.

In only very few studies have all the components of the energy budget been measured independently. It is usual to find that one or more of the major components has been estimated 'by difference' in order to produce a balanced budget. Because they form a balanced equation, determinations of any four of the five main components – ingestion (R), faecal loss (F), excretion (U), metabolism (M) and production (P) – allow the fifth to be estimated by difference. All errors associated with the determinations of the measured components will, however, become a pooled error in the component estimated by difference. Further problems may arise because it is not possible to determine all the components simultaneously. For example, production is ideally studied as the change in energy content in the fish body with time, but because the energy content can only be determined after the fish has been slaughtered, it is not possible to obtain information about the energy content of a particular fish both before the start of, and on completion of, an experiment. In this instance the researcher must rely on information obtained from sampling different individuals, and care must therefore be taken to ensure that the fish sampled before the start of the experiment are as similar as possible to those used in the experiment, with respect to genetic background, nutritional history and physiological status. Research is currently being directed at the problem of the estimation of the energy content of live fish, and methods being investigated include tomography, the measurement of electrical conductivity and the examination of the correlation between morphological characters and energy status.

Chapter seven

Ingestion, absorption and excretion

7.1 INTRODUCTION

Whilst ingestion (R) represents the input term in the energy budget equation and both faecal losses (F) and excretion (U) are debit terms, the processes involved in ingestion, absorption and excretion are so inexorably linked that they merit treatment as an integrated whole.

Ingestion refers to the act of taking materials into the body, so by strict definition the term could be used to encompass all the processes involved in feeding, digestion and the absorption of the digested nutrients. In the current work, however, ingestion will be defined in the more limited sense as being confined to food consumption (R).

Once ingested the food is digested, and the majority of the digested nutrients are absorbed over the gut wall into the bloodstream. Unabsorbed material is passed through the gut and defecated (F). Some workers opt to classify faecal losses as part of excretion, i.e. that which is discharged from the animal body, but the majority tend to apply the term excretion to the discharge of products resulting from post-absorptive metabolic processes. In the present context the term excretion will be used to refer to the discharge of the nitrogenous end products of protein metabolism (U).

7.2 FACTORS INFLUENCING INGESTION (R)

The majority of biological traits, including physiological rates, are size dependent. The relationship between a biological variable and body weight can usually be described by an equation of the form:

$$f(W) = a\, W^{b} \tag{7.1}$$

where the biological variable or function, $f(W)$, increases allometrically with increasing body weight, W. The terms a and b are constants, with b representing the exponent relating $f(W)$ to W (the scaling or weight exponent). The relationship between ingestion (food intake) and body weight can be described by an equation of this type, and the weight exponent, b(1), is almost invariably found to be less than 1 (Table 7.1). For the majority of fish species, maximum rates of ingestion have been found to scale in proportion to body weight raised to the power 0.6–0.8, and, as a rule of thumb, a weight exponent, b(1), of approximately 0.75 can be assumed.

Large fish consume more food than do small fish, but it is common practice to express food intake, or rates of ingestion, in relative terms, i.e. food intake per g body weight, or as percent body weight consumed per day. When ingestion is expressed in these terms it is clear that relative food intake declines with increasing body weight (Table 7.2), and relative food intake will scale in proportion to body weight raised to approximately −0.25.

Water temperature will have a major influence on the amount of food consumed by a fish. When fish are given access to an unlimited supply of food an increase in temperature will initially lead to increased rates of ingestion. Food consumption will then peak at some intermediate temperature before declining precipitously as the temperature approaches the upper thermal tolerance limit of the species (Fig. 7.1).

Table 7.1 Weight exponent in the relationship between body weight and food intake in a range of fish species. Information about size range and feeding methodology is also given

Species	Size range (g)	Feeding regime	Weight exponent
Sockeye salmon, *Oncorhynchus nerka*	4–216	To satiation, 3 times daily	0.657
Rainbow trout, *Oncorhynchus mykiss*	3–1300	To satiation, 3 times daily	0.830
Brown trout, *Salmo trutta*	5–300	To satiation, 4 times daily	0.761
Arctic charr, *Salvelinus alpinus*	10–700	*Ad libitum*	0.651
Largemouth bass, *Micropterus salmoides*	8–150	To satiation, twice daily	0.711
Sea bass, *Dicentrarchus labrax*	2–180	To satiation, 3 times daily	0.727
Plaice, *Pleuronectes platessa*	0.7–9.2	*Ad libitum*	0.697

Table 7.2 Food intake of different sizes of brown trout, *Salmo trutta*, at two water temperatures. Food intake data are expressed in either *absolute* (mg dry wt) or *relative* (% body wt) terms. Based on Elliott (1975), *Freshwater Biol.*, **5**, 287–303

Fish weight (g)	Food consumed per fish per day			
	10 °C		15 °C	
	mg dry wt	% BW	mg dry wt	% BW
5	64	5.3	135	11.3
10	108	4.5	230	9.6
50	367	3.1	791	6.6
100	621	2.6	1346	5.6
200	1068	2.2	2290	4.8
300	1429	2.0	3125	4.3

Rates of ingestion may be affected by levels of dissolved oxygen, with both food intake and growth being depressed as oxygen content of the water declines. For example, the appetite of rainbow trout, *Oncorhynchus mykiss*, appears to be suppressed once oxygen saturation falls below approximately 60%.

Measurements of food consumption are often characterized by large inter- and intra-individual variations. When fish are fed in groups, social interactions and the establishment of dominance hierarchies can lead to the suppression of food intake and growth of some individuals even though the fish appear to have been fed to satiation or in excess. The effects of such interactions are, however, more marked when food supply

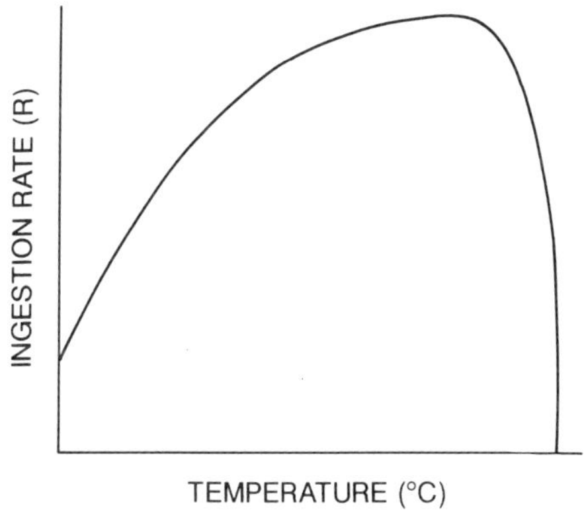

Fig. 7.1 Influence of water temperature on rates of ingestion of fish.

is restricted or the food resource is defensible. Under such conditions the socially dominant individuals may secure first access to the food resource and consume greater amounts of food than subordinate fish.

The fact that social interactions may have a marked influence on food intake has been implied from the results of experiments carried out on rainbow trout fed according to different restricted feeding regimes (0.5, 1.0 or 1.6% of tank biomass). Food was provided for 60 min each day and the amounts of food consumed by each fish in the group were recorded on a number of occasions. This allowed both interindividual and intra-individual variations in food intake with time to be examined. For fish fed according to the most restricted feeding regime, those fish which ate most consumed over six times as much food as the fish which ate least. Moreover, the fish which ate most displayed relatively little day-to-day variation in food intake, apparently being able to secure adequate amounts of food during each daily 60 min feeding bout. On the other hand, the fish which, overall, consumed small amounts of food showed marked day-to-day variations in food intake, sometimes managing to consume quite large amounts, and on other days obtaining very little food at all. For groups of fish fed at the highest level, the fish which ate most consumed only about 2.5 times as much food as those which ate least. In addition, increasing the food supply led to a marked drop in the intra-individual variability in the amounts of food consumed on different days, with this reduction in variability being most marked for those fish which ate the least amount of food.

These results suggest the following interpretation. When food supplies are limited, some individuals, the social dominants, monopolize the resource and consume the majority of the food. These fish are able to consume adequate amounts of food each time it is presented. The remain-ing fish are prevented from obtaining an equal share of the food supply owing to the activities of the dominant individuals, and feeding by the subordinate individuals may be more opportunistic. On some occasions they may obtain a reasonable share of the amount of food supplied, but at other times they may obtain little or nothing at all. Thus, the overall 'average' food intake by the subordinates may be considerably lower than that of the dominants, and the day-to-day variability in the amount consumed will be large. As food supply is increased, the resource may be less strongly defended, or may become indefensible, which would give the subordinate individuals a greater opportunity to feed. Under such circum-stances the dominants would still obtain adequate supplies of food and show relatively little day-to-day variability, but there would be expected to be marked changes for the subordinates. An increased food supply would be expected to lead to an overall increase in the total amounts of food consumed by these fish, and their day-to-day variability in consumption would be expected to be reduced.

The effects of social interactions described above refer to a case in which the amounts of food supplied were manipulated, but social interactions may also come into play when the temporal distribution of a given food amount is manipulated. For example, it might be expected that if a given amount of food is provided at a point source over a prolonged period of time, then a single, or small number of individuals would be able to defend and monopolize the resource and thereby exclude the remaining members of the group. Under these circumstances one, or a few, fish would consume most of the food supplied and the remaining fish would obtain little or no food at all. If, however, the same amount of food were supplied over a very limited time span, then resource defence and monopolization would become more difficult and the food resources would be expected to be more evenly distributed amongst the individuals making up the entire fish group. This has been confirmed in laboratory studies carried out on groups of zebrafish, *Brachydanio rerio*, supplied with *Daphnia pulex* prey at a point source. When prey were supplied at a rate of 1 per min for a period of 5 h, one individual in the group of six fish tended to defend the feeding station and this individual consumed about 70% of the prey. When the same number of prey (300) were presented over much shorter time intervals (3 or 10 min), only about 30% of the prey were consumed by the dominant individuals and the majority of the fish within the group managed to obtain access to the food. When food was supplied during a very restricted time interval the dominant individuals were not always the most successful in obtaining food, but as the time over which food was supplied increased, then the more successful were the dominant fish in obtaining greater proportions of the total food supplied. Thus, resource defence and monopolization was possible when small amounts of food were provided at regular intervals over a protracted time period, but when there was a temporal 'clumping' of prey supply, the dominant individuals were unable to defend the food source and other fish within the group were able to obtain quite large proportions of the food supplied.

Suppression of food intake through the effects of social interactions is not, however, the only cause of the observed variability, because large day-to-day differences in consumption have been observed when individual fish have been studied. In most cases the variability in food consumption will be seen as a day or two of heavy feeding followed by days on which the fish feed little or not at all.

In addition to short-term (day-to-day) variability in consumption, there may also be longer-term variations in rates of ingestion that cannot be directly attributable to factors such as seasonal changes in water temperature. Juvenile Atlantic salmon, *Salmo salar*, for example, display seasonal differences in food intake that do not appear to be linked to changes in water temperature. In these fish, rates of food consumption may be greater at times of the year when daylength is increasing (spring) than when it is decreasing (autumn). Rates of ingestion (and growth) may, however, also show long-term variations even when the fish appear to have been kept under uniform environmental conditions (constant light and temperature). For sexually mature fish, variations in food intake may be linked to different stages in the reproductive cycle, with mature fish feeding little during the spawning season. Sexually immature fish also display long-term variability in food consumption and growth, with there being a two-to-threefold difference between peaks and troughs, so factors other than those linked to the reproductive cycle must be important. The potential roles of environmental (other than temperature) and endogenous factors on the long-term regulation of food consumption and growth of fishes have, however, been relatively little studied.

7.3 LOSSES IN FAECAL MATERIAL (F)

Part of the food consumed by the fish will pass through the gastrointestinal tract without being digested and absorbed. In other words, part of the ingested food will be lost as faeces. The faeces comprise not only unabsorbed food, but also mucus and cells sloughed from the walls of the gut, some digestive enzymes, some bile components and bacteria derived from the gut microflora. If it is assumed that mucus, dead cells, and bacteria make only a minor contribution to the total faecal material, then a faecal analysis can provide a reasonably accurate estimate of how much of the nutritional content of the food has been absorbed by the fish.

Absorption efficiency (AE), which is also known as digestive efficiency or the digestibility of food nutrients, is a measure of the proportion of the food energy, or nutrient, content absorbed by the fish. Absorption efficiency can, therefore, be defined as follows:

$$\mathrm{AE} = 100\,(R - F)/R \qquad (7.2)$$

where R is the energy, or nutrient, content of the food and F represents faecal losses. In this definition of absorption efficiency there is no attempt made to distinguish between faecal losses accruing from unabsorbed food and those derived from other sources, such as cellular breakdown products. Rarely, attempts are made to correct for the faecal components derived from non-food sources, and in such cases 'true' absorption efficiency is given by:

$$\text{'true' AE} = 100\,[R - (F - F^{\mathrm{P}})]/R \qquad (7.3)$$

where F^{P} is the non-food component of the faeces. Corrections for the faecal components derived from cellular and bacterial sources have most usually been made when the aim of an experiment has been the determination of the digestibilities of different protein sources in foods. Protein digestibility is usually determined as the efficiency with which the nitrogen content is absorbed. The faecal nitrogen component that is not derived from unabsorbed proteins in the food is known as metabolic faecal nitrogen. Metabolic faecal nitrogen usually amounts to approximately 100–200 mg N per 100 g dry diet consumed. Under most circumstances, failure to take account of metabolic faecal nitrogen will lead to the estimate of absorption efficiency being 2–3% points lower than the 'true' value. When diets of low protein content are fed, however, metabolic faecal nitrogen may constitute a major part of the total nitrogen content of the faeces. Under these circumstances the digestibility of the protein source may be considerably under-estimated if the contribution of the metabolic faecal component is not corrected for in the calculations of absorption efficiency.

Some authors have referred to absorption efficiency as *assimilation efficiency* (AsE), but the latter term is usually taken to represent the food energy or nutrients remaining after both faecal and nitrogenous losses have been accounted for. In other words, assimilation efficiency is usually defined as:

$$\text{AsE} = 100\,[R - (F + U)]/R \qquad (7.4)$$

where U is the loss via the products of nitrogenous excretion, and the other terms are as described above. Thus, care must be taken in choice of terminology if confusion is to be avoided.

Determination of absorption efficiency

A number of techniques have been developed in an attempt to determine the efficiency with which fish digest and absorb their food. These can be broadly divided into *direct* and *indirect* methods. The direct method requires both knowledge of the total amount of food consumed by the fish and the collection of all the faeces produced. This is usually impracticable, and most estimations of absorption efficiency have been made using indirect methods.

Indirect methods involve the inclusion of an inert marker in the diet, and absorption efficiency is calculated as:

$$\text{AE} = 100 - 100\,([XA/XB] \times [YB/YA]) \qquad (7.5)$$

where XA and XB are the concentrations of the inert marker in the food and faeces, respectively, YA is the nutrient concentration in the food and YB is the concentration of the nutrient in the faeces. Because the calculation of absorption efficiency is based upon the ratios of nutrient to marker in the food and faeces, total food consumption need not be known and there is no necessity to collect all the faeces produced.

In order for the indirect method to give accurate results a number of conditions must, however, be fulfilled. The marker used must be inert. It should not interfere with the normal processes of feeding, digestion and absorption, and should, itself, not be broken down and absorbed during passage through the gastrointestinal tract. The food containing the marker must be fed over a sufficiently long time period to enable representative sampling of faeces to be made, i.e. samples collected should be totally free of contamination by faeces produced as a result of consumption of previously unmarked food. An ideal marker would also be easily analysed in food and faeces, and the marker should be transported through the gut at the same rate as the food components.

Most of the markers used (chromic oxide, titanium dioxide, acid-insoluble ash) either are not normal components of the natural food of fishes or are present in such small quantity that they cannot be used for the estimation of the efficiency with which natural prey are digested and absorbed. Most studies have therefore been conducted with fish fed on artificially prepared diets to which the marker has been added.

Chromic oxide is the most commonly used marker and it is incorporated into the food at inclusion levels of 0.25–1.5% on a weight basis. The marker imparts a green colour to the food and this may present some acceptability problems with some species. Chromic oxide may also cause some disturbance to digestive function and there may be some degree of separation of the marker from the food during passage through the gut. Thus, it is advisable to pool samples of faeces collected at different times in an attempt to overcome potential problems caused by differential egestion of marker and the other faecal components.

Acid-insoluble ash has the advantage of being a natural component of the food, but in order for it to be used as a marker, the level of inclusion in the food should be 1% or over. It is, however, cheap and easy to analyse, but there may be some slight loss of this marker during the processes of digestion and absorption.

Cellulose, lignin, chlorophyll, ash and chitin are all natural products and they have been suggested as being suitable markers for use in digestibility studies conducted on various animal species. All of these components may, however, be digested and absorbed to some extent, and extreme caution should be exercised in the interpretation of the results obtained in experiments where these substances have been used as markers.

Assuming that a suitable marker can be found, there may be further problems associated with the type of sampling procedure adopted for the collection of faeces. Collection by siphoning or netting the faeces directly from the water is cheap and simple, but there may be leaching of some components from the faeces if they are left in the water for protracted periods of time. Loss of nutrients through leaching will obviously lead to an over-estimate of absorption efficiency. Several collection systems have been designed in order to ensure that the faecal contact with water, and risk of leaching, is kept to a minimum.

Faecal contact with the water can be excluded by collecting the faeces by anal suction, stripping or intestinal dissection. Intestinal dissection, which requires fish to be killed, is expensive and, unless care is taken to collect intestinal contents from only the hindmost part of the gut, the material may not be truly representative of the faeces. Both anal suction and stripping are best conducted on large fish, and both methods permit only periodic sampling of material. If faeces are collected by stripping there is the possibility of contamination with urine and surface mucus, and there is also the risk of including a small proportion of incompletely digested food in stripped samples. All of these contaminants would lead to an underestimation of absorption efficiency.

Some studies have been conducted in which comparisons have been made between the estimates of absorption efficiency obtained using different methods of faecal collection, and examples of the results of this type of study are given in Table 7.3.

Factors influencing faecal losses

How large a proportion of the nutrient content of the food is lost in the faeces is influenced by a wide range of factors, but food type and composition are particularly important. For carnivorous fish consuming natural prey, absorption efficiency will usually be found to be within the range 70–95%. Plant material is digested and absorbed much less efficiently,

Table 7.3 Effects of faecal collection methods on estimations of the digestibility (absorption efficiency) of different food components

Method	Digestibility (%)		
	Dry matter	Protein	Lipid
Dissection (removal of faeces from rectum)	80	90	97
Anal suction	79	90	97
Netting/siphoning*	84	95	97
Stripping†	73	78	62

*Leaching of nutrients from the faeces may give an over-estimation of digestibility.
†Contamination of faeces with incompletely digested food may have led to an underestimation of digestibility.

Table 7.4 Nutrient digestibilities (absorption efficiencies) for a number of natural food organisms consumed by fish species. Note that in ecological studies it is common to use the expression absorption efficiency, whereas in studies of fish nutrition it is more usual to refer to nutrient digestibility, or digestibility coefficient

Food type	Fish species	Absorption efficiency (%)	
		Protein	Energy
Detritus	Cichlids	~45	~40
Macroalgae	Gobiids, pomacentrids, cichlids, stichaeids	45–80	30–70
Aquatic macrophytes	Cichlids, cyprinids	60–80	30–60
Oligochaetes	Salmonids, percids, cyprinids, pleuronectids	80–95	70–85
Crustaceans	Salmonids, percids	~90	80–88
Fish	Percids, gadoids	85–98	80–97

and herbivorous fish may often absorb their food with an efficiency of 40–80% (Table 7.4). For farmed fish that are fed artificially formulated feeds, the composition of the diet will be of overriding importance for the efficiency with which it is digested and absorbed. Dietary composition and digestibility will be affected both by the choice of ingredients and by the processing to which the ingredients have been subjected (Table 7.5). These factors can result in large differences in digestibility, both between feed ingredients and between different types of complete feed formulations.

The efficiency with which food is digested and absorbed can also be influenced by feeding rate. In other words, an increase in the rate of ingestion may result in reduced absorption efficiency. For example, the energy loss in the faeces of brown trout, *Salmo trutta*, amounted to 25% of

Table 7.5 Nutrient digestibilities of a number of commonly used feed ingredients, for tilapia, *Oreochromis niloticus*, and channel catfish, *Ictalurus punctatus*

Feed ingredient	Digestibility coefficient (%)			
	Tilapia		Channel catfish	
	Protein	Energy	Protein	Energy
Fishmeal	87	80	87	85
Poultry by-products	74	59	74	67
Soya meal	91	57	77	56
Wheat middlings	76	58	84	60
Corn (maize) meal	83	76	66	59

Table 7.6 Digestibilities of different food components in small (S, 20 g) and large (L, 600 g) rainbow trout, *Oncorhynchus mykiss*, held at various water temperatures

Temp. (°C)	Digestibility (%)							
	Protein		Lipid		Carbohydrate		Energy	
	S	L	S	L	S	L	S	L
7	73	92	80	93	43	60	46	76
11	91	95	85	91	57	62	72	75
15	92	92	86	92	57	58	70	75

the ingested energy when fish were fed maximally on *Gammarus*, but for trout given a reduced food supply faecal losses were reduced to approximately 15% of ingested energy. Similarly, when fish are fed on formulated pellet feeds, an increase in ingestion leads to an increased nutrient loss in the faeces, and there may be a negative linear correlation between feeding rate and absorption efficiency. This effect of feeding rate on absorption efficiency will rarely be of major proportions, with efficiency decreasing by about 10% as feeding rates increase from low levels to maximum rates.

Absorption efficiency has also been reported to be influenced by factors such as water temperature, fish size and the type of feeding regime employed. In some studies, absorption efficiency has been reported to be reduced at low water temperature. For rainbow trout the effect of water temperature on absorption efficiency has been found to be dependent upon the size of fish used in the experiments, with small fish digesting and absorbing food less efficiently than larger fish at low temperatures (Table 7.6).

7.4 PRODUCTS OF NITROGENOUS EXCRETION (*U*)

When lipids and carbohydrates are used as respiratory substrates in aerobic metabolism, they are completely oxidized to carbon dioxide and water, but when proteins are utilized as an energy source, only the carbon chains of the constituent amino acids enter the energy production cycle. The nitrogen (amino) fractions of the amino acids that remain when protein is used as a respiratory substrate cannot be metabolized and must, therefore, be excreted. The chemical processes by which the amino groups are removed from the amino acids are known as transamination and deamination, and there are several enzymatic pathways that can lead to the liberation of ammonia from amino acids. Deamination also occurs when amino acids absorbed from the protein fraction of the food are used for the synthesis of energy-storage products such as lipids and glycogen.

The amino acid deamination pathways can be divided into two major groups. The first type of pathway is able to produce ammonia from a variety of amino acids by transfer of amino groups to a common acceptor, which then becomes the substrate for deamination. Indirect production of ammonia from a range of amino acids may occur either by transdeamination or via the process of purine nucleotide cycling. In transdeamination the amino group of the amino acid is transferred to ketoglutarate by enzymes known as amino-transferases, and the glutamate that results from this series of reactions is deaminated by the enzyme glutamate dehydrogenase. The process of purine nucleotide cycling involves the initial transfer of amino groups from amino acids to oxaloacetate, and this is accomplished by amino-transferase enzymes. The aspartate that is produced is then deaminated via fumarate and malate, by a series of enzymes that use purine nucleotides as intermediate catalysts (Fig. 7.2). The second type of pathways contain a number of specific enzymes, which produce ammonia from single amino acids by direct deamination. Of the different enzymatic pathways present in fish species, the glutamate transdeamination pathway is considered to be the most important.

Thus, deamination of amino acids leads to the release of amino groups that cannot be recycled through other metabolic processes. The amino groups must be excreted, and this excretion of amino nitrogen constitutes an energy loss for the fish. In the majority of teleost fish species, the excretion of waste nitrogen occurs, predominantly, at the gill surface, with the nitrogenous excretion across the gills usually amounting to 6–10 times that in the urine. Ammonia and ammonium are the main excretory products in fish and many other aquatic animals. The majority of the ammonia/ammonium is produced in the liver of the fish and is then transported in the blood system to the gills for excretion to the surrounding water. Some ammonia/ammonium is also produced in the gills, kidneys and muscles, with the production in the muscle tissue appearing to be greatest when the fish are either exposed to hypoxic conditions or are swimming at high speed.

Ammonia binds to a proton in aqueous solution to produce ammonium ions. The fraction of the 'total ammonia' (ammonia + ammonium) found as ammonia increases with increasing pH and temperature, but decreases in solutions of high ionic strength. There are a number of possible mechanisms of ammonia/ammonium excretion at the gills, and those that appear to be most important are passive diffusion and ion-exchange mechanisms. Ammonia/ammonium is very water soluble and both ammonia and ammonium ions exist in aqueous solutions at physiological pH values. Both forms may be expected to diffuse passively across biological membranes down their respective partial pressure gradients, so the diffusion of ammonia and ammonium ions may represent two independent, but parallel, pathways for nitrogenous excretion at the gills.

Ammonia is a small lipophilic molecule that can diffuse easily, and rapidly, across biological membranes. Diffusion will be aided by the fact that the ammonia concentration in the blood plasma will usually be several-fold (up to 40 ×) higher than that in the surrounding water. There is little doubt, therefore, that diffusion of ammonia represents a major pathway for the branchial excretion of waste nitrogen.

Ammonium ions, on the other hand, are larger than molecules of ammonia, and they also carry a positive charge. Biological membranes are relatively impermeable to cations, so the contribution made by the diffusion of ammonium ions to the total branchial excretion of nitrogenous waste products may be relatively low. It seems probable that the excretion of ammonium is strongly coupled to the movement of other cations, with ammonium serving as a counter ion in the uptake of sodium. According to this mechanism, a transepithelial exchange involving sodium uptake and ammonium ion efflux would aid the fish in performing ionic regulation and assist in nitrogenous excretion.

The relative contributions made by the different mechanisms to the branchial nitrogenous excretion are not known with any degree of certainty, but it has been suggested that the diffusion of ammonia may account for 60% or more of the efflux. Approximately 20% of the excretion may occur via ion-exchange mechanisms, and the remainder would be accounted for by the passive diffusion of ammonium ions across the branchial membrane.

The advantages of ammonia/ammonium over other products of nitrogenous excretion such as urea and uric acid, can be summarized as follows. The costs of synthesis are minimal compared with those required for the synthesis of urea and uric acid. Ammonia is a relatively small molecule which diffuses readily across the gill membrane from the blood to the surrounding water, and when waste nitrogen is excreted in the form of ammonium this ion can be exchanged for sodium, and thereby facilitate ion uptake.

Other end products of nitrogen metabolism, e.g. urea and creatine, are produced by fish, but generally in lesser quantity than ammonia/ammonium. These nitrogenous end products may be excreted either in the urine, through the skin or via the gills. It is not unusual for fish to excrete over 80% of their nitrogenous wastes as ammonia/ammonium, but the proportion of nitrogen excreted as urea and other metabolic end products may reach 30–40% in some marine species. Thus, the impression is gained that marine species tend to excrete a greater proportion of their

a)

Transdeamination

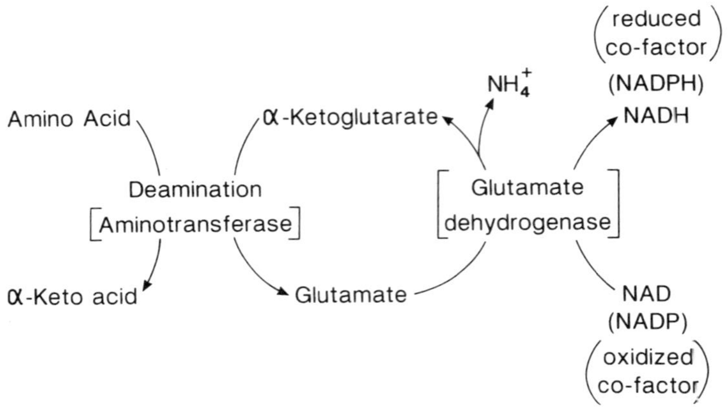

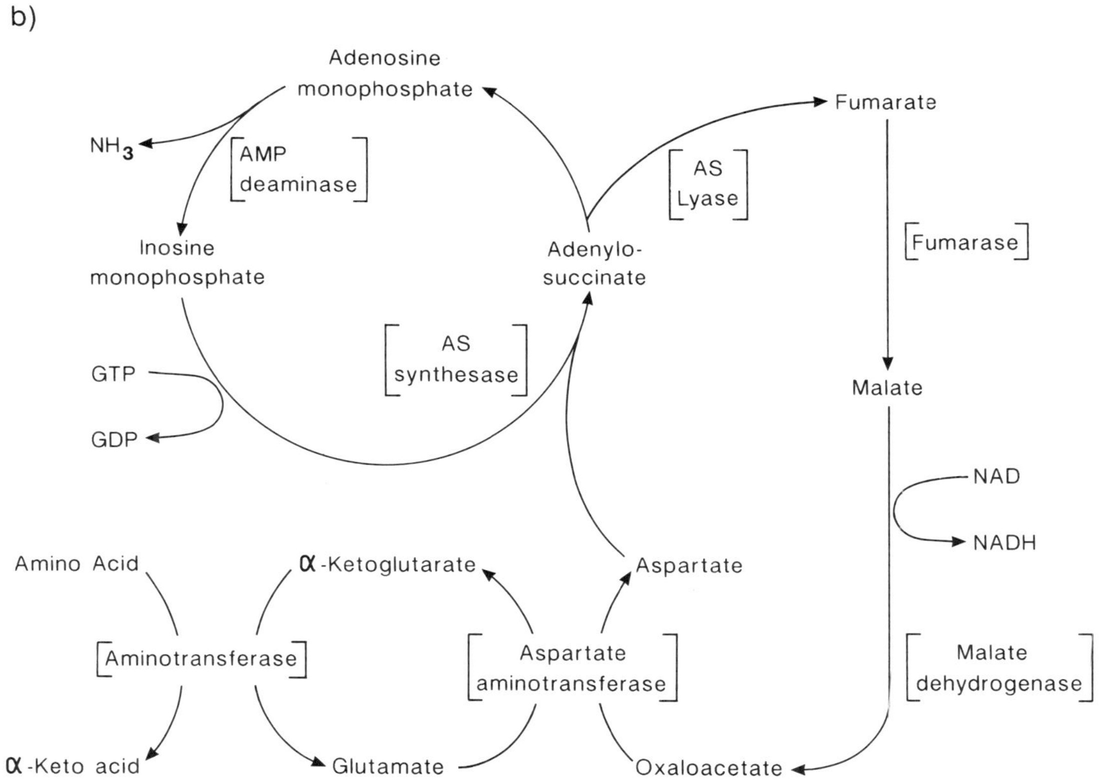

Fig. 7.2 Deamination pathways of amino acids leading to the release of amino-nitrogen. (a) Transdeamination, (b) purine nucleotide cycling. NAD, Nicotinamide-adenine dinucleotide.

nitrogenous wastes as urea, creatine and purines than do freshwater fish species.

The majority of teleost fish species lack some of the key enzymes of the ornithine–urea cycle, and have reduced ability to synthesize urea via the pathway which is the main route of ureogenesis in mammals. In most teleosts, therefore, formation of urea probably occurs via the purine pathway, with uricolysis being the main source of urea production (Fig. 7.3). Some air-breathing teleosts appear to have the full complement of enzymes for production of urea via the ornithine–urea cycle, and, in addition, teleosts that inhabit bicarbonate-rich, 'soda' lakes may excrete the vast majority of their nitrogenous wastes in the form of urea.

For fish living in soda lakes the high concentrations of bicarbonate present may create problems for respiratory exchange, acid–base balance and the processes normally involved in the excretion of nitrogenous waste products. As a consequence, fish inhabiting these environments have developed special adaptations for meeting this series of problems.

For fish living in more benign environments, the excretion of carbon dioxide over the gill tissue may assist in creating a micro-environment suitable for the rapid passive diffusion of ammonia from the gills to the surrounding water. The excretion of carbon dioxide will lead to the acidification of the thin 'boundary layer' of water close to the gills:

$$H_2O + CO_2 = H_2CO_3 = H^+ + HCO_3^- \qquad (7.6)$$

and boundary layer acidification promotes the formation of ammonium ions from ammonia:

$$NH_3 + H_2O = NH_4^+ + OH^- \qquad (7.7)$$

and this, in turn, will lead to the maintenance of the diffusion gradient for ammonia across the gills.

For fish living in soda lakes (pH 10, 60 mM bicarbonate), the acidification of the boundary layer may be much reduced, or prevented, owing to the buffering effect of the bicarbonate. At high pH, most of the excreted ammonia would remain as ammonia, rather than being converted to ammonium ions, leading to a reduction in the diffusion gradient across the gills. In addition, high bicarbonate concentrations in the external environment will lead to bicarbonate moving down its concentration gradient from the water and into the fish. Thus, the fish inhabiting soda lakes face the dual problem of increased bicarbonate loading and reduced diffusion gradients for ammonia excretion. The problem is solved through the production of urea, these fish relying almost exclusively upon urea as the nitrogenous excretory product. Production of urea via the ornithine–urea cycle requires the input of amino nitrogen and carbon dioxide, so the employment of this excretory pathway

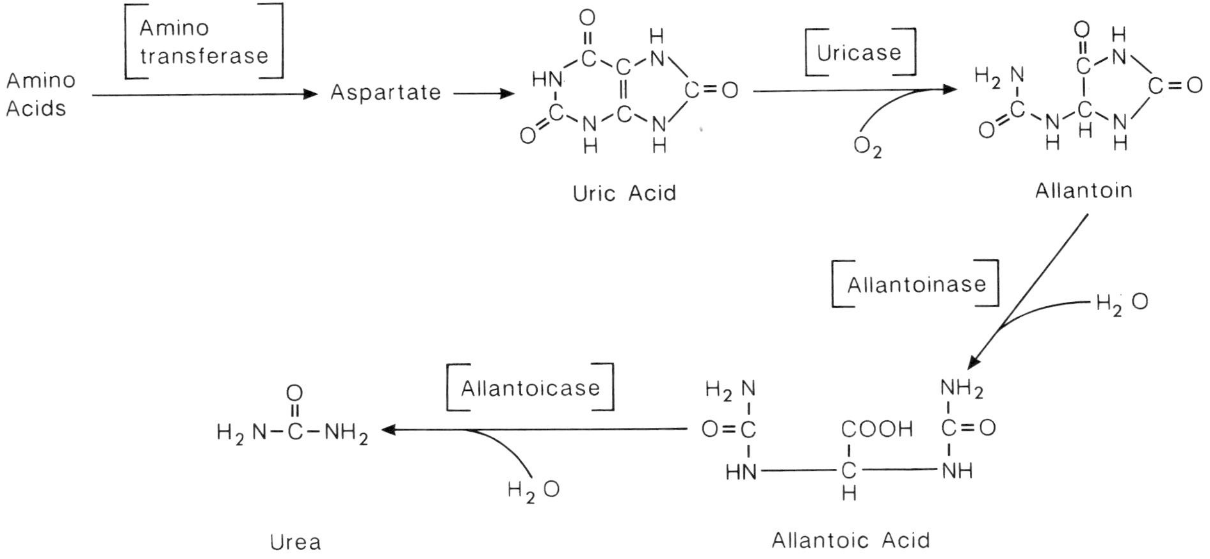

Fig. 7.3 Pathway for the production of urea via uricolysis.

maintains nitrogenous excretion by consuming amino nitrogen, and assists in the maintenance of acid–base balance by consuming bicarbonate equivalents.

Cartilaginous fish also possess the full complement of enzymes required for production of urea via the ornithine–urea cycle, and this is the predominant route for ureogenesis in these species. In the cartilaginous fish, however, urea is employed as an organic osmolyte rather than being a nitrogenous excretory product. This leads to there being a number of differences in the pathways of urea synthesis between the ureosmotic cartilaginous fish and ureotelic species. The most significant difference is that involved in the initial step of ammonia assimilation for urea synthesis. In the cartilaginous fish, ammonia is first combined with glutamate, in the presence of the enzyme glutamine synthesase, to produce glutamine. The glutamine is subsequently utilized as the substrate for the synthesis of carbamoyl phosphate, the reaction being catalysed by glutamine-dependent carbamoyl phosphate synthesase. The resulting carbamoyl phosphate then enters the ornithine–urea cycle. Thus, the production of carbamoyl phosphate involves glutamate–glutamine cycling which is dependent upon an input of ammonia sequestered from the blood. This sequestration of ammonia from the blood for urea synthesis in these ureosmotic fish leads to there being an overall conservation of ammonia-nitrogen within the body (Fig. 7.4). In the ureotelic animals, on the other hand, glutamate, rather than glutamine, acts as the intermediary ammonia donor for the synthesis of carbamoyl phosphate. The entry of the carbamoyl phosphate into the ornithine–urea cycle results in the production of urea, which is ultimately excreted (Fig. 7.4).

Factors influencing rates of excretion

The protein fraction of the food is usually digested and absorbed very efficiently, but if the amounts of amino acids absorbed are in excess of immediate requirements they will be deaminated and the carbon skeletons used for the synthesis of storage products. The resulting amino groups are excreted. This is usually defined as the *exogenous* fraction of nitrogenous excretion because the waste nitrogen arises directly from the deamination of amino acids absorbed from the food. Since *exogenous excretion* is considered to result from the direct deamination of amino acids ingested and absorbed from the food, the exogenous component of nitrogen excretion will be influenced by factors such as feeding rate, the protein content of the food and the amino acid composition of the diet with respect to levels of essential and non-essential amino acids.

If fish are fed protein- (nitrogen-) free diets they will continue to excrete some nitrogenous products, and these arise as the result of the catabolism

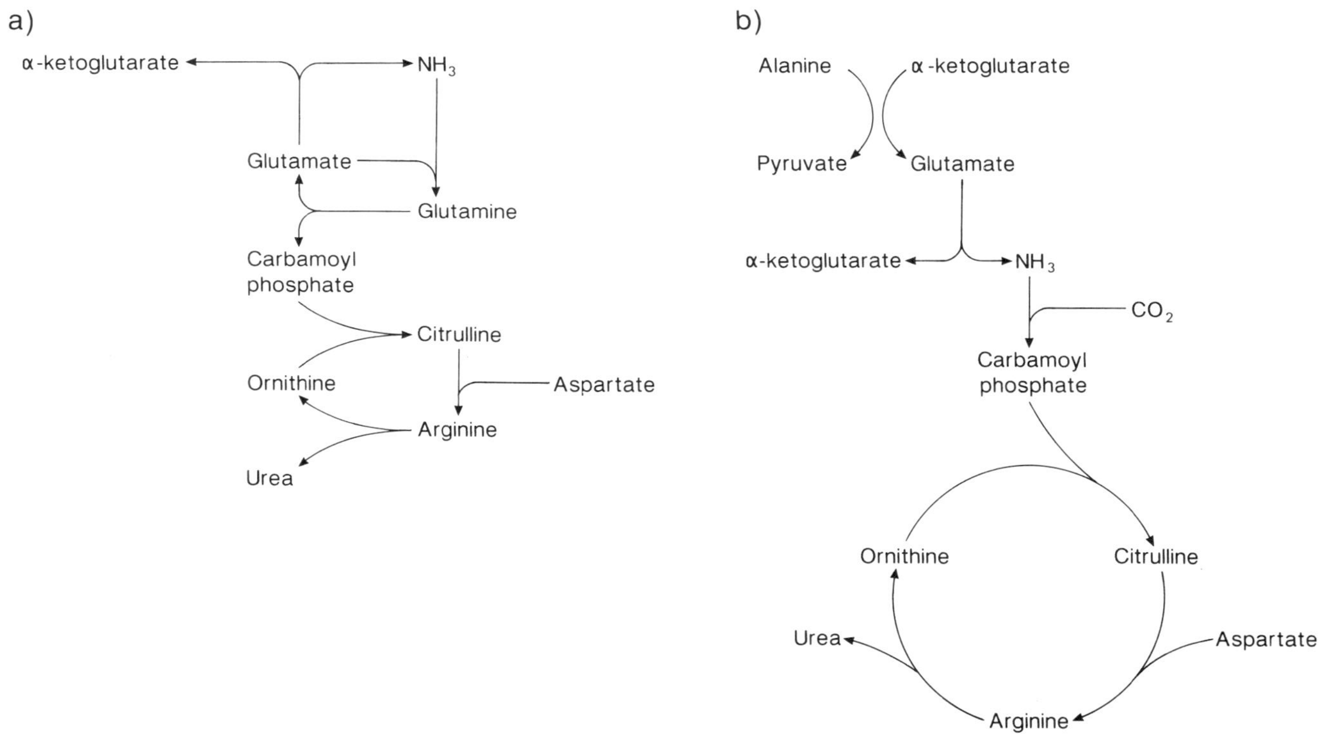

Fig. 7.4 Urea production via the ornithine–urea cycle in (a) ureosmotic and (b) ureotelic animals.

of body tissue proteins. Thus, *endogenous nitrogen excretion* is defined as the nitrogenous excretory products resulting from the transamination and deamination of the amino acids which arise as the result of the turnover and breakdown of tissue proteins. Because the vast majority of the amino acids released as a result of protein breakdown are reutilized in synthesis (usually over 90%), rates of endogenous nitrogen excretion are generally quite low. In practice, it is often extremely difficult to get fish to accept protein- (nitrogen-) free diets, so endogenous nitrogen excretion is often estimated as the rate of excretion of a fish which has been deprived of food for a number of days.

The differentiation between endogenous and exogenous excretion leads to the patterns of nitrogen flow through the fish being divisible into distinct routes depending upon the particular source of the nitrogenous excretory products, but, in practice, it is impossible to distinguish between the endogenous and exogenous components of nitrogenous excretion in an actively feeding and growing fish.

When endogenous nitrogenous excretion has been estimated by examining the excretory rates of fish deprived of food for a number of days, the two factors that have been found to have the greatest influence on rates of excretion are fish size and temperature (Table 7.7). The relationship between size and the rate of nitrogenous excretion can be described by an allometric function, and weight exponents of 0.47–0.95 have been found in experiments conducted with a range of fish species.

Ingestion of food will lead to an increase in the rate of nitrogenous excretion, with excretion peaking some hours after the completion of consumption of the meal. Most of the increase in nitrogenous excretion

Table 7.7 Effects of body weight and water temperature on rates of nitrogenous excretion in some fish species

Species	Body weight (g)	Temperature (°C)	Nitrogenous excretion (mg N d^{-1})	
			Ammonia	Total N
Bluegill sunfish,	50	26		7.7
Lepomis macrochirus	144	26		13.7
Goldfish,	1	20	0.11	0.14
Carassius auratus	10	20	0.75	1.00
Rainbow trout,	75	13		12
Oncorhynchus mykiss	129	13		17.5
	900	10	28	
	900	15	50	
	900	20	77	

will be due to a marked rise in ammonia/ammonium excretion, with the rate of urea production and excretion being little affected by the ingestion of a meal. At its peak, the rate of nitrogenous excretion of a recently fed fish can be several-fold higher than that of an unfed fish. This exogenous excretion will then slowly decline until the endogenous level of excretion is reached after a period of several hours, or perhaps days. The length of time required for the rise and fall of the exogenous excretion will be determined by the size of the meal, its composition and water temperature. The composition of the food will have a marked influence on the exogenous excretion, with factors such as the protein:energy ratio and the balance of amino acids in the diet being of particular importance.

Exogenous excretion may be expected to be high when the fish are fed on diets rich in protein and, also, when the balance of the amino acids is unsuitable for the promotion of protein synthesis and growth. For a number of carnivorous fish species, 40–60% of the available dietary energy should be provided as protein if good growth is to be ensured. When fish are fed such diets 30–50%, or in some cases more, of the nitrogen absorbed from the food may be lost in the form of exogenous excretion.

It is difficult to provide accurate information about the energy losses that accrue as a consequence of the excretion of nitrogenous wastes, because these will vary with both the amount and the composition of excretory products. It has been suggested that a conversion factor of approximately 25 kJ per g N excreted may give a realistic estimate of the energetic losses associated with nitrogenous excretion in fish. Proteins comprise approximately 55% carbon, 7% hydrogen, 20% oxygen, 2% sulphur and 16% nitrogen, so for each gram of protein that is used as respiratory substrate, 0.16 g of nitrogen will be excreted. Therefore, for each gram of protein metabolized there will be an energy loss of approximately 4 kJ (0.16×25), which amounts to about 17% [$(0.16 \times 25/24) \times 100$] of the energy content of the protein. Similarly, when proteins provide the carbon skeletons for the synthesis of lipids and carbohydrates, the deamination of the amino acids will also lead to the loss of 4 kJ per g protein utilized. On the basis of this it can be calculated that the exogenous excretion will amount to approximately 5% of the ingested energy when the diet contains 50% protein, and 50% of the nitrogen absorbed from the food is lost in the form of ammonia.

A few experimental studies have been carried out in which attempts have been made to determine the energetic losses accruing from the excretion of waste nitrogen, and the results suggest that the losses may be equivalent to 4–15% of ingested energy depending upon feeding regime, food composition and the conditions under which the fish are held.

Chapter eight

Respiration and metabolism

8.1 INTRODUCTION

Losses due to metabolic demands usually constitute a large proportion of the energy budget of a fish. These energy costs can be partitioned into the minimal costs required for maintaining basic bodily functions, M_M, those associated with activity, M_A, and those related to the digestion, absorption and processing of food, M_F. Whilst this partitioning may be relatively easy to grasp from a theoretical standpoint, it is far from easy to quantify the different components experimentally.

In work with mammals, metabolism may be recorded directly as 'heat loss', or indirectly by the simultaneous monitoring of oxygen consumption and carbon dioxide production. The indirect method, involving the calculation of the respiratory quotient (RQ), allows heat production to be estimated relatively easily. In studies with fish, on the other hand, it is usual to measure oxygen consumption, but not carbon dioxide production, so that the estimation of heat production requires assumptions to be made about the types of respiratory substrates being metabolized.

Based upon the assumption that protein and lipid are the primary respiratory substrates in fish, an oxycalorific coefficient of 19.4 kJ per l oxygen has been recommended, but it must be re-stressed that the use of this oxycalorific coefficient assumes that metabolism is occurring aerobically.

8.2 COSTS OF MAINTAINING BODILY FUNCTIONS, M_M

The 'minimal metabolism' represents a measure of the minimum rate of heat production free from the effects of muscular activity, consumption of food and its subsequent processing. In experimental studies, these variables have been controlled with varying degrees of success, and a number of terms have arisen to describe the types of measurements that approximate to the 'minimal metabolism'.

A wide range of methodologies have been used in attempts to determine the 'minimal metabolism' of fish species, and the terminology has not been satisfactorily standardized. Terms such as basal, standard, resting and low routine metabolism have been used as synonyms, but distinct definitions have also been given to each of these terms (Table 8.1). Problems arise, however, because there is a lack of consistency in the way each of the terms has been defined by different workers, and no universally acceptable set of definitions has yet been presented.

In studies with mammalian species the measurement of basal metabolism requires a rigidly defined set of conditions to be fulfilled, and it is open to question whether the term basal metabolism should be used when referring to measurements of 'minimal' metabolic rates of fish species. There is a good case to be made for the introduction of a new term – *fasting metabolic rate* – for describing estimates of minimal metabolism made on fish, because, in practice, most estimates of the 'minimal metabolism' in studies with fish have been obtained using experimental protocols that are similar to those used for the estimation of fasting metabolic rates in domestic animals. In fish studies, however, the metabolic rate recorded has usually been termed the *resting, resting routine,* or *low routine metabolic rate.* All of these terms presuppose that measurements are being made on a quiescent fish, in a post-absorptive condition, but it is recognized that

Table 8.1 Definitions of terminology used in describing metabolic rates (oxygen consumption) of fish species

Nutritional status	Metabolic rate	Activity level
Post-absorptive	Basal; Standard*	Complete rest: zero activity
	Resting; Low routine Resting routine; Fasting	Spontaneous swimming movements
Daily feeding	Routine; Intermediate	Normal activity without stress
	Active	Maximum sustained activity
	Hyper-active	Burst performance; Escape response

*Note that standard metabolism is estimated by extrapolation from the relationship between swimming speed and oxygen consumption.

measurements include some increase over the true minimum owing to the metabolic costs of low levels of spontaneous activity (Table 8.1). Thus, on the grounds of both the promotion of standard terminology and for simplicity, it may be of value to replace the plethora of current terms used in fish studies when referring to 'minimal metabolism' with *fasting metabolic rate.*

A case can, however, be made for retaining the term *standard metabolism* in fish studies, provided that the term is only used when referring to estimates of 'minimal metabolism' obtained by specific means. It has become usual to calculate *standard metabolism* from data obtained on swimming fish. A relationship between swimming speed and the metabolic rate of the fish is established, and then the relationship is extrapolated to zero velocity. The estimate of the metabolic rate at zero swimming speed is defined as the *standard metabolism* of the fish. Whilst the validity of these extrapolations can be questioned on both statistical and physiological grounds, the method is useful for obtaining estimates of the 'minimal metabolism' of active species which do not remain quiescent in respirometry chambers.

A prerequisite in all studies designed to measure 'minimal metabolism' is that the animals be in a post-absorptive state, i.e. the animals must have been fasted to ensure that the effects of previous meals, and food processing, are negligible. There have been few studies carried out on fish species in order to quantify the time required to reach a post-absorptive state, and a fasting period of 48–72 h has often been arbitrarily adopted prior to the conducting of measurements of 'minimal metabolism'. Whilst a fast of this duration may be adequate to ensure a return to the post-absorptive condition in many cases, it certainly does not have general application. The time required for the ablation of the effects of previous meals on metabolic rate is highly variable, and may be governed by factors such as meal size and composition, and by water temperature. It may, however, be generally assumed that the time taken to reach the post-absorptive state is related to the retention time of food in the digestive tract. In other words, some elevation in metabolic rate due to the effects of feeding is observed for slightly longer than the time during which products of digestion are being transported across the gut mucosa into the bloodstream.

Experimental procedures that have been employed for the estimation of the 'minimal metabolism' of fish species are many, and various, and it is difficult to give recommendations for a standardized method. On the other hand, an experiment conducted according to the following procedure would be generally accepted as giving a close approximation to the 'minimal metabolism'. Fish should be fasted for some time (48–72 h) prior to being introduced into the respirometry chamber. Following a period of acclimatization (often overnight), measurements can commence. Metabolic rate should be monitored for a period of several hours (an 8–12 h monitoring period during the course of the working day is usual, but longer would be preferable) before the fish is removed from the respirometer and is replaced by the next subject. The respirometer should be located in a laboratory shielded from disturbance and the chamber exposed to no more than dim light. Oxygen concentrations should be maintained well above limiting levels, both during the acclimatization period and when metabolic rate measurements are being made. At the termination of the experiment, records should be examined and those showing either a decline in metabolic rate with time (due to continued subsidence of the effects of previous meals) or marked fluctuations (due to uncontrolled activity) should be discarded.

Fish size and metabolic rate

Large fish generally consume more oxygen than small fish, but on a unit-weight basis, small fish will consume more oxygen than larger conspecifics. This allometric relationship can be described as

$$M_M = a(2)\, W^{b(2)} \tag{8.1}$$

where M_M is the rate of oxygen consumption (ml O_2 per fish per hour) (energy expenditure), W is body weight (in g), and a(2) and b(2) are constants. There are several hundred studies in the literature reporting the effects of body size on rates of oxygen consumption in fish, and the majority of the estimates of the weight exponent, b(2), lie within the range 0.65–0.9. Results of a recent analysis suggest that a value of 0.86 for the weight exponent may be appropriate for many species.

The decline in oxygen consumption per unit body weight that accompanies increases in body size has been attributed, in part, to ontogenetic changes in the relative sizes of different body organs. For example, as the fish grow, highly metabolically active tissues, such as the gut and liver, will gradually represent a lower proportion of the body weight, whereas the relative size of the swimming musculature will tend to increase. In addition to these ontogenetic changes in organ sizes there is also evidence to suggest that the 'metabolic intensities' of the different tissues tend to decline as fish increase in size and/or age.

The reported values of the weight exponent are, however, sufficiently disparate to imply that interspecific differences may exist. There is also evidence to suggest that there may be intraspecific differences depending upon measurement conditions, such as water temperature and season during which the experiments are carried out. It has been reported, for example, that the weight exponent may decline with increasing test temperature, and this has been attributed to the differential effects of temperature on the 'minimal metabolism' of fish of different sizes.

Temperature effects on metabolic rate

When measurements of 'minimal metabolism' are made on fish it is important to ensure that the fish are acclimated to the temperature at which the measurements are to be made. If fish are transferred from one temperature to another, several days, or even weeks, may be required before the fish become acclimated to the new conditions. Thus, the results of metabolic measurements made immediately after transfer ('acute' response) will usually be found to differ from those obtained from experiments carried out on acclimated fish. For example, if a fish held at low temperature (5 °C) is transferred to a higher temperature (25 °C), the metabolic rate will usually be observed to increase abruptly ('acute' response) (Fig. 8.1). If the fish is then held at this higher temperature for a prolonged period (days or weeks) the metabolic rate may gradually decline and then stabilize at a new level – in other words, the fish becomes acclimated to the new temperature conditions. On transferring the warm-acclimated fish to cooler water, an 'acute' response will again be observed, before the fish gradually becomes re-acclimated to low-temperature conditions (Fig. 8.1).

These effects of temperature on metabolism can be displayed graphically by plotting rate–temperature curves. In many instances where an increase in temperature has been found to lead to an increase in the rate of oxygen consumption, a simple exponential (semilogarithmic) transformation comes close to linearizing the data:

$$\ln M_M = \mathrm{k} + \mathrm{c}\,T \qquad (8.2)$$

where T is temperature, and c represents the temperature coefficient (Fig. 8.2). Thus, when starting with fish acclimated to low temperature (5 °C) (point A in Fig. 8.2), an abrupt increase in temperature will result in a marked increase in metabolic rate ('acute' response, denoted by the filled circles). If the fish are then held for a prolonged period at high temperature (25 °C) metabolic rate may gradually fall (to point D in Fig. 8.2) as the fish become acclimated to the higher temperature. Exposing these warm-acclimated fish to water at low temperatures will result in an 'acute' response (denoted by filled squares). Acclimation of fish for a prolonged period to some intermediate temperature (15 °C) would be expected to result in a metabolic rate (point C in Fig. 8.2) lying between those recorded for fish acclimated to low and high temperatures.

Thus, when the metabolic rates of the fish are recorded at their respective temperatures of acclimation (test temperature = acclimation

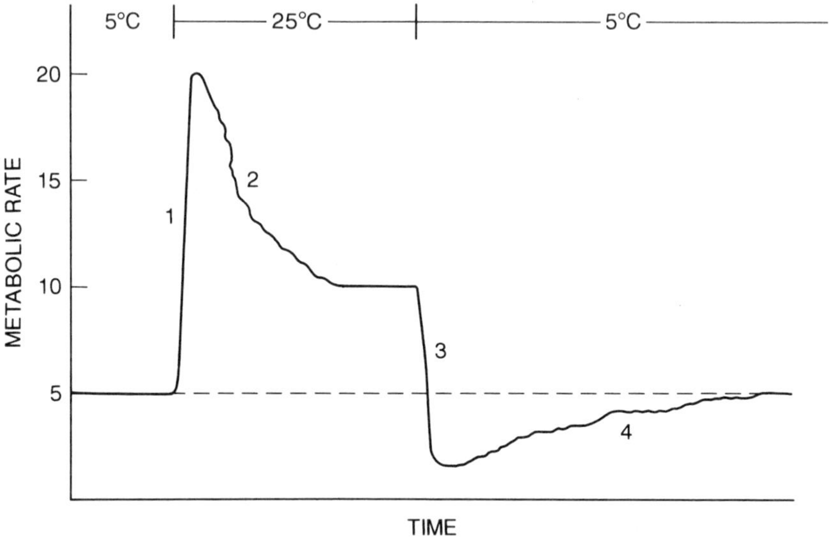

Fig. 8.1 Schematic figure showing the time course of thermal acclimation and metabolic compensation following transfer between different temperature conditions. (1) 'Acute' response following temperature increase; (2) acclimation to new temperature conditions resulting in a partial compensation of metabolic rate; (3) 'acute' response following return of acclimated fish to low-temperature conditions; (4) re-acclimation of fish to low temperature accompanied by an increase in metabolic rate.

temperature) the rate–temperature curve will be that denoted by the dashed line A–C–D in Fig. 8.2. The situation described here is one in which 'partial' compensation has occurred. In other words, metabolic rates of fish acclimated to 15 and 25 °C are higher than those of fish acclimated to 5 °C, when all are tested at their respective acclimation temperatures. It is also possible to envisage situations under which 'complete' compensation may occur, and under these circumstances the rate–temperature curve would follow the dashed line indicated by A–B–E in Fig. 8.2. When rate–temperature curves are plotted the temperature coefficient is usually found to lie within the range 0.05–0.1, which means that metabolic rates increase 1.65–2.7-fold for every 10 °C rise in temperature. A coefficient of 0.07 implies that a 10 °C rise in temperature leads to a doubling of metabolic rate.

An expression known as the Q_{10} has often been used to describe the effects of temperature upon the rates at which physiological processes proceed. The Q_{10} is defined as the increase in the rate of a physiological process resulting from a 10 °C increase in temperature. Thus, if the rate

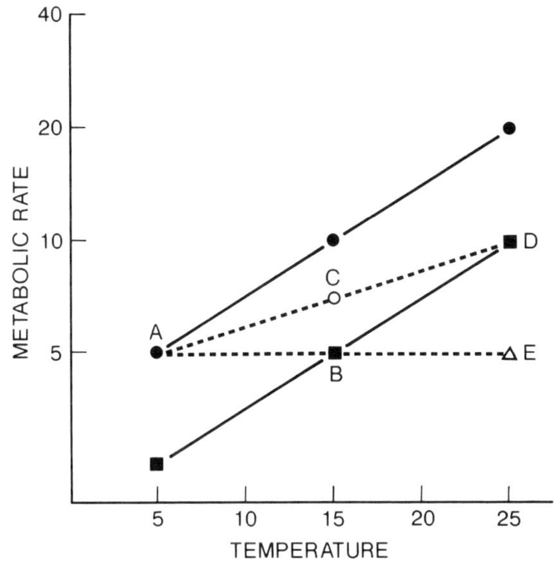

Fig. 8.2 Rate–temperature curves showing some possible effects of temperature on the metabolic rates of fish. Solid lines indicate 'acute' responses to different temperature conditions. Dashed lines represent two possible types of thermal compensation, observed when 'acclimation' temperature = test temperature. A–C–D indicates partial compensation, whereas A–B–E shows the response of a fish showing complete metabolic compensation to different temperature conditions. Note vertical scale.

doubles, the Q_{10} is 2, if it triples the Q_{10} is 3, and so on. The Q_{10} can be calculated from the following formula:

$$Q_{10} = (V_2/V_1)^{10/[T(2)-T(1)]} \qquad (8.3)$$

where V_1 and V_2 are the rates of the physiological process measured at temperatures $T(1)$ and $T(2)$, respectively. The Q_{10} bears a simple relationship to the temperature coefficient (c) in the equation $\ln M_M = k + cT$, in that $\ln Q_{10} = 10c$, under conditions where the temperature coefficient (c) remains constant over a wide temperature range.

The data represented by the dashed line A–C–D in Fig. 8.2 can be used to calculate a Q_{10} value: $Q_{10} = (10/5)^{10/[25-5]} = (2)^{0.5} = 1.41$. This represents the effects of temperature upon the metabolic rates of fish acclimated to their respective test temperatures (i.e. test temperature = acclimation temperature), but Q_{10} has also been used to describe 'acute' temperature responses. In the latter case, the data presented in Fig. 8.2 would yield a Q_{10} value of: $Q_{10} = (20/5)^{10/[25-5]} = (4)^{0.5} = 2$. Thus, whilst the Q_{10} has been widely used as a convenient measure of the temperature effects upon physiological processes, it is essential that the measurement procedures be clearly stipulated if confusion is to be avoided.

Further problems may arise because the Q_{10} does not remain the same over the entire temperature range, but tends to vary with temperature. Q_{10} is often found to be high at low temperatures, and to decrease as temperature increases. Rates of physiological processes may also become independent of temperature over part of the temperature range of the test species, with a Q_{10} close to 1 often being recorded in the mid-part of the range. Thus, it has become conventional to give information about the temperature range over which the Q_{10} was determined, e.g. a Q_{10} determined over the temperature range 5–15 °C would be denoted as $Q_{10(5-15)}$.

8.3 ENERGY EXPENDITURE ASSOCIATED WITH SWIMMING, M_A

Swimming is a very energy-demanding activity, and the energy expenditure of fish swimming at high speed may rise to 10–15 times that of a fish at rest, but this is maintained for only a very short period of time. The factorial aerobic scope for activity, defined as the rate of oxygen consumption of an actively swimming fish divided by the consumption of a fish at rest, is more usually found to be within the range 3–7, with the majority of species lying towards the bottom end of the range.

Rates of oxygen consumption are usually found to increase exponentially as swimming speed increases (Table 8.2). Information about the

Table 8.2 Effects of swimming speed and water temperature on rates of oxygen consumption (metabolic rates) of some fish species. Relationships between swimming speed and oxygen consumption are linearized by semilogarithmic transformation: $\log M = \log m + d A$, where M is oxygen consumption (mg O_2 kg^{-1} h^{-1}), A is swimming activity (speed) (body lengths s^{-1}), and m and d are constants

Species	Temperature (°C)	log m	d
Flounder *Platichthys flesus*	10	1.79	0.20
	15	1.97	0.21
Dab, *Limanda limanda*	5	1.77	0.21
	15	1.96	0.21
Sockeye salmon, *Oncorhynchus nerka*	10	1.78	0.28
	20	2.08	0.22
Haddock, *Melanogrammus aeglefinus*	5	1.81	0.26
Cod, *Gadus morhua*	15	2.12	0.21

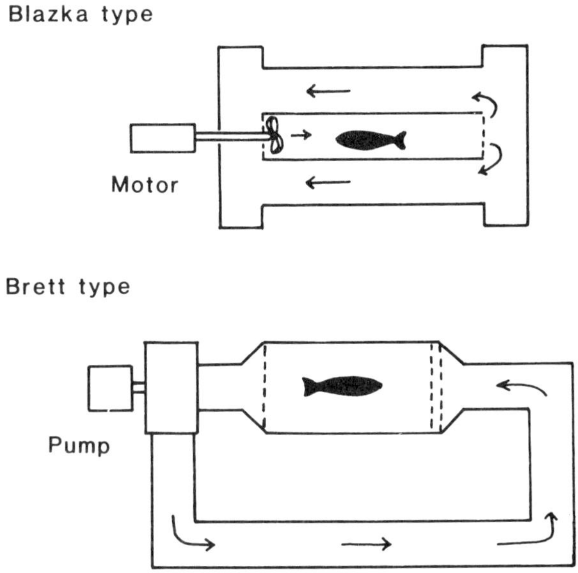

Fig. 8.3 Schematic diagrams of two types of respirometry chambers used for studying the effects of swimming activity on metabolic rates (rates of oxygen consumption) in fish species.

oxygen consumption of swimming fish may be obtained from experiments conducted on fish that are voluntarily active or from those in which fish are forced to swim against a water current. Most measurements have been made using tunnel respirometers in which fish are forced to swim against currents of known speed (Fig. 8.3). The advantage of this type of respirometer is that the activity level of the fish can be accurately controlled. Other types of swimming respirometer rely upon the fish being voluntarily active. Respirometers may be designed to take advantage of the optomotor response (the tendency of some fish species to hold station against a moving background), but all respirometers that rely upon the fish being voluntarily active must incorporate some form of activity counter.

To what extent measurements of oxygen consumption give a true reflection of the energetic costs of swimming activity is open to question, since anaerobic metabolism, with the production of lactate, increases as levels of swimming activity rise. In some species, such as flatfishes, gadoids and typical 'sit-and-wait' predators, anaerobic respiration begins in the swimming muscle at relatively low activity levels (0.5–2.0 body lengths per second, $BL.s^{-1}$), whilst in salmonid species and actively foraging species, higher swimming speeds (2.0–5.0 $BL.s^{-1}$) are reached before the muscles begin to respire anaerobically and lactate begins to accumulate. There also appears to be a distinct size effect since anaerobic metabolism is induced at a lower relative swimming speed (expressed as $BL.s^{-1}$) in large than in small fish.

Terminology of swimming activity in fish

A distinction can be made between burst performance swimming, and swimming at slower speeds that can be maintained for much longer periods of time. Consequently, it has become usual to classify swimming activity into a number of categories – sustained, prolonged and burst swimming. This classification not only defines the duration over which the fish can perform at a given level, but also gives an indication as to the respiratory substrates and metabolic pathways employed to provide the energy required for the different types of swimming.

The term *sustained swimming* performance is applied to those swimming speeds which can be maintained for long periods without resulting in muscular fatigue. Sustained swimming can, theoretically, be maintained indefinitely, but for practical purposes sustained swimming speeds are defined as those that can be maintained for a period of 200 min or longer. Metabolism during sustained swimming is aerobic, and there is no accumulation of lactate. The activities that would be encompassed by the term sustained swimming include station holding in mid-water, foraging at low swimming speed and long-distance migration.

Prolonged swimming is of shorter duration (20 s–200 min) than sustained swimming, and results in fatigue. The fish will be said to be fatigued when it collapses, and is no longer able to maintain the given swimming speed. The energy supply for prolonged swimming may be provided from both aerobic and anaerobic metabolism. A special category of prolonged swimming is the *critical swimming speed*. The critical speed is a useful term that can be used for comparing the swimming abilities of different sizes or species of fish. Critical swimming speeds are determined under rigidly controlled laboratory conditions, in which the fish are forced to swim against water currents in a flume apparatus. The fish are subjected to stepwise increases in swimming speed until fatigue occurs, and the critical swimming speed is calculated from the maximum speed achieved prior to fatigue. For example, if a fish can maintain a swimming speed of 50 $cm.s^{-1}$ for 60 min, but fatigues after 30 min at a speed of 70 $cm.s^{-1}$, then the critical swimming speed lies between 50 and 70 $cm.s^{-1}$. In this example, the critical swimming speed would be calculated as $50 + [(70 - 50) \times (30/60)] = 60$ $cm.s^{-1}$, where 50 $cm.s^{-1}$ is the swimming speed at the last completed step, $(70 - 50)$ is the incremental increase in speed ('step size'), and 30/60 is the time into the last step at which fatigue occurred (30 min) divided by the prescribed step duration (60 min). Because both the duration and velocity increment of each step will influ-ence the critical swimming speed obtained, it is important that this infor-mation be given when presenting the results of studies of this type.

The highest swimming speeds of which fish are capable are termed *burst swimming* speeds. These high speeds can be maintained for only a very short period of time, and fish usually swim at burst speeds for periods of less than 20 s. The energy for burst swimming is provided, predominantly, by anaerobic metabolism. Burst swimming is characterized by an initial acceleration phase followed by a sprint phase of steady swimming. Steady swimming refers to swimming in a given direction at constant speed, whereas swimming that involves either acceleration or direction change is usually termed unsteady swimming.

Thus, when fish are forced to engage in strenuous, rapid swimming the energy requirements of the muscles are greater than those that can be supplied by aerobic metabolism. The immediate energetic needs for short bursts of muscular activity are met from stores of high-energy phosphate compounds (creatine phosphate), but these are exhausted rapidly, and the majority of the extra energy requirement for burst swimming is usually derived from the anaerobic metabolism of muscle glycogen, with the formation of lactate. Since anaerobic metabolism is much less efficient than aerobic metabolism, the respiratory substrates present in the muscle are used relatively rapidly. This, in turn, leads to lactate accumulation and the cessation of swimming activity.

Sustained, aerobic versus exhaustive swimming activity

Several physiological responses distinguish strenuous exercise from steady-state aerobic swimming. Muscle ATP and creatine phosphate levels are not depleted in fish swimming aerobically, and, whilst some lactate may be generated during steady-state swimming, the rates of production and oxidation appear to come into equilibrium. Thus, elevations of muscle and blood lactate levels are of minor proportions in fish swimming at moderate speeds. During exhaustive exercise, on the other hand, there is a rapid depletion of muscle ATP, creatine phosphate and glycogen reserves, with anaerobic glycolysis leading to a large build-up of lactate in the white muscle. Some of the lactate may be released from the muscle, leading to a substantial rise in blood lactate levels, but blood lactate levels never reach those seen in the white muscle. The release of lactate into the blood may give rise to a metabolic acidosis, with a concomitant fall in blood pH. By way of contrast, blood pH is little altered during the performance of sustained aerobic swimming activity.

Both sustained aerobic, and strenuous exhaustive, swimming result in increased water influx over the gills of freshwater fish. This is presumably brought about as the result of the increased lamellar recruitment that occurs in exercising fish. Lamellar recruitment, which results in an increase in gill area for oxygen uptake and gaseous exchange, will also lead to an unavoidable increase in the area over which there is easy entry of water from the external environment and into the blood. The increase in water influx is compensated by an increase in urine production.

Swimming at relatively moderate speeds does not result in changes in plasma electrolyte levels, and there are only minimal changes in ion exchange over the gill epithelium. By contrast, exhaustive exercise causes profound disturbances to ionic and osmotic balance. The increase in intramuscular lactate results in a net flow of fluid from extracellular to intra-cellular compartments. In other words, there is a decrease in plasma water and an increase in plasma ionic concentrations. In addition, the release of lactate, and the accompanying general acidosis and fall in blood pH, may be expected to influence the oxygen-transporting capacity of the blood.

A major difference between the physiological responses invoked by sustained aerobic, and strenuous, exercise is the extent to which catecholamines are mobilized. Aerobic exercise at moderate swimming speeds does not lead to a marked increase in circulating levels of catecholamines over those seen in resting fish, but following the imposition of a heavier swimming load there may be a several-fold increase in blood catecholamine levels. Combinations of declining blood oxygen content and the acidosis resulting from lactate accumulation may be the stimuli promoting catecholamine release. One of the major effects of catecholamine release is the offsetting of the possible negative influence of a general acidosis on blood oxygen transport capacity. The liberation of the catecholamines into the blood leads to the maintenance of the internal pH of the red blood cells at a 'normal' value, despite the general blood acidosis. Consequently, the Root shift is not observed and there is no marked oxygen offloading from the haemoglobin. The red blood cell pH is maintained via the activation of an Na^+/H^+ exchange mechanism in the cell membrane, and this mechanism results in extrusion of protons from the red blood cell (Fig. 8.4). Because the membrane of the red blood cell is relatively impermeable to protons, the main way in which the red blood cell pH could be disturbed is via H^+ generation resulting from the hydration of carbon dioxide. In the blood plasma the reaction $HCO_3^- + H^+ = H_2O + CO_2$ is not catalysed by carbonic anhydrase, so the reaction proceeds relatively slowly. Consequently, the rate of proton transfer from the red blood cell to the plasma via the Na^+/H^+ exchange mechanism is potentially faster

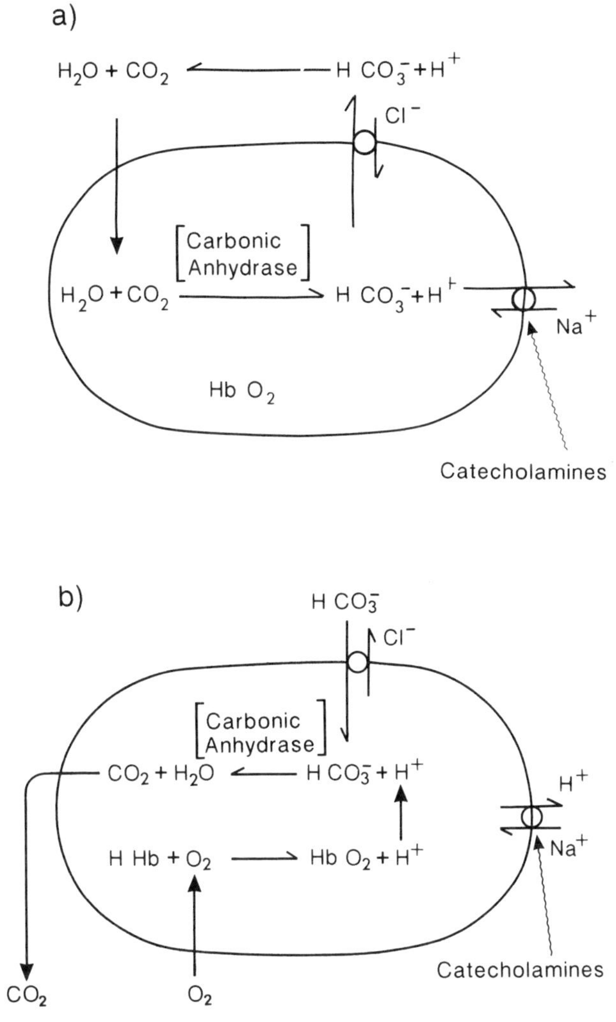

Fig. 8.4 Ionic and gaseous exchanges over red blood cell (RBC) membranes during metabolic acidosis. (a) Na^+/H^+ exchange stimulated by catecholamine release regulates intracellular pH during acidosis by transporting protons (H^+) out of the cell. (b) Carbon dioxide (CO_2) excretion during stimulation of RBCs by catecholamines. Protons (H^+) required for bicarbonate (HCO_3^-) dehydration are released from haemoglobin (Hb) during oxygenation at the gills.

than the rate at which protons can be regenerated due to the influx of carbon dioxide from the plasma.

When the blood passes through the gills and the haemoglobin is oxygenated, there is a release of protons from the haemoglobin. These protons may be used to drive bicarbonate dehydration, and thereby result in the production of carbon dioxide, which can be excreted. The red blood cells contain carbonic anhydrase and the reaction proceeds, therefore, at the rapid catalysed rate (Fig. 8.4). Consequently, during the transit of the blood through the gills, there can be some excretion of carbon dioxide even though the ion-exchange mechanisms of the red blood cell membrane have been influenced by catecholamine stimulation. Nevertheless, bicarbonate (HCO_3^-) dehydration in the red blood cells, and carbon dioxide excretion over the gills, may be initially reduced compared with the situation in resting fish. This, in turn, results in an accumulation of carbon dioxide in the blood, which can exacerbate the fall in blood plasma pH during strenuous exercise. It is possible that the inhibition of carbon dioxide excretion is only of relatively brief duration. The elevation in the plasma carbon dioxide content will lead to an increase in the blood-to-water gradient for carbon dioxide diffusion over the gills, and thereby promote an increased rate of diffusion. Thus, the increase in the plasma diffusion gradient resulting from increased plasma levels of carbon dioxide may, to a large extent, offset the reduced role of the red blood cells in the dehydration of bicarbonate, and lead to a relatively rapid restoration of carbon dioxide excretion to normal levels.

The release of catecholamines that occurs during the course of a strenuous swimming bout has a relatively small effect upon heart rate, with beat frequency only increasing 15–20% over that seen in resting fish. The catecholamines may, however, have a significant effect upon cardiac output (heart rate × stroke volume) via influences on the stroke volume of the heart. Catecholamine release also has a marked influence on the patterns of blood flow through the gills, brought about by effects upon the gill vasculature. The blood vessels of the gills appear to be very sensitive to catecholamines, and stimulation leads to a pronounced vasodilation. This results in a reduction in the gill resistance to blood flow, and a consequent recruitment of additional respiratory lamellae.

Strenuous swimming activity can only be maintained for short periods of time, and it is obvious that the physiological state of a fish after completion of an exhaustive bout of swimming is very different from that of the fish prior to the start of the exercise. Following exhaustive exercise, there will be a period of recovery in which there will be a restoration of physiological status to the pre-exercise condition.

Part of the lactate that accumulates during strenuous exercise can be re-oxidized to glycogen following the cessation of activity. Obviously, this requires the utilization of oxygen, and the oxygen consumption of a fish in the period following a bout of strenuous exercise is greater than that of a fish at rest. This excess post-exercise oxygen consumption over resting levels, defined as the total amount of oxygen used in recovering from exercise, is usually termed the *oxygen debt.* The oxygen debt is not only the result of the re-oxidation of lactate, but there are also costs associated with the restoration of ionic and acid–base balance, and with the recovery of ATP, creatine phosphate and internal oxygen stores. The processes involved in the recovery of ATP and creatine phosphate represent the *alactacid oxygen debt,* and these occur rapidly, to be followed by a second, slower component (*lactacid oxygen debt*) which includes the oxygen consumed in association with the re-oxidation of lactate. The alactacid oxygen debt may be repaid within a period of a few minutes to hours, but several hours will usually be required for repayment of the lactacid oxygen debt. Consequently, during recovery from strenuous activity, the rate of oxygen consumption may remain elevated above resting levels for a period of several hours. Only part of the lactate produced anaerobically during burst swimming is re-converted back to glycogen, because some of the lactate is used as respiratory substrate to provide the ATP to fuel the re-conversion and restoration processes (Fig. 8.5). In addition, following a period of strenuous exercise, lactate may be a major respiratory substrate, being used in preference to other substrates, so that the major flux of lactate may not be directed towards the regeneration of glycogen stores.

In mammals, some of the lactate released into the bloodstream during the course of strenuous exercise is re-converted into glycogen via the Cori cycle. This involves the uptake of the circulating lactate by the liver, where a small proportion is metabolized to fuel the conversion of the major part of the lactate to glucose. The glucose is, in turn, transported to the muscles to restore glycogen levels. Re-conversion of lactate to glycogen may also occur in the muscles themselves. In fish, there is some release of lactate into the bloodstream during bouts of exhaustive exercise, but the majority appears to be retained in the white, anaerobic swimming muscle tissue. Thus, the Cori cycle may only play a minor role in lactate re-conversion during post-exercise recovery in fish. Further evidence for this comes from the fact that the capacity of the liver for synthesis of glucose from lactate may be relatively limited. The bulk of the lactate produced during strenuous exercise appears, therefore, to be metabolized within the white muscle, either as a respiratory substrate resulting in ATP production, or for the regeneration of the glycogen stores.

Recovery from strenuous swimming activity in fish also involves the re-establishment of acid–base and ionic balance, and the restoration of fluid volume. During exhaustive exercise the disturbances to acid–base balance result from carbon dioxide accumulation and the release of protons into the bloodstream from the muscles. Consequently, post-exercise recovery requires adjustments in carbon dioxide excretion and ion-exchange mechanisms in the gill epithelium. The increase in muscle lactate levels during

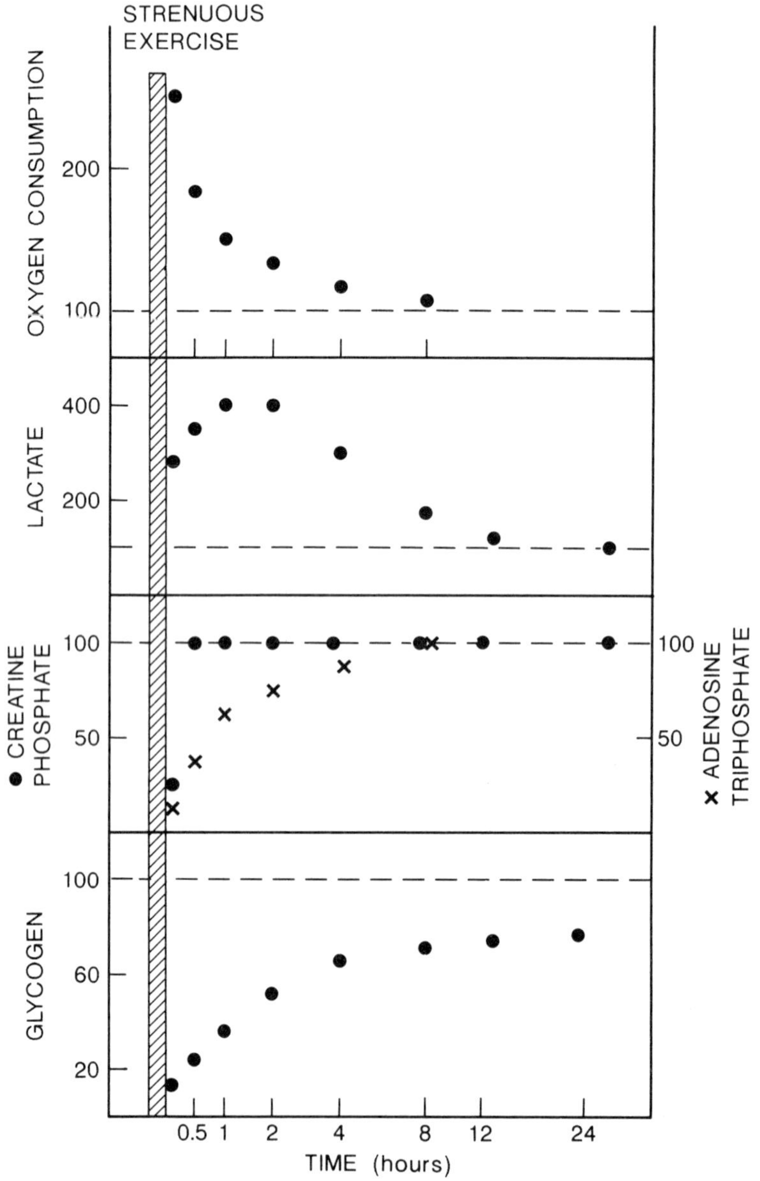

Fig. 8.5 Time course illustrating the effects of a bout of strenuous swimming (hatched region) on rates of oxygen consumption and metabolic indicators of exhaustive exercise, in the immediate post-exercise period and during recovery. Values for oxygen consumption and metabolic indicators presented in the figure are given as percentages of levels observed in resting fish. Note horizontal scale.

exercise results in considerable water movements between extracellular and intracellular compartments due to osmotic effects, and these disturbances to fluid balance are also corrected during post-exercise recovery. These corrections involve the transport of ions across the cell membranes, and the metabolic costs associated with these transport processes may be considerable. Thus, increased ion transport during post-exercise recovery may also impose metabolic costs that make a large contribution to the maintenance of the high rates of oxygen consumption observed following a bout of exhaustive exercise.

The high requirement for oxygen during the process of recovery from strenuous exercise necessitates that the fish maintain relatively high levels of gill ventilation following the cessation of the swimming bout. Because the oxygen content in the arterial blood leaving the gills will be close to normal during the recovery period, it is unlikely that a low level of blood oxygen is the stimulus for the continued high level of gill ventilation during this period. It seems more likely that the acidosis, resulting from carbon dioxide and proton accumulation in the blood, is the primary stimulus driving the post-exercise hyperventilatory response.

8.4 ACTIVITY LEVELS AND 'FIELD METABOLIC RATES'

Studies conducted under controlled laboratory conditions in which fish are forced to swim at a constant speed against a water current give no information about energy expended on swimming activity by fish in the wild. Attempts have been made to collect information about activity levels of fish in nature using various forms of telemetric equipment. The simplest telemetric method involves the fixing of a small radio transmitter to the fish, enabling the position and movements of the fish to be monitored at regular intervals. The fish is captured, tagged, released and then movements monitored over the course of the subsequent few days or weeks by taking positional recordings at short intervals (usually a few minutes between measurements). Thus, by taking a series of measurements it is possible to build up a picture of the activity pattern and movements of the fish within the water body. This method has, however, a number of shortcomings, and energy expenditure estimates based upon information obtained in this type of study are of limited accuracy. For example, in calculating the distance swum it is assumed that the fish swims in a direct line between two points, such that the estimate of the distance swum by the fish is a minimum estimate. It is also assumed that the fish swim at constant speed. On the basis of this information about the distances swum at given swimming speeds, the energy expenditure of the fish is estimated by extrapolation of data about swimming speed–

metabolism relationships obtained in the laboratory. This method of calculation will result in an underestimation of the energy expenditure by the fish on activity, and estimates obtained must, therefore, be considered as being minima.

Attempts have also be made to estimate activity levels by monitoring changes in a range of physiological parameters, such as ventilation frequency, heart rate and muscular contractions in the swimming muscle. It is, however, no simple task to use physiological measurements in the estimation of energy expenditure for activity, since each method has specific weaknesses and shortcomings. For example, both ventilation frequency and heart rate are influenced by a range of factors, so a change in these parameters need not necessarily imply that the swimming activity of the fish has changed. In addition, a number of physiological changes may be brought into play to meet the increase in tissue oxygen demand brought about by increased activity. The physiological response to increased activity may not only be an increase in ventilation frequency and heart rate, but there will usually also be an increase in stroke volume. In practice, the increase in cardiac output (heart rate × stroke volume) that occurs with increasing activity is more often the result of quite large changes in stroke volume than of marked increases in heart rate. For example, although there are close correlations between heart rate and locomotor activity in some species, the heart rate of the sea bass, *Dicentrarchus labrax*, appears to be relatively constant over a comparatively wide range of activity levels. Thus, the fish can swim at a range of different swimming speeds without there being any detectable change in heart rate, and the increased oxygen required by the muscles during swimming activity must, therefore, be provided by invoking physiological changes other than marked increases in heart rate.

At high swimming speed some species of fish may switch from the usual form of gill ventilation to ram ventilation, and ventilation frequency will, under these circumstances, become uncoupled from the overall activity level of the fish. Thus, there need not necessarily be close correlations between ventilation frequency and swimming activity.

Extreme caution must, therefore, be exercised if information about heart rates and ventilation frequency is used to estimate energy expenditure associated with various levels of activity. Some of the difficulties inherent in using information of this type can be overcome by monitoring the activity of the swimming muscles directly, but this method, too, is not completely devoid of problems. Different muscle fibre types are activated at different swimming speeds. This can lead to either changes in, or an uncoupling of, the relationship between the electrical activity recorded in the muscle tissue and swimming speeds of the fish, making accurate assessments difficult.

Estimates of the energy expenditure of fish in the wild have been made for a small number of species. Despite the limitations of the different methods employed, there appears to be a consensus that relatively little time and energy are expended upon swimming activity. Under normal circumstances, approximately 10%, and seldom more than 20%, of the total energy expenditure will be used for swimming. It must be stressed, however, that the species studied (brown trout, rainbow trout, cod, *Gadus morhua*, largemouth bass, *Micropterus salmoides*, pike, *Esox lucius*) can be considered as being relatively sedentary, and a larger proportion of the total energy expenditure would be expected to be devoted to swimming in highly active pelagic species.

8.5 EFFECTS OF FEEDING ON METABOLIC RATE, M_F

The metabolic rates of fed fish are higher than those of fish deprived of food, and regular provision of food can lead to rates of metabolism being maintained well above those recorded for unfed fish. Fed fish may be more active than unfed fish and this may contribute to the elevated metabolic rate. In addition, feeding fish may use energy in searching for food, in the capture of prey, or in the general excitement of a feeding frenzy. Whilst the activities associated with food capture may lead to some elevation in energy expenditure, such activities cannot be the major cause of the increase in metabolism because the highest rates of metabolism are usually observed several hours after the cessation of feeding.

Thus, following the ingestion of food there is an increase in the metabolic rate, which is not the direct result of increased activity associated with food capture, and the question of why this should occur is one that has received considerable attention. The 'causes' of the post-prandial increase in metabolic rate are many, but some of the components may contribute to the increase in energy expenditure to a greater extent than others.

In fish, the effects of feeding on metabolic rate have been studied by examining the post-prandial increase in the rate of oxygen consumption. The rate of oxygen consumption increases steeply shortly after the ingestion of a meal, and has usually been found to peak at a level 2–3 times the pre-feeding level. The peak usually occurs within a few hours after the end of a meal, and the metabolic rate then gradually declines to the pre-feeding level (Fig. 8.6). The duration over which the effect can be observed is affected by a number of factors including meal size and composition, with an increase in meal size leading to a prolongation of the time the metabolic rate remains elevated above the pre-feeding level. For fish fed a single meal, the duration of the effect of feeding on metabolic

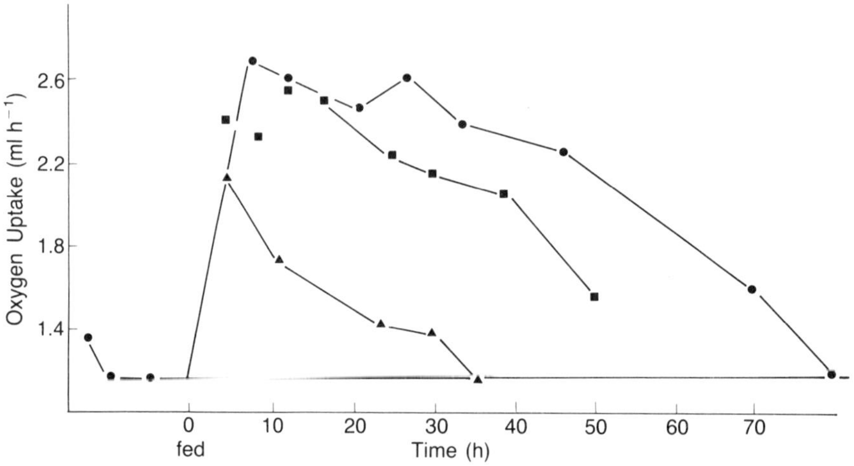

Fig. 8.6 Effects of feeding meals of different sizes on rates of post-prandial oxygen consumption in the plaice, *Pleuronectes platessa.* The example shows the time course of changes in oxygen consumption of an individual fish following the feeding of meals consisting of different amounts of a fish paste diet: triangles, 0.25 ml (1.4 kJ); squares, 0.5 ml (2.8 kJ), circles, 1.0 ml (5.6 kJ). Reproduced with permission from Jobling and Davies (1980), *J. Fish Biol.*, **16**, 629–38.

rate can be markedly affected by temperature. For example, when plaice, *Pleuronectes platessa,* were fed standard-sized meals under different temperature conditions, rates of oxygen consumption remained elevated above pre-feeding levels for approximately 51, 35 and 26 h at 10, 15 and 20 °C respectively. There appear to be clear links between the duration of the effect and the rate of food passage through the gut. Some elevation in oxygen consumption is usually observed for a period several hours longer than the time required for the digestion and absorption of the meal.

The post-prandial increase in metabolic rate results from the energy requirements for the digestion, absorption and storage of nutrients, for the deamination of amino acids and synthesis of excretory products, and for the biosynthesis, turnover and deposition of tissue components, but the relative contributions of each of these processes are exceedingly difficult to evaluate with any degree of accuracy.

Following the ingestion of a meal there is a marked increase in the motor activity of the gastrointestinal tract. Results from studies in which inert bulk has been introduced into the gut suggest, however, that the contribution of the mechanical component to the post-prandial increase in metabolic rate is small. Processing of the food in the gut also involves the digestion and absorption of nutrients, both of which may involve some energy expenditure. Energy will be required for the synthesis and secretion of digestive enzymes, and for the active transport of the products of digestion across the gut mucosa, but the energetic costs associated with these processes remain largely unknown. Nevertheless, the increase in oxygen consumption following the ingestion of a meal is probably best regarded as being a post-absorptive phenomenon, because intravenous injection of nutrients is known to give rise to a marked increase in metabolic rate.

Early studies with mammalian species suggested that there was a close link between the amount of protein fed and the post-prandial increase in metabolic rate. This gave rise to the suggestion that the effect was associated with some aspect of protein metabolism. Work with fish has provided evidence that infusion of amino acids directly into the blood stream leads to a marked increase in metabolic rate. Rates of nitrogenous excretion may increase markedly following feeding. Consequently, it has been suggested that the post-prandial increase in metabolism is the result of the energetic costs linked with amino acid deamination and the synthesis of the nitrogenous products of excretion. There are, however, not always close links between post-prandial metabolic rates and rates of nitrogenous excretion, so this cannot provide the sole explanation for the phenomenon.

Tissue constituents are in a dynamic state and there is a continuous turnover of body proteins, involving both protein synthesis and protein breakdown. Infusion of essential amino acids into the blood has been shown to lead to an increase in both rates of protein synthesis and metabolic rates, and a cause and effect relationship has been suggested. If ingestion of food results in an increase in protein turnover, this would imply that there is an increase in energy expenditure required to maintain the status quo (i.e. metabolic costs increase without there being any net gain of body protein), but when animals are fed they also grow, and the nutrient energy supplied in the food is deposited as protein and lipid. The rate of retention of protein in the body tissues is a net quantity representing the difference between the rate of synthesis and the rate of protein degradation, and it is easy to see that the deposition of 1 g protein as tissue growth will require considerably more than 1 g protein to be synthesized. Results from research with mammals suggest that protein deposition efficiency is often within the range 30–60%, and the few results available for fish species suggest that 2–3 g protein may be required to be synthesized per g protein deposited as tissue growth. Tissue growth is, however, more than just the deposition of protein, and the energetic expenditure associated with the synthesis of lipids must also be considered. There is a great deal of uncertainty involved in the estimation of the energetic costs associated with the synthesis and deposition of tissue biomolecules, but it has been suggested that 60–80% of the increase in metabolic expenditure associated with the ingestion of food may be the direct result of increased protein and lipid synthesis.

The synthesis and deposition of tissue proteins and lipids represents an accretion of new tissue, or body growth, but the ingestion of food will also lead to the production of intermediate storage products that are ultimately completely oxidized. Temporary storage of nutrients as glycogen or fat imposes an energetic cost due to the fact that 'indirect' oxidation, involving 'substrate cycling', is less efficient than would be oxidation of the nutrient directly. A large proportion of the substrate transformations occur in the liver, and the liver shows a rapid metabolic response to large, short-term changes in nutrient supply.

Results from the earliest studies carried out on the effects of feeding on metabolic rate attest to the large contribution made by the liver to the post-prandial increase in metabolism. There are marked increases in rates of protein synthesis in the liver shortly after the ingestion of a meal, and the response to feeding appears to occur earlier in the liver than in other tissues. This can be taken as providing evidence that the costs of protein synthesis, along with those of intermediate storage of nutrients and lipid synthesis, may be the major causal factors of the post-prandial increase in metabolic rate.

8.6 EFFECTS OF PREVIOUS NUTRITIONAL HISTORY ON METABOLISM

The metabolic rates of fish, and other animals, are markedly influenced by feeding, with food intake leading to a post-prandial increase in oxygen consumption that may last for several hours or days. This is a short-term effect of feeding on metabolic rate, but results of studies carried out on domestic animals suggest that the *fasting metabolic rate* ('minimal metabolism') may be affected by the previous nutritional history of the animal. In other words, estimates of fasting metabolic rate are often found to be higher for growing animals that have been well fed, than for animals that have been subjected to long periods of undernutrition or starvation. These differences may be the result of the effects of different feeding levels, or 'planes of nutrition', on the biochemical composition of the body tissues, changes in the relative sizes of different organ systems and influences on the rates of synthesis and turnover of various tissue components.

The majority of the energy expended as fasting metabolism can be viewed as that required to meet two types of function. The first can be considered as 'service' functions, such as the expenditures of the respiratory and circulatory systems in meeting the costs associated with supplying the various body tissues with a supply of oxygen and nutrients. The second type of function relates to 'cellular maintenance' functions, which would include the costs for ion transport and those associated with the synthesis and turnover of the biochemical constituents of the cells. Service functions may account for 35–50% of energy expenditure, and cellular maintenance functions for 40–55%, but the biochemical and metabolic events contributing to each set of functions may not be mutually exclusive.

The most important processes contributing to tissue energy demand are the turnover of cellular proteins and lipids, and ion transport, the latter requiring substantial ATPase enzyme activity. Other energy-consuming processes contributing to energy demand include those related to the turnover of nucleic acids and those required for the synthesis of nitrogenous excretory products, but these additional costs are probably of relatively minor proportions.

The enzyme Na^+,K^+-ATPase, which extrudes 3 Na^+ from the cell whilst transporting 2 K^+ into the cell, is the most important of the ATPase enzymes involved in cellular transport processes. Movement of both Na^+ and K^+ occur against their respective concentration gradients at the cost of one high-energy phosphate bond. Activity of the Na^+,K^+-ATPase enzyme, also known as the 'sodium pump', is responsible for the active transport of substrates, the maintenance of cellular homeostasis, and is involved in the generation and maintenance of membrane potentials. In studies with mammals, it has been estimated that Na^+,K^+-ATPase enzyme activity accounts for 20–40% of energy expenditure, and there can be no doubt that a considerable proportion of the fasting metabolic rate is associated with ion transport via the sodium pump.

A second major component of energy expenditure of cells is that associated with the turnover of cellular constituents, with the energetic costs of protein turnover appearing to be of greatest importance. Protein turnover involves both the synthesis and degradation of proteins, and there may be metabolic costs associated with both processes. The energetic costs of protein synthesis are often assumed to be 4–5 kJ per g protein synthesized, leading to estimates that protein synthesis can account for 12–25% of the energy expenditure in mammals. Protein degradation may occur both via processes that are dependent upon, and via those that are independent of, the utilization of ATP, but the overall costs of protein degradation are not known with any degree of certainty. Nevertheless it is clear that the energetic costs of protein turnover require considerable metabolic expenditure, and these costs may account for approximately 20–25% of ATP utilization.

The protein synthetic capacity of a cell or tissue can be estimated by examining the numbers of ribosomes present. This assumes that the synthetic capacity per ribosome is more or less constant. In practice, the synthetic activity of ribosomes is not constant, but can vary from minute to minute. These changes can exert a short-term control over rates of protein synthesis. Long-term adaptations of synthetic capacity to changes in food supply and nutritional status are, however, brought about more by changes in total ribosome numbers, than by individual activity. Because about 80% of the RNA content of a cell is ribosomal RNA (15% transfer RNA, 5% messenger RNA), determination of the RNA concentration within a cell will provide an estimate of ribosomal numbers and, thereby, give an indication of the protein synthetic potential.

Changes in the nutritional status of an animal lead to changes in tissue RNA concentrations, and these changes can take place gradually over periods of several hours, days or weeks. In animals that are deprived of food for prolonged periods, there will be a gradual decline in tissue RNA concentration. In other words, food deprivation leads to a reduction in ribosomal numbers and a fall in protein synthetic capacity. This decline in protein synthetic capacity that accompanies starvation, or poor food availability, may give rise to an overall reduction in turnover, and the net result may be a reduction in metabolic activity with a conservation of energy reserves.

Re-feeding of a previously starved or undernourished animal leads to a gradual increase in tissue RNA concentrations, a restoration of protein synthetic capacity, and an increase in rates of protein turnover and energy expenditure. Thus, increased food availability may lead to both increased synthetic capacity and an increase in protein turnover in the tissues. In other words, rates of turnover and energetic expenditure may increase, leading to a rise in the metabolic costs of maintenance. Consequently, the energetic expenditure required for the maintenance of the status quo does not appear to be constant, but varies with the previous level of food supply and the long-term nutritional status of the animal.

Many fish species are capable of surviving long periods of food deprivation, and during starvation both metabolic rates and rates of protein synthesis are known to decline. The fall in protein synthesis is the result of both a reduction in ribosomal activity and a decrease in ribosomal numbers. Following re-feeding, there is an increase in both ribosomal numbers and activity level leading to an increase in rates of protein synthesis. Thus, in fish, as in mammals, rates of protein synthesis and tissue turnover appear to be influenced by food availability, and this would be expected to have consequences for rates of energy expenditure. Some evidence for an adjustment of energy expenditure related to available food resources comes from the results of a growth trial conducted with brown trout. Groups of fish that were fed on low rations continued to grow despite the fact that the amount of food supplied appeared to be below the maintenance energy requirement of groups of 'well-fed' fish. Thus, it was suggested that levels of metabolic expenditure became down-regulated under conditions of limited food supply, resulting in energy conservation, and allowing growth to continue despite reduced nutritional input.

Chapter nine

Production and growth

9.1 INTRODUCTION

Ingested energy that is neither lost as faecal or excretory products, nor used for metabolism, is available for growth. Growth can take the form of either somatic or reproductive growth. Within the framework of the energy balance equation, growth is defined as an increase in the energy content of the fish body.

In studies with fish, growth has usually been recorded in terms of weight gain, and it has often been assumed that an increase in body weight is synonymous with an increase in energy content. In other words, it has been assumed that the composition of fish tissue is constant and that a change in weight will accurately reflect changes in the energy content of the body.

The composition of the fish body is not, however, constant, and the relative proportions of protein, lipid, carbohydrate and water making up a given weight gain will vary with feeding conditions and levels of food supply. Lipid deposition tends to increase with increasing food supply, such that the bodies of fish fed maximally will generally contain proportionately more lipid than those held on restricted rations. Similarly, fish tend to deposit increasing amounts of body fat as they grow older, so the percentage body lipid content is usually found to be higher in large than in rapidly growing young fish.

Deposition of 1 g lipid (38 kJ) leads to a weight increase of 1 g, whereas deposition of 1 g protein (24 kJ) also leads to the deposition of 3–4 g water. In other words, the deposition of 1 g protein will lead to a gain of 4–5 g tissue weight. Thus, changes in weight may not always give an accurate reflection of growth in terms of energy gain. Failure to take into account the changes in body composition that may occur under different rearing conditions can result in incorrect conclusions being drawn about the efficiency with which ingested energy has been deposited as body tissue.

Results of experimental studies on growth must be presented with reference to some definite framework of time, so that it becomes appropriate to discuss growth in terms of a *growth rate.* If the fish has increased in weight from W_1 to W_2 during the time period between t_1 and t_2 then the weight gain ($W_2 - W_1$) is an expression of absolute growth, and weight gain with time ($W_2 - W_1/t_2 - t_1$) is known as the *absolute growth rate.* Results of growth studies are, however, much more frequently expressed in terms of relative growth, such as weight gain relative to the initial body weight of the fish: $W_2 - W_1/W_1(t_2 - t_1)$. A special case of relative growth occurs when reference is made to growth at a particular instant in time rather than over a protracted time interval. This is given the name the *instantaneous growth rate* (g), and multiplying by 100 gives the *specific growth rate* (G), which is the most frequently used expression of growth in experimental studies. In practice, specific growth rate (G) is calculated as:

$$G = [(\ln W_2 - \ln W_1)/(t_2 - t_1)] \times 100 \qquad (9.1)$$

where W_1 and W_2 are body weight at the start and end of the growth period, respectively, and $t_2 - t_1$ is the length of the period in days.

9.2 BASIC RELATIONSHIPS BETWEEN FEEDING AND GROWTH

Under any given set of circumstances it might be expected that the more food a fish consumes, the more rapidly it will grow, with the upper limit to the rate of growth being determined by the maximum amount of food the fish will, or can, consume. The overall relation of growth rate to food consumption has been determined for several species of fish and this *growth–rations* curve is a fundamental relationship in the understanding of the influence of environmental factors on growth.

An example of a growth–rations curve is given in Fig. 9.1(a), and it can be seen that the relationship between growth rate and food consumption (ingestion rate) is not a simple linear one, but is curvilinear. Starting at zero ingestion ($R = 0$), where the fish are starving and losing weight (note the negative intercept on the growth axis), the curve rises steeply to cross the point of zero growth rate ($G = 0$) at the maintenance ration. R_{maint}, the maintenance ration, is defined as the amount of food required to be ingested by the fish in order to maintain bodily functions without either loss or gain in weight. With increasing levels of ingestion the curve begins to flex, and it plateaus at the point of maximum food consumption and maximum growth rate.

The relationship between growth (G) and ingestion (R) is curvilinear, rather than linear, so the drawing of a tangent through the origin will

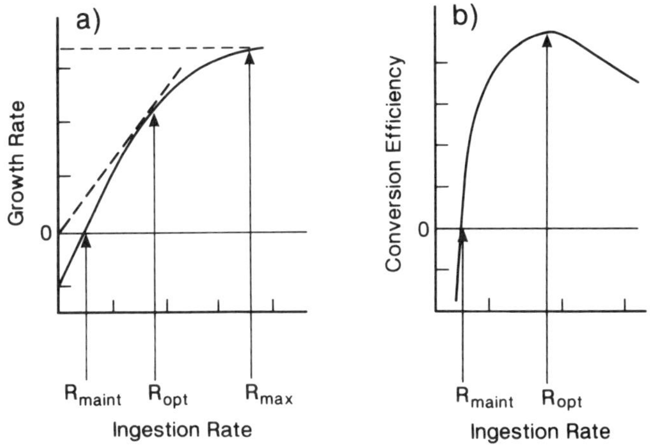

Fig. 9.1 The relationships between rates of ingestion, growth rates and food conversion efficiency for fish species. (a) A growth–rations curve illustrating maintenance (R_{maint}), optimum (R_{opt}) and maximum (R_{max}) ingestion rates. (b) Effects of ingestion rate on food conversion efficiency, illustrating that best food conversion is achieved when ingestion is at an intermediate level (R_{opt}). R_{maint}, maintenance ration, is the ingestion rate required for fish to maintain body weight. R_{opt}, optimum ration, is the ingestion rate at which gross conversion efficiency (weight gain per unit food intake) is maximized. This is indicated in (a) by the point at which the dashed line running from the origin strikes the curve at a tangent. R_{max}, is the maximum ration. The horizontal dashed line in (a) indicates the growth rate, G_{max}, to be expected at R_{max}.

touch the curve at the point where the ratio of growth to food consumption is maximal (Fig. 9.1(a)). That is, the tangent will touch the curve at the point where the growth per unit food intake is greatest, and conversion efficiency (G/R) is maximized (Fig. 9.1(b)). Therefore, it is obvious that maximum growth and maximum conversion efficiency are not achieved under the same conditions of feeding, and this fact should be borne in mind when experimental protocols are being designed.

Several attempts have been made to describe the curvilinear relationship between growth and ingestion in mathematical terms, enabling comparisons between growth studies to be made. One of the simplest forms of expression involves a logarithmic transformation of the ingestion and growth data leading to the relationship being described as:

$$\ln (G + 1) = \mathrm{A} + \mathrm{B} \ln (R + 1) \qquad (9.2)$$

where A and B are the intercept and slope of the regression, respectively. A value of 1 is added to growth and ingestion rates to remove negative and zero values.

Growth rate may also be plotted as a function of ingestion rate using a Monod threshold-corrected hyperbolic equation, assuming substrate-limited growth. According to this threshold-corrected hyperbolic growth equation, the growth rate (G) for any given ingestion rate (R) can be calculated as:

$$G = [G_{max}\,(R - R_{maint})]/[(R_m - R_{maint}) + (R - R_{maint})] \quad (9.3)$$

where G_{max} is the maximum growth rate, R_{maint} is the maintenance ration and R_m is the rate of ingestion that results in the fish showing growth rates half of the maximum.

9.3 LABORATORY METHODS FOR GROWTH STUDIES

When conducting laboratory studies on fish energetics and growth, the first objective will usually be to obtain information about the maximum growth rates of the fish under different environmental conditions. Fish will, therefore, often be fed according to regimes that aim to ensure that rates of ingestion are maximized, and the feeding options open to the experimenter are feeding to excess and satiation feeding.

Feeding to excess

This can be employed both when natural prey and when formulated pellets are used as food. When pelleted feeds are used, the food can be provided from automatic feeders timed to distribute portions of food at short intervals. This should ensure that food is available more or less continuously, and therefore give the fish the opportunity to consume maximum rations and grow maximally under the given set of experimental conditions. The disadvantages of this method are that it is wasteful of food and, unless care is taken, the excess food may disintegrate and lead to a deterioration of water quality in the experimental tanks. Furthermore, although the fish may be feeding maximally, it is impossible to determine a value for this maximum rate of feeding without applying special techniques.

Quantitative determinations of the gut content of fish can be made using X-radiography, and application of X-radiographic techniques allows estimations of food intake to be made on groups of fish that have been fed in excess. Prior to the start of a measurement period the fish are continuously provided with food distributed using automatic feeders. A measurement period starts when the normal food is replaced by food of the same nutritional composition, but formulated to contain a low concentration of X-ray-dense particles. Following completion of a feeding period the fish are

X-ray photographed, the X-ray plates are developed and the numbers of X-ray-dense particles present in the guts of the fish are counted. By reference to previously calculated regressions between particle numbers and weight of food, the amount of food consumed by the fish can be estimated. The method has the advantages that fish need not be deprived of food prior to the start of a measurement period, disturbance of the fish can be kept to a minimum during feeding and the method is also non-invasive, which allows repeated measurements to be made on individual fish.

If the X-radiographic method is to be used for the assessment of food intake, several criteria must be fulfilled. Firstly, it is clear that the X-ray-dense particles must be retained in the gastrointestinal tract of the fish for some time, and not be rapidly passed through the gut and lost in the faeces. Any defecation of the X-ray-dense particles in the time between the start of a feeding period and the time of X-ray photographing would lead to an underestimation of food intake. Secondly, it is obvious that, if repeated measurements of food intake are to be made on individual fish, all particles from previous meals must have been evacuated from the gut before any additional measurements can be undertaken. Because several days may be required before all the X-ray-dense particles consumed in one feeding bout are evacuated, there must be a reasonably long time interval between successive measurement periods. Provided these limitations are recognized, the X-radiographic method is a valuable tool for the study of food intake and feeding behaviour of fish exposed to a variety of conditions.

Satiation feeding

This is defined as the maximum amount of food a fish will consume, and can therefore only be achieved by having food continuously available. Some species of fish can be trained to operate demand-feeders that release food when they perform some specific act such as attacking a trigger or breaking a light beam. Demand feeding should, in theory, satisfy the criteria for satiation feeding. Not all the food released by the feeder will, however, be consumed by the fish, and when demand feeders have been used in experimental studies considerable food wastage has often been observed. Thus, unless the amounts of food wasted are recorded, the use of demand feeders does not permit the food consumption of the fish to be monitored with any degree of accuracy.

There are, therefore, problems in providing conditions that fulfil the requirements for satiation feeding, and 'satiation feeding', as practised in laboratory growth and energetics studies, usually represents a compromise. The term has been used to describe feeding regimes employed in a wide range of experimental protocols. Most usually, 'satiation feeding' has come to mean the maximum amount of food a fish will consume when presented with food two or three times per day (Table 7.1).

It is questionable how accurately this approximates to true satiation feeding, but it is suggested that under the experimental conditions most often employed with salmonids (temperature 10–15 °C; fish weight 10–50 g) the estimate obtained is close to maximum ingestion. For example, the feeding of small (10 g) rainbow trout more frequently than twice per day at 10 °C gave no significant increases in rates of food intake and growth. Similarly, in a comprehensive study of the effects of feeding frequency, fish size and temperature on food intake in brown trout, it was found that feeding the fish two or three times per day was adequate to ensure the consumption of maximum rations under normal rearing conditions.

Even though the presentation of food to the fish two to three times per day should be sufficient to ensure maximum rates of ingestion, the experimenter will be faced with the problem of deciding when to present food in order to achieve maximum feeding. Fish may not show equal propensity to feed at all times of the day and, when allowed to feed voluntarily, fish of many species display peaks and troughs of feeding activity during the course of a 24 h cycle.

Some species are most active at night, others during the day and several predatory fish species feed most actively in the periods around dusk and dawn. The fish may not only display differences in feeding activity at different times during the course of the 24 h cycle, but the times of day at which the fish feed most avidly may also change with season. For example, during spring and autumn both rainbow trout and Arctic charr, *Salvelinus alpinus*, show peaks of feeding activity around dusk and dawn, but during the winter months the majority of food ingestion occurs during the hours of darkness. In summer, feeding activity is greatest during the daylight hours. By contrast, juvenile Atlantic salmon appear to feed almost exclusively during the hours of daylight, irrespective of season.

Restricted feeding

If the aim of an experiment is the investigation of a growth–rations relationship, fish will have to be fed on a range of rations varying from zero (starvation) up to maximum. In other words, various restricted feeding regimes will have to be employed. This will usually entail the feeding of a fixed amount of food to the fish at some predetermined time of the day. Within medical and agricultural sciences, the effects of different feeding and rearing practices on a number of physiological variables have been investigated, but there is comparatively little information available for fish species. There is, however, evidence that both growth and fat deposition may be influenced by the time of day at which food is provided to the fish. Thus, by arbitrarily adopting a given feeding routine (feeding once or twice per day at fixed times), the experimenter may inadvertently impose a growth restriction on the fish, whilst at the same time influencing patterns of fat deposition.

When fish are held together in groups interactions between individuals will often lead to the establishment of social (dominance) hierarchies. Once established, a hierarchy can remain stable for an extended period of time, with the social rank of the dominant fish being reinforced by performance of threat displays or overt acts of aggression (chase and bite) against subordinate individuals. One consequence of the social hierarchy is that resources (e.g. space, feeding sites and food supplies) are not equally divided between all members of the group, and it is usually the dominant individuals that secure first access to any limited resource. Thus, when food supply is restricted, the food will not be equally apportioned and this may lead to large differences in food consumption and growth between individual fish within the group (Chapter 7). Acts of aggression may increase in frequency during feeding, and severe restriction of food supply will also, generally, lead to increases in agonistic interactions between the fish as they compete for the limited resource. Thus, it is clear that the establishment of social hierarchies may influence the results obtained in a feeding trial, and the effects of hierarchy formation are likely to become increasingly important with decreasing food supply.

Growth–rations relationships are traditionally studied by supplying groups of fish with different amounts of food, and then the relationship between the biomass change of the group of fish as a whole and the food supply is examined. Thus, the relationships are determined from bulk biomass changes. The possible effects of hierarchy formation on food intake and growth of individual fish preclude the use of individual growth data in the examination of the growth–rations relationships. Consequently, the determination of growth–rations relationships is a very resource-consuming process, requiring access to large numbers of fish and extensive rearing facilities.

An alternative to group rearing is to work at the level of the individual fish, in which each fish is held isolated and fed according to a predetermined feeding regime. The following of the feeding and growth of individuals has advantages over the investigation of growth–rations relationships using groups of fish and bulk changes methodology, but relatively few studies have adopted the individual approach. When individual rearing is employed it is essential that sufficient numbers of fish be held at each feeding level (a minimum of five may be considered satisfactory, but more would be preferable) and the time required for the individual feeding of large numbers of fish may be considerable. Thus, time constraints may be one reason why the individual approach has not proven to be particularly popular. In addition, fish of many species do not adapt well to isolation and may either cease feeding when held out of contact with conspecifics, or require considerable time to become thoroughly acclimatized to this type of rearing condition.

Chapter ten

Environmental factors and growth

10.1 INTRODUCTION

In its simplest form, the energy balance equation can be written as $E(\mathrm{P}) = E(\mathrm{In}) - E(\mathrm{Out})$, where $E(\mathrm{P})$ is energy available for growth, $E(\mathrm{In})$ is ingested energy and $E(\mathrm{Out})$ represents the energy losses due to metabolic and other processes. Consequently, any abiotic factors that affect rates of food consumption and metabolism may be expected to have profound influences on the growth of fish.

Under laboratory or commercial hatchery conditions it is possible to perform environmental manipulations that can lead to significant improvements in both growth rates and food utilization efficiencies. For example, growth rates of sockeye salmon, *Oncorhynchus nerka*, held at the optimum temperature for growth (15 °C) could be improved from $1.4\%.d^{-1}$ to $2.4\ \%.d^{-1}$ by making adjustments to water flow rates, photoperiod and salinity. When fish were held in water of isosmotic salinity under long-daylength conditions, not only did food intake and growth rates increase, but food utilization efficiencies were improved at all ration levels.

In practice, it is extremely difficult to conduct experiments which can provide sufficient information to allow a detailed analysis of the influences of the simultaneous manipulation of several environmental factors. Consequently, it has been usual to consider the effects resulting from the manipulation of a single environmental variable under otherwise constant conditions, and there are very few instances where the influences of environmental factor interactions have been examined.

10.2 EFFECTS OF TEMPERATURE ON GROWTH

Temperature, which is known to influence both ingestion (Chapter 7) and metabolism (Chapter 8), will also affect growth rates. It is, however, important to distinguish between the effects of temperature *per se*, and the effects on growth induced by the interactions between temperature and limited food supply.

Under conditions of unlimited food supply, an increase in temperature will lead to an increase in food intake, but at high temperatures there will be an abrupt decline in rates of ingestion (Fig. 10.1(a)). Metabolic rate, on the other hand, usually increases with increasing temperature (Fig. 10.1(a)). The changes in food intake and metabolic rate brought about by changes in temperature are shown in Fig. 10.1(a), and the difference between the two lines represents the resources available for growth. Figure 10.1(b) shows that growth increases with increasing temperature, peaks and then declines at high temperature. The temperature at which growth is maximized is called the *optimum temperature for growth*, and it

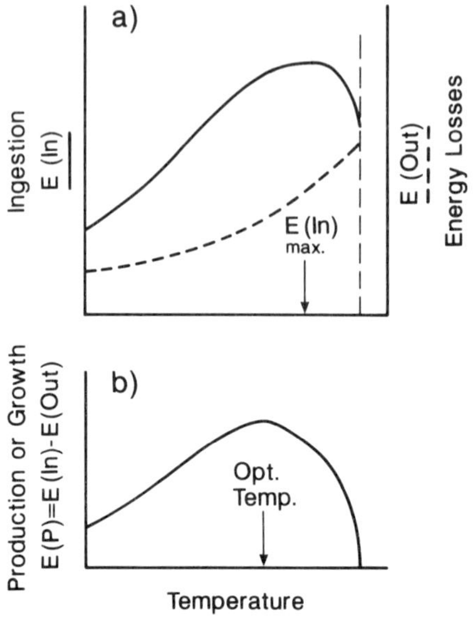

Fig. 10.1 Effects of temperature on (a) rates of ingestion and metabolism, and (b) the consequences of these relationships for resources available for growth. Energy available for growth, *E*(P), is calculated as ingested energy, *E*(In), minus energetic losses, *E*(Out). Note that the temperature at which growth is maximized (Opt. Temp.) is a few degrees lower than the temperature at which the rate of ingestion is highest (*E*(In) max.). The vertical dashed line in (a) indicates the maximum temperature tolerated by the species.

should be noted that this optimal temperature is a few degrees lower than the temperature at which food intake is greatest.

The term optimum temperature for growth should only be used when describing the results obtained in studies in which fish have been fed maximally, because restricted feeding will have a marked influence upon the growth rate observed at any given temperature.

There are large interspecific differences in temperature optima. The majority of salmonids, for example, have temperature optima within the range 12–17 °C (Table 10.1), whereas many cyprinids have optima of 20 °C or higher. In addition, there may be ontogenetic changes in temperature optima for a given species, with larvae and juveniles often having a higher optimum temperature for growth than larger conspecifics. Thermal optima for growth may also be influenced by photoperiod (daylength). For example, 0-group hybrid bass (female *Morone saxatilis* × male *M. chrysops*) displayed best growth at 27.9 °C on increasing photoperiods, whereas the temperature optimal for growth was reduced to 25.7 °C when fish were held on simulated autumnal photoperiods, with decreasing daylength.

Once the thermal optima for growth have been determined, attention is likely to be directed towards the investigation of conditions that result in the most efficient utilization of food. This involves the construction of growth–rations curves for a range of temperatures, a very resource- and time-consuming task. Consequently, there are very few complete records of these relationships, but some work of this nature has been carried out for salmonid and other commercially farmed species. An example of growth–rations curves for different temperatures is given in Fig. 10.2. The drawing of tangents through the origin to touch the various growth–

Table 10.1 Temperature optima for growth and the effects of body weight (*W*, g) on the specific growth rates (*G*) of different salmonid species. Relationships between body weight and growth rate are linearized by plotting data on double logarithmic axes: $\ln G = \ln a + b \ln W$, where a, the growth coefficient, is the growth rate of a fish weighing 1 g and b is a weight exponent. Temperatures indicated are the, presumed, optimal temperatures for growth of the different species

Species	Temperature (°C)	Growth coefficient, a	Weight exponent, b
Sockeye salmon, *Oncorhynchus nerka*	15.5	5.42	−0.40
Rainbow trout, *O. mykiss*	17	6.86	−0.32
Pink salmon, *O. gorbuscha*	15	9.78	−0.45
Brown trout, *Salmo trutta*	13	2.79	−0.32
Arctic charr, *Salvelinus alpinus*	14	8.54	−0.35

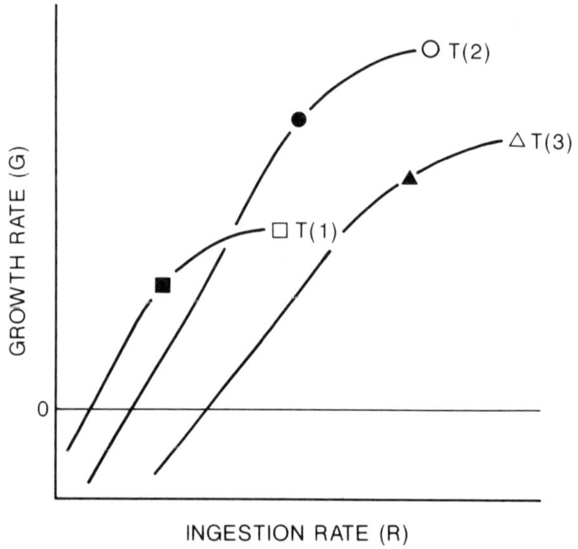

Fig. 10.2 Growth–rations curves for fish reared at different temperatures, ranging from low, T(1), to high, T(3). Open symbols indicate combinations G_{max}:R_{max} at the three different temperatures, whereas filled symbols indicate G_{opt}:R_{opt} combinations.

rations curves gives estimations of the growth rates (G_{opt}) and ration levels (R_{opt}) at which food utilization (G/R) is best at the different temperatures. An examination of the growth–rations curves reveals a number of important points with respect to food utilization efficiency. Firstly, rates of feeding that result in maximum growth are not those that give rise to the most efficient utilization of food. Secondly, the growth–rations curves for different temperatures are not superimposed upon each other, nor do they necessarily run parallel to each other (Fig. 10.2). In other words, the curves are displaced in relation to each other, as a result of the fact that the amounts of food required to support metabolism and maintain a given rate of growth increase with increasing temperature. One consequence of the displacement of the growth–rations curves is that the values of G_{opt}:R_{opt} differ between temperatures. In other words, the feeding conditions giving best food utilization vary with temperature.

The plotting of values of G_{opt} and R_{opt} together with those for maximal growth (G_{max}) and rations (R_{max}) reveals that G_{opt} rises steadily from a position well below G_{max} at low temperatures to be almost superimposed on G_{max} at high temperatures (Fig. 10.3). From this it follows that R_{opt} gradually approaches the maximum ration (R_{max}) as temperature increases towards the upper tolerance limit for growth (Fig. 10.3).

In the previous paragraphs discussion has been directed towards the

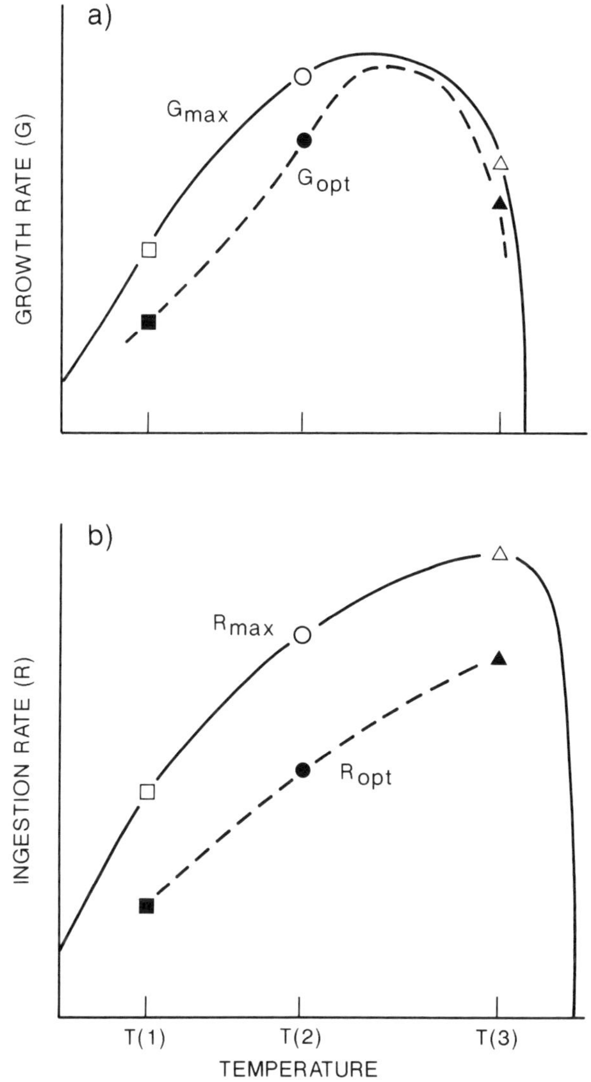

Fig. 10.3 Summary configurations illustrating the effects of temperature on (a) maximum (G_{max}) and optimum (G_{opt}) growth rates, and (b) maximum (R_{max}) and optimum (R_{opt}) rates of ingestion. Points given in the figure (open and filled symbols) refer to points indicated on the growth–rations curves presented in Fig. 10.2.

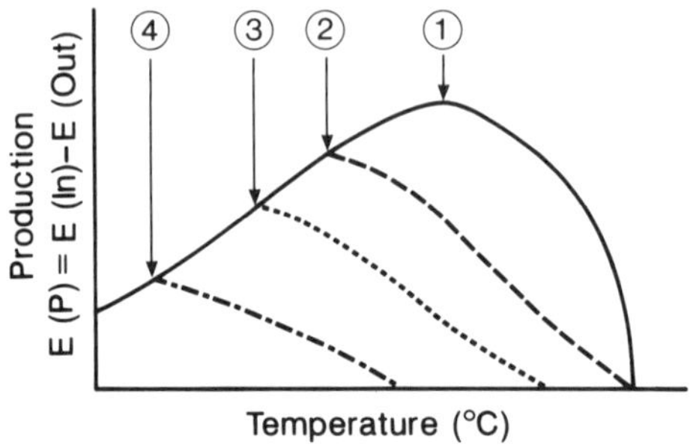

Fig. 10.4 Influence of temperature upon growth (production) of fish held under different conditions of food restriction: curve (1), unlimited food supply; (2), slight restriction; (3), moderate restriction; (4), low level of food supply. Arrowheads (1–4) indicate the temperatures at which growth rates will be greatest under the different regimes.

determination of the conditions under which food conversion is most efficient, and this required the plotting of growth–rations curves for a range of temperatures. Growth–rations curves also permit the investigation of growth–temperature interactions for any given ration level. Under conditions of restricted, as opposed to unrestricted, feeding there will be changes in the relationships between temperature and growth rate, and as food supply becomes more and more restricted, the best rates of growth will be observed at lower and lower temperatures (Fig. 10.4). The most obvious reason for this can be seen by considering the effects of temperature on metabolic rate. With increasing temperature there is an increase in metabolic rate and, consequently, the amount of food required to hold body weight constant (maintenance ration) increases with a rise in temperature (Fig. 10.2). Thus, for any given size of ration the 'scope for growth' (energy available from the food minus the maintenance requirement) decreases as temperature is increased. Consequently, as the ration size is gradually reduced, the scope for growth is greatest at progressively lower temperatures, and the marked increase in energy requirements for maintenance that accompanies rising temperature accounts for the fact that fish lose weight when fed low rations at high temperatures.

10.3 EFFECTS OF SALINITY ON GROWTH

The osmotic and ionic concentrations of fish body fluids differ from those of the surrounding medium, so the fish is required to expend a certain amount of energy in order to meet the metabolic costs of ionic and osmotic regulation. Consequently, it has been hypothesized that if the external environment were manipulated to ensure the reduction of the metabolic costs of iono- and osmoregulation to a minimum, then the growth and food utilization of the fish would be improved. Metabolic costs would be expected to be minimized when fish are held in isionic and isosmotic media, and it has been observed that the abilities of a number of freshwater fish species to osmoregulate decline as the osmotic concentration of the external medium becomes hyperosmotic to the body fluids.

The results of some experimental studies indicate that when salmonids, and some other euryhaline species, are held in media that are isosmotic with their body fluids, their metabolic rates are lower than those of fish held in either fresh water or seawater. These studies would seem to provide evidence that the energetic costs of iono- and osmoregulation are lowest in environments where gradients between blood and water are minimal. Other workers have failed to confirm these results and consider that the metabolic costs required for iono- and osmoregulation (perhaps as little as 1–2% of total metabolism) are so small that any reductions would be almost undetectable, and would certainly not be of the magnitude required to result in any growth advantage.

There have been a number of attempts to test the validity of the 'isosmotic hypothesis' by conducting growth trials on fish held under different salinity conditions, but the results have been somewhat equivocal. The majority of the experiments have been conducted on small salmonids (prior to parr–smolt transformation), and in a number of studies it has been found that rearing salinity has little influence upon growth at the lower end of the range (fresh water to 15‰). At higher salinities, however, growth has been found to be depressed (Fig. 10.5). In very small fish (initial weights 0.5–1.0 g) of some salmonid species, e.g. rainbow trout and chinook salmon, *Oncorhynchus tshawytscha*, there may be a steady decrease in growth rate as the salinity of the rearing water is increased from fresh water to brackish water. At the same time, metabolic rates, monitored as rates of oxygen consumption, may increase as salinity is raised.

The growth reductions observed at higher salinities need not merely be a reflection of changes in the metabolic costs of iono- and osmoregulation with increased salinity. Since growth is a net result reflecting the difference between energy provided by ingestion of food and metabolic expenditure, any influences of changes in salinity on food intake, digestive and absorptive processes would also lead to effects upon growth. Thus, a consideration of growth responses alone is insufficient to provide an adequate analysis of the effects of changing salinity.

Growth–rations curves can be constructed from the data collected from a series of growth experiments carried out on juvenile Atlantic salmon

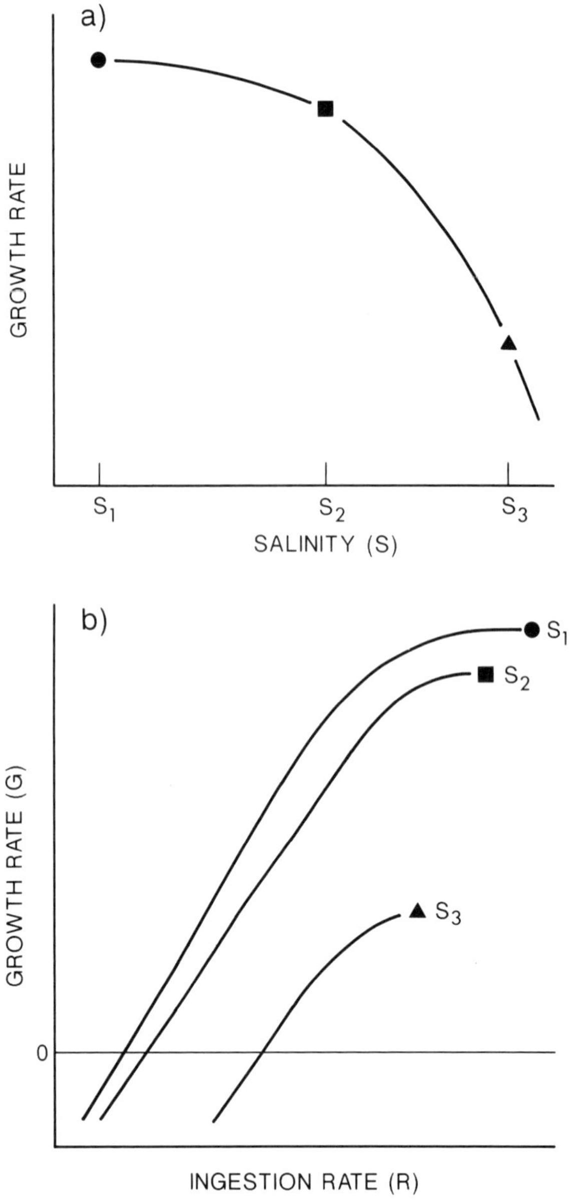

Fig. 10.5 Effects of different salinities (from low, S_1, to high, S_3) on (a) growth rates and (b) growth–rations curves. Note that the points plotted in (a) denote the points corresponding to G_{max}:R_{max} at the respective salinities, as indicated on the growth–rations curves shown in (b).

(40–60 g) reared in a range of salinities at 10 °C. There were minimal differences in the growth–rations curves for the fish held in fresh water and in salinities of 10 and 20‰. The growth–rations curve of the fish held in full-strength seawater was, however, shifted to the right and downwards in comparison with those of the other groups (Fig. 10.5). In other words, the salmon held in seawater displayed reduced growth and poorer food utilization efficiency than fish held in fresh and brackish water. In addition, the maintenance ration increased from 1.3% $BW.d^{-1}$ in fresh water to almost 3% $BW.d^{-1}$ for the fish held in seawater.

Studies of food intake and growth have also been carried out on Arctic charr (150 g) held at a range of salinities, under constant temperature conditions (8 °C). Experiments have been conducted at different seasons of the year. During winter (short daylength) the fish exposed to fresh and brackish water (20‰) fed and grew well, but charr held in seawater fed little and many of the fish lost weight. By contrast, when the experiment was repeated during the summer months (long daylength), feeding and growth was good in fish exposed to all salinities. These differences are probably related to the seasonal changes in hypo-osmoregulatory ability that have been observed in anadromous Arctic charr.

In winter the charr are to be found in fresh water, hypo-osmoregulatory ability is poor and exposure of the fish to full strength seawater (especially at low temperature) may result in some mortality. During the course of the spring, the hypo-osmoregulatory ability of the fish improves and the charr appear to be able to tolerate exposure to full-strength seawater without experiencing any osmoregulatory problems. The fish migrate to the sea, where they spend the summer months, and then return to fresh water during the early autumn.

The seasonal alterations in hypo-osmoregulatory ability, salinity tolerance and the ability to maintain growth in seawater appear to be influenced by changes in daylength, with the increasing daylength of spring triggering the physiological changes required for improved performance in seawater. Thus, the effects of salinity changes may differ depending upon the time of year at which the experiments are carried out. It is clear that the observed responses to changing a single environmental factor (salinity) are not simple, but can be the result of a series of complex interactions with other variables (daylength, temperature).

10.4 PHOTOPERIODIC EFFECTS ON GROWTH

Photoperiod is known to influence both feeding and growth of fish, but, because natural seasonal variations in daylength follow similar cycles to temperature, the effects of the two variables may be confused. Thus, for fish in the wild it may often be difficult to distinguish between the effects of temperature and photoperiod on feeding and growth responses. In some species, however, there may be anomalies that are attributable to photoperiodic, rather than temperature, effects. For example, under natural conditions of photoperiod and temperature, some salmonid species display a rapid increase in food intake and growth during the spring, as daylength increases, even though water temperature may still be quite low. The reverse trend may be observed during autumn as daylength becomes shorter: food intake and rates of growth decline despite the fact that water temperatures remain reasonably high.

For fish that feed visually, long periods of daylight may be expected to increase the length of time that the fish can forage and capture prey. Under such conditions, the effects of long daylengths on feeding and growth would be related to the time available for feeding. Thus, there might be expected to be a relationship between the number of daylight hours ('feeding opportunity'), the amount of food consumed and growth. A relationship of this kind has been reported for juvenile Atlantic salmon during the first summer of life: the greater the total number of daylight hours experienced, the larger was the size of the fish at the end of the summer.

The situation is, however, not as simple as this, because even under conditions when the fish are given equal opportunities to feed (e.g. fed to satiation once or twice per day, or provided with food for the same number of hours per day), there may be differences in growth of groups of salmon reared under differing photoperiods. The exposure of fish to extended daylengths during the autumn and early winter has been found to enhance growth, and this appears to be related to the stimulation of the brain–pituitary axis, resulting in increased somatotrop and thyrotrop activity. Thus, photostimulation may result in an increased production of somatotropin (growth hormone), elevated activity of the thyroid gland and, possibly, increased production of anabolic steroids, leading to an overall enhancement in rates of growth. Beneficial effects of long daylength on growth have also been observed for species other than salmonids. For example, growth rates of channel catfish, *Ictalurus punctatus*, were almost 20% higher when fish were held under a regime of long daylength (14L:10D) than when held on short days (10L:14D).

Changes in photoperiod, with respect to both the direction and the rate of change in daylength, may have profound influences on the growth of a range of fish species. For example, the effects of increasing and decreasing photoperiods on food intake, growth and food utilization efficiency have been studied for hybrid bass, *Morone saxatilis* × *Morone chrysops*, held at a range of water temperatures. At all test temperatures, rates of food intake and growth were greater, and food conversion more efficient, on

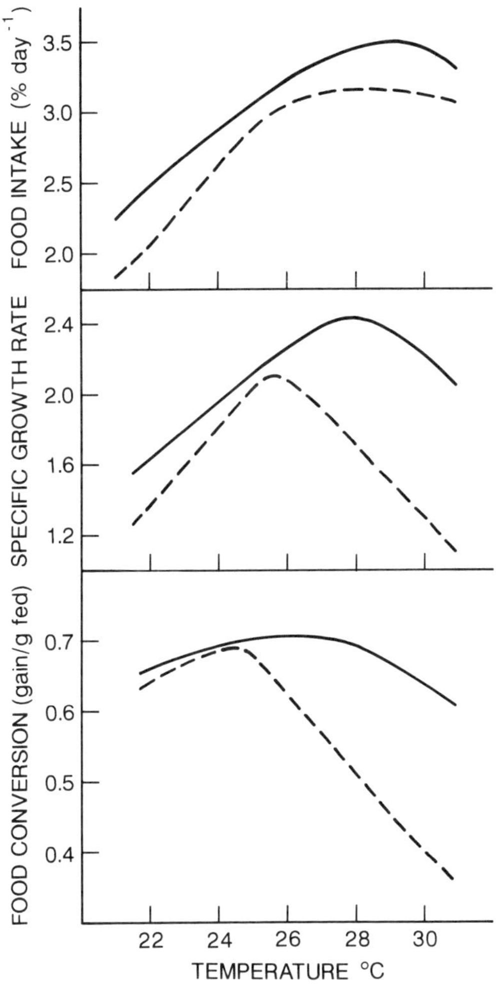

Fig. 10.6 Influence of photoperiod (daylength) on the relationships between temperature and food intake, growth rate and food conversion in hybrid bass (*Morone saxatilis* × *M. chrysops*). Solid curves indicate relationships obtained when daylengths were increasing, and broken curves show results obtained on decreasing daylengths. Based upon data presented in Woiwode and Adelman (1991), *Trans. Am. Fish. Soc.*, **120**, 217–29.

increasing than decreasing photoperiods (Fig. 10.6). There was, however, found to be a photoperiod–temperature interaction, with the optimum temperature for growth being approximately 2 °C lower on the decreasing than on the increasing photoperiod. When fish were fed on submaximal rations, growth was also found to be more rapid on the increasing than on the decreasing daylength regime.

Whilst both constant long, and increasing, photoperiods may have a growth-stimulatory effect, fish species seem to be more sensitive to changing photoperiods than to constant photoperiods. Thus, increasing daylengths appear to produce the greatest endocrine-mediated photostimulation of growth. The growth of the green sunfish, *Lepomis cyanellus*, has been studied using both constant (either long or short daylength) and variable (gradually increasing or decreasing daylength) photoperiods at different seasons of the year. Experiments were conducted under constant temperature conditions. Higher growth rates were obtained for fish held on increasing (8L – 16L), than decreasing (16L – 8L), photoperiod, irrespective of the time of year. In addition, greater growth was observed for fish held on constant long (16L) than constant short (8L) photoperiods, but overall enhancement of growth was greatest on the increasing photoperiod. In this series of experiments, weight gains were greatest during the spring and lowest during the autumn for all the photoperiod regimes studied, suggesting that the effects of photoperiod may have been superimposed upon an innate (endogenous) cyclic growth pattern. Endogenous growth patterns have been reported for a number of species of fish when held under constant conditions of photoperiod and temperature for protracted periods of time, and it is possible that the responses of the fish to changing daylength are of primary importance in the synchronization of an endogenous rhythm of food intake and growth with the prevailing environmental conditions.

10.5 OXYGEN AVAILABILITY AND GROWTH

The metabolic rates of well-fed fish are higher than those of starving fish, and rates of oxygen consumption increase as the feeding conditions of the fish are improved (Chapter 8). When the level of dissolved oxygen in the water is low, food intake may be suppressed (Chapter 7), and this is thought to be due to the fact that reduced oxygen availability would be unable to support the high energy demands of a well-fed fish. A reduction in food intake at low levels of dissolved oxygen would obviously have consequences for growth, and attempts have been made to determine whether there are critical levels of dissolved oxygen at which food intake and growth become affected.

In studies designed to determine critical levels of dissolved oxygen it has been common practice to express levels of dissolved oxygen in terms of concentrations ($mg.l^{-1}$ or ppm). A number of authors have stated that an oxygen concentration of 5 $mg.l^{-1}$ is the critical level for food intake and growth of a range of fish species, but an examination of the literature demonstrates that this is incorrect. For example, the minimum oxygen concentration required for the promotion of maximal growth in coho salmon, *Oncorhynchus kisutch*, may range between 5–8 $mg.l^{-1}$, depending upon rearing conditions, and it has been reported that food intake and growth of rainbow trout may become depressed if oxygen concentrations fall below 6–7 $mg.l^{-1}$.

It is open to question whether oxygen concentration is the appropriate criterion to be used when making recommendations about required levels of dissolved oxygen. Whilst the oxygen concentration determines the volume of water that must be pumped over the gills in order for the fish to obtain a given amount of oxygen, the rate at which oxygen will diffuse from the water to the blood will be dependent upon the oxygen tension. Thus, oxygen tension is also an important factor determining the performance of the fish. The degree of oxygen saturation of the water can be taken as giving an indication of the oxygen tension. Consequently, it may be equally, if not more, appropriate to use oxygen saturation (% saturation), rather than concentration ($mg.l^{-1}$), as a criterion when examining critical oxygen requirements for growth.

These different ways of expressing levels of dissolved oxygen can lead to problems when attempts are made to compare results obtained under different experimental conditions. For example, fully saturated fresh water has a dissolved oxygen concentration of 12.72 $mg.l^{-1}$ at 5 °C, so a 5 $mg.l^{-1}$ critical level of dissolved oxygen would represent 39% saturation ((5/12.72) × 100) at this temperature. The same critical concentration would, however, be equal to 61% saturation ((5/8.24) × 100) at 25 °C. Thus, extreme caution should be exercised in both the presentation and interpretation of results collected under different conditions. Because both oxygen concentration and tension may be expected to influence fish feeding and growth performance in a complex fashion, it is perhaps simplistic to expect that a single critical level of dissolved oxygen will be found to be applicable under all rearing conditions.

When rainbow trout were reared under different conditions of dissolved oxygen at 15 °C, food intake was reduced when the oxygen saturation of the water fell below 60%. Growth and food conversion efficiency, however, appeared to be affected when saturation fell below 70% (concentration 7 $mg.l^{-1}$). Rates of food intake and growth of channel catfish held at 26–27 °C, were lower for groups of fish reared in water at 60% oxygen saturation (concentration 4.9 $mg.l^{-1}$) than for those held in fully saturated water. These two sets of data not only indicate the problems inherent in presentation of results of experiments designed to determine critical levels of dissolved oxygen, but also suggest that oxygen saturation may be of greater importance than concentration in limiting the growth

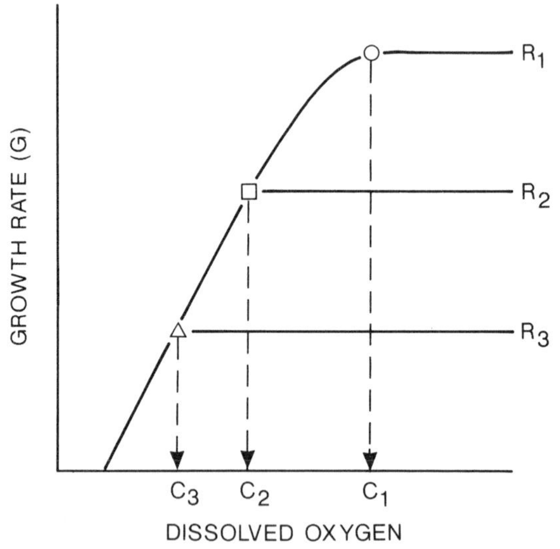

Fig. 10.7 Influence of level of available dissolved oxygen upon the growth rates of fish fed at different feeding rates, ranging from high (R_1) to low (R_3). The critical levels of dissolved oxygen required for the maintenance of maximum growth responses at each feeding rate are indicated by C_1, C_2 and C_3 for the high, intermediate and low rations, respectively.

performance of the fish. Data relating to the critical dissolved oxygen levels for growth in other species of fish are sparse, but from the evidence available it appears that critical levels of oxygen saturation are usually 50–70%, with salmonids being towards the top end of the range.

Under conditions of restricted feeding and reduced food intake, the demand for oxygen will be lower than under conditions of unrestricted feeding. Consequently, the level of dissolved oxygen that is critical for the support of growth gradually falls as levels of feeding become more and more restricted. For example, when channel catfish were fed unrestricted rations, a dissolved oxygen saturation of approximately 60% was found to be critical, but when the fish were fed on a fixed, restricted ration (3% $BW.d^{-1}$) there were no significant differences in growth between groups of fish held in 60% and fully saturated water. Growth was, however, depressed in groups of fish reared in water at 35% oxygen saturation. Thus, as rations become more restricted the levels of dissolved oxygen that impose a limit upon growth are gradually shifted to lower saturation percentages (Fig. 10.7).

Chapter eleven

Biotic factors and growth performance

11.1 INTRODUCTION

Within fish farming it is the desire of the producer that the fish grow rapidly and make efficient use of the feed provided. Whilst this requires that the fish be provided with both correctly formulated feeds and a good rearing environment, there are also numerous biotic factors that can influence both feeding responses and growth performance. For example, growth will be poor if food is presented suboptimally in either time or space, or if feeding responses are depressed as a result of the fish being stressed by the presence of dominant conspecifics. Thus, in order to ensure that the fish are enabled to display rates of growth that approach the physiological potential, it is essential to have information about the ways in which various biotic factors can influence the feed intakes and growth performances of individual fish maintained within larger groups.

11.2 FISH SIZE AND GROWTH

Size-effect studies have been conducted on a range of fish species, but the majority of work has been carried out on salmonids in order to obtain information that can be used in standardizing hatchery feeding and growth practices. The results of some of these studies will be considered in the discussions that follow, but before considering the experimental results, some attention should be given to the theoretical background.

Within the simplest representation of the energy balance equation $E(\text{In}) = E(\text{Out}) + E(\text{P})$, both $E(\text{In})$ and $E(\text{Out})$ can be expressed relative to fish body weight in the form of allometric equations:

$$E(\text{In}) = \text{a}(1)\ W^{\text{b}(1)} \text{ and } E(\text{Out}) = \text{a}(3)\ W^{\text{b}(3)} \quad (11.1,\ 11.2)$$

such that:

$$E(P) = a(1)\ W^{b(1)} - a(3)\ W^{b(3)} \tag{11.3}$$

It has earlier been stated (Chapter 7) that ingestion scales in proportion to body weight raised to the power 0.75, and the weight exponent for the minimal metabolism is approximately 0.86 (Chapter 8). Thus, the plotting of ingestion and metabolism against body weight on double logarithmic axes will produce two straight lines that gradually approach each other as body weight increases. The two lines will eventually intersect at the point where ingestion exactly balances metabolism. In other words, at this point all the ingested energy is required for meeting the metabolic costs of maintaining the body, and no energy is available for growth. The point of intersection will occur at what is, theoretically, the maximum body size of the fish, because at a larger body size the fish would be unable to ingest sufficient food to meet maintenance requirements.

Rates of growth are usually presented in terms of relative growth, such as a percentage increase in body weight per day (% $BW.d^{-1}$). It should be obvious that any growth increment (in real terms) is relatively smaller for a large fish. For example, if a 10 g fish gains 0.2 g in weight during the course of a day the growth rate will be 2% $BW.d^{-1}$, but if a 50 g fish gained the same amount of weight the growth rate would be only 0.4% $BW.d^{-1}$.

The factors discussed in the foregoing paragraphs are important determinants governing the form of relationships between growth rates and fish body size. From the above, it should be apparent that specific growth rate (G) will decline with increasing fish weight. The relationship between growth rate and body size may be described by an allometric function:

$$G = a(4)\ W^{b(4)} \tag{11.4}$$

where b(4) is the weight exponent (in this case a negative exponent) and a(4) represents the growth rate of a fish weighing 1 g. A number of studies have been carried out in which the effects of body weight on growth have been investigated, and there is information available for several salmonid species (Table 10.1). Values for the weight exponent are often close to −0.35, and are usually found to be within the range −0.3 – −0.45. The value of the intercept a(4) in the above equation is dependent upon factors such as the inherent growth characteristics of the species and water temperature, but it is clear that for any given species the value of the intercept will be greatest under those conditions that promote maximum rates of growth (Table 10.1).

It must be stressed that in the discussion of the relationship between growth rate and body size it has been presupposed that fish were feeding at maximal rates, such that growth was not being limited by food supply.

The interrelationships between the growth–rations curves for fish of different sizes show that there is a decline in maximum growth rate with increasing fish size, but at ration sizes below the maximum the growth rates of larger fish can be greater than those of small fish (Fig. 11.1(a)). It is important to note that ration sizes are expressed in proportion to fish weight and, therefore, represent different absolute weights of food. It should also be clear that the 'body size effect' on growth rate will not be observed when experiments are conducted under conditions of food limitation (Fig. 11.1(b)).

11.3 BEHAVIOURAL FACTORS AND GROWTH

Within any group of fish, there will always be some that grow faster than others and, with time, size differences will tend to become more and more exaggerated. Some of the observed differences in growth rates may be due to behavioural factors, and when behavioural elements apply, it has often been found that the largest fish grow fastest and the smallest fish are growth depressed. This has given rise to the concept of the size-related dominance hierarchy, in which the largest fish become socially dominant and secure first access to resources (food, breeding sites, mates). Smaller individuals become subordinate and obtain a disproportionately low share of available resources.

During the initial formation of a dominance hierarchy the social status of different individuals may be determined as the result of overt aggressive interactions. Once established, a hierarchy can remain relatively stable for some time, with the social status of individuals changing little. The position of an individual within the hierarchy is continually reinforced, but this rarely occurs by direct physical contact and fighting. Most often the dominant fish resort to threat displays, or posturing, in order to intimidate subordinate individuals and, in this way, the risk of incurring physical damage is much reduced. Restatement of social status and maintenance of social position can also achieved with the aid of water-borne chemical agents (pheromones). In some species, for example, exposure of a fish to water that has previously held a dominant individual is sufficient to cause cessation of feeding and induce submissive behavioural posturing.

It might be expected that the effects of behavioural interactions on food intake and growth would be most obvious under conditions of restricted food supply. Under such conditions competition for the limited resource would be expected to be heightened (Chapter 7). When food supply is severely restricted, there may be marked increases in agonistic interactions between the fish as they compete for access to food. In experiments where

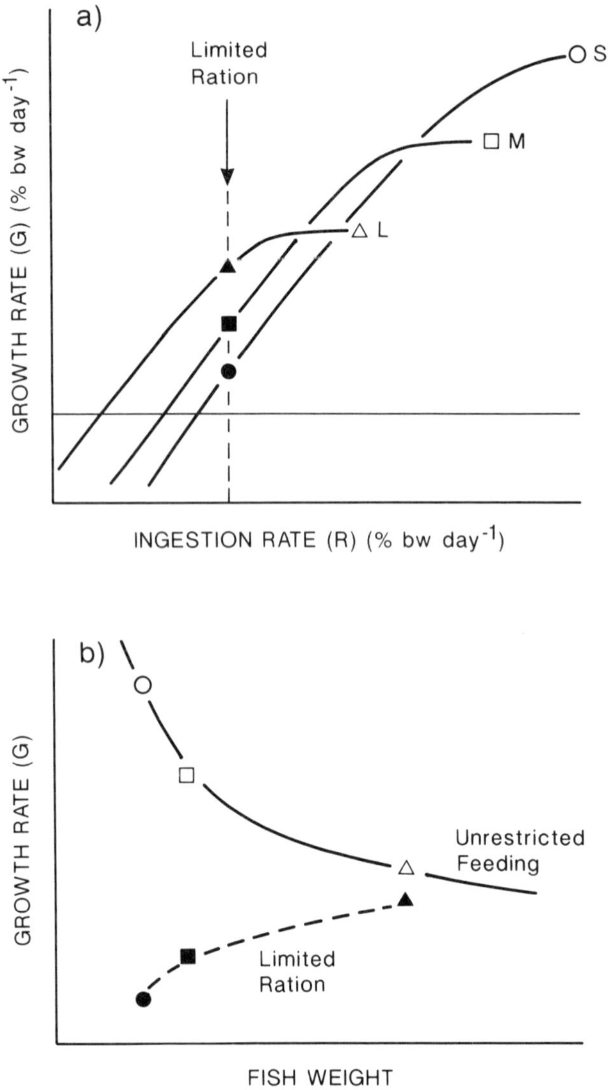

Fig. 11.1 The effects of fish body size on rates of growth illustrated by (a) growth–rations curves for fish of different sizes (S, small; M, medium; L, large) and (b) the relationships between growth rate and body size for fish fed on either unrestricted or restricted rations. Note that both growth rates and ingestion rates are presented as *relative rates,* i.e. as a percentage of the fish body weight per day.

the behavioural interactions between the fish have been observed closely it has often been found that the large, dominant individuals consume a disproportionately large percentage of the food. It is not, however, always the most overtly aggressive individuals that consume the most food and grow fastest. Very aggressive individuals may expend so much time and energy engaging in agonistic encounters that this reduces the amount of time devoted to foraging and feeding activities. In the long term, this may be detrimental to growth. Thus, the fastest-growing fish within a group may be either those that seek to avoid conflict, or those that use less time-consuming and physically demanding methods (e.g. threat displays and posturing) than fighting in order to confirm dominance status and resolve disputes.

Even when the total food supply appears to be well in excess of that required to ensure good growth of all fish within the group, and direct competition for food should be minimized, behavioural interactions between the fish may still have a strong influence on food intake and growth. For example, if the food is unequally distributed in space, certain fish may establish territories that ensure that they monopolize the food to the detriment of other individuals. This type of behaviour has been observed under aquaculture conditions when food has been supplied from automatic feeders at distinct feeding stations. In channel catfish ponds, for example, it has been observed that certain fish, which have usually been found to be the largest individuals, establish territories that include the feeding stations. The remainder of the fish are excluded from the area until the dominants cease feeding and move off to occupy positions in other parts of their home range.

Similar behaviour amongst salmonids may result in the exclusion of some fish from gaining access to food under conditions where the food is distributed at a point source using automatic feeders timed to deliver portions of food at predetermined time intervals. For example, in a study of the feeding behaviour of Atlantic salmon in a sea cage it was found that a small proportion (25%) of the fish obtained two-thirds of the food provided. Furthermore, subordinate fish would not attempt to feed if a dominant individual was near by, even though the dominant was not actively engaged in feeding. Thus, the subordinates appeared to be intimidated from feeding by the close presence of a dominant fish. It is generally believed that the feeding response, food intake and growth of subordinate fish is improved if the food is widely spread over the water surface, rather than being delivered at a point source.

When salmonids are reared in small groups, the results of behavioural interactions between fish will often be manifested by the establishment of clearly defined territories. The behavioural patterns observed may include the patrolling and defence of the territorial boundaries, involving attacks on intruders. As the number of fish within the group becomes larger, incursions over territorial boundaries may increase in frequency and the territory holder will be required to devote an increasing amount of time and energy to territorial defence. The costs incurred in defending the territory will approach and, at some point, eventually exceed the benefits accruing from holding the territory. At this point, territorial behaviour is no longer profitable and the territory should be abandoned. The exact point at which any given territory should be abandoned will be dependent upon the interplay between the costs of defence and the perceived quality of the territory with respect to defensible resources, such as food, hiding places and breeding sites. Thus, as group size increases, the behaviour of the fish will change and when the size of the group becomes sufficiently large the fish may form either distinct schools, or looser aggregations and shoals. These changes in behaviour that occur with increasing group size may be expected to have consequences for feeding and growth, and some attention has been given to the investigation of the effects of group size (stocking density) on growth under fish farming conditions.

Stocking density and growth

Although there have been many experiments performed in order to investigate stocking density effects, it is frequently difficult to distinguish the separate effects of stocking density *per se* from those due to changes in one or more abiotic factors. Thus, whilst there have been several reports indicating that rates of growth become reduced when fish are held at high stocking densities, the reduction can, in many cases, be traced to limitations caused by deteriorations in water quality parameters such as a decline in dissolved oxygen levels or an increase in the level of nitrogenous waste products. However, even when such limitations are removed, there appears to be an optimum density range for some species of fish when fed unrestricted rations.

When held in isolation, fish of some species do not feed and grow well, and they seem to require the social stimulation arising from being reared with conspecifics. Rates of feeding and growth may remain low in groups reared at very low stocking densities, and then increase with increasing stocking density, but under highly crowded conditions growth rates will again decline. For example, growth rates of Arctic charr have been found to be low when fish have been stocked at densities below 10 $kg.m^{-3}$. The growth of individual fish improves quite markedly as stocking densities are increased, and growth rates appear to reach a maximum when fish are stocked at 40–60 $kg.m^{-3}$. Relatively good rates of growth are, however, maintained at stocking densities in excess of 80–100 $kg.m^{-3}$, but there is a decline in growth as stocking densities become extremely high.

When examining the effects of stocking density upon growth, it is important to distinguish between the influence of density upon the growth of individual fish and the effects on the biomass gain of the group as a whole. Thus, whilst the growth rates of individual fish may be maximized at a particular stocking density, the biomass production (individual weight gain × numbers of fish) of the group as a whole is likely to be greatest at a different level of stocking. For example, numerous studies have been carried out with rainbow trout in order to define the optimal stocking densities to be used in commercial production units. Within the range of stocking densities tested (5–100 $kg.m^{-3}$), it has generally been found that the growth rates of the fish decline as stocking density increases, but the greatest production of fish biomass occurs at an intermediate stocking density. Similarly, in studies conducted on Atlantic salmon parr, the highest rates of individual weight gain have often been found at the lowest stocking density tested, but overall production may be greatest in the groups of fish stocked at the highest density. Food utilization efficiency, on the other hand, is often best at an intermediate density, showing that the highest values for individual weight gain, biomass production and food conversion efficiency may occur at different stocking densities.

Similar effects of stocking density on individual growth rates and production of fish biomass have been reported for some non-salmonid fish species. For example, results from a growth trial conducted to examine the aquaculture potential of the pacu, *Colossoma mitrei,* indicated that stocking density had a marked influence on growth and production. Fish (about 50 g body weight) were stocked at initial densities of 5, 11 and 15 $kg.m^{-3}$ and growth was examined in a 3 month trial. At the end of the growth trial the fish stocked at the lowest density were the largest, there having been a 25% difference in weight gain between the fish stocked at the lowest and highest densities. Stocking densities increased during the course of the growth trial, and were at harvest within the range 16–41 $kg.m^{-3}$. Thus, production of biomass had differed markedly between the groups of fish stocked at the different densities, and had been 10.5, 18 and 25 $kg.m^{-3}$ for the low, intermediate and high stocking densities, respectively. These results demonstrate that although an increase in stocking density had led to a reduction in the average growth rates of the fish, overall rates of biomass production were greatest at the highest stocking density tested.

In addition to interactions with other fish, there may be behavioural interactions with environmental factors and these can have consequences for growth. For example, it has been demonstrated that available space can affect growth. When female guppies, *Poecilia reticulata,* were grown in aquaria of differing sizes, the greatest growth was achieved by fish having access to the greatest space. In addition, if fish were switched from a small to a large aquarium there was an increase in growth. Similarly, when four green sunfish were reared together in an aquarium of 4 l volume the fish grew faster than did individual fish limited to a 1 l volume each. Thus, it seems that the rate of growth may be related to the total space available in relation to the size of the fish. In the case of the green sunfish, however, the results could have arisen as a consequence of growth enhancement due to social interactions between the fish. Nevertheless, the 'space effect' may be important, and when rainbow trout fry were reared at constant stocking densities (number of fish per m^3 water volume or number of fish per m^2 tank bottom) it was found that lowest mortality and best growth occurred when the total amount of space was increased in relation to the size of the individual fish.

11.4 SUSTAINED EXERCISE AND GROWTH

In connection with the farming of salmonid species, attempts have been made to avoid exposing the fish to strong water currents in the rearing units. The rationale behind this is that the high levels of energy expenditure required by the fish in swimming, to hold station against the current, would lead to reduced rates of growth and/or increased food costs. Results of a number of experimental studies have not borne out these expectations, and there is an increasing body of evidence suggesting that growth rates of salmonids increase when the fish are forced to swim at moderate speeds for prolonged periods of time.

There is no denying that the energy costs of swimming are high, and increased demands for oxygen during swimming have been confirmed in studies carried out on a range of fish species (Chapter 8). Studies on the relationship between swimming activity and metabolic rate have usually been conducted on individual fish, and, when not forced to swim against a current, these fish generally show only low levels of spontaneous activity. When groups of fish are held in respirometers, however, levels of spontaneous activity may increase and this can lead to marked, though transient, increases in metabolic rate. These increases in activity levels appear to be associated with bouts of aggressive behaviour, which occur at irregular intervals. When groups of salmonid fish are exposed to water currents they show a series of behavioural changes. The fish orientate against the current, may form schools, and levels of aggression appear to be markedly reduced compared with those seen in groups of fish held in standing water. The net result may be that the overall energy expenditures of groups of fish swimming for prolonged periods at moderate speed, and those exhibiting high levels of spontaneous activity at irregular intervals, differ very little from each other.

Thus, levels of energy expenditure may differ little between groups of fish exposed to water currents and those reared in standing water, but this cannot explain why the exercised fish show more rapid rates of weight gain. The improved weight gain in the exercising fish does not, primarily, appear to be due to increased food consumption, but is more the result of these fish displaying better food conversion (Fig. 11.2). In other words, fish forced to swim at moderate speeds for prolonged periods show a greater weight gain per unit food consumed than do conspecifics reared in standing water. The bases of this improved food conversion must lie in the behavioural and physiological changes brought about by inducing the fish to swim for prolonged periods of time.

It is known that prolonged swimming at moderate speeds can give rise to marked changes in the swimming muscles of salmonid fishes. Since these muscles represent 50–60% of the body weight in salmonids, any changes here will have a considerable influence on the growth of the body as a whole. Prolonged exercise induces marked increases in rates of protein synthesis, turnover, and deposition in muscle tissue, and can also lead to a dramatic increase in muscle fibre size (hypertrophy). Rates of protein synthesis, breakdown and deposition are increased in all tissues during exercise, but the most marked changes occur in the swimming

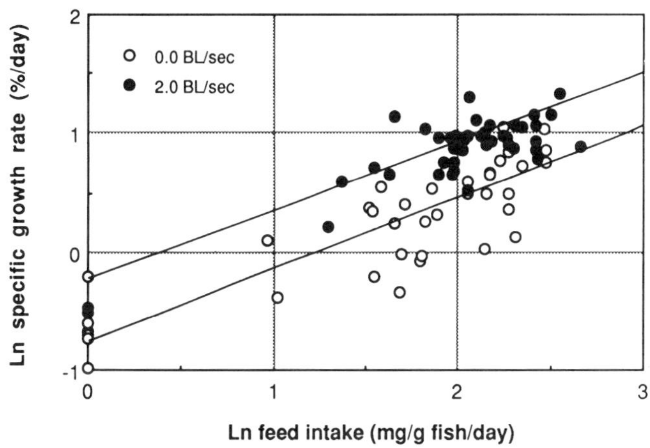

Fig. 11.2 Effects of different levels of sustained exercise (fish reared in standing water or forced to swim against a water current for a period of 9 weeks) on the relationship between rates of food intake and growth of juvenile Arctic charr, *Salvelinus alpinus*. Each point indicates an individual fish. Filled circles indicate exercised fish forced to swim against a water current (initial speed 2 $BL.s^{-1}$). Note that exercised fish show higher growth rates, at a given level of food intake, than do the controls. Reproduced with permission from Christiansen and Jobling (1990), *Can. J. Zool.*, **68**, 2185–91.

muscles. Tissues such as gill and heart have inherently high rates of protein turnover, and protein deposition efficiency is low. In these tissues there may be only relatively small changes in rates of protein synthesis and breakdown in response to exercise. On the other hand, protein turnover in muscle is low and deposition efficiency high. Protein deposition efficiency in the white swimming muscle may be over 70%, which is much higher than the 40–50% often recorded for the efficiency of protein deposition in the body as a whole. Exercise leads to marked increases in both protein synthesis and breakdown in the swimming muscle, and these changes are accompanied by a substantial increase in protein deposition. These disproportionately large effects of exercise on the swimming muscles may lead to both increased body protein growth and an overall improvement in whole-body protein deposition efficiency in exercised fish compared with controls reared in standing water.

Exposure of salmonids to water currents leads to behavioural changes, including a reduction in the frequency of aggressive encounters. Reduced aggression may indicate that the influences of dominance hierarchies are less strong in exercised than in unexercised salmonids, and that exposure of the fish to currents may result in there being a more even distribution of food among the fish making up the group. The coefficient of variation (CV = (SD/mean) × 100) may be used as an index of interindividual variability, and examination of results obtained in studies with salmonids suggests that variability in food acquisition is lower in fish exposed to water currents than those held in static water (Table 11.1). Furthermore, this more even distribution of food is reflected in the rates of growth of the exercised fish being more homogeneous than those of the unexercised

Table 11.1 Interindividual variability in food intake and growth of juvenile Atlantic salmon, *Salmo salar*, and Arctic charr, *Salvelinus alpinus*, reared under different exercise regimes. Water currents were equivalent to swimming speeds of 0, 1.0, 1.5 and 2.0 BL.s^{-1}. Table entries are expressed as the coefficient of variation (CV = (SD/mean) × 100), which is used as the index of variability. Data are from Christiansen and Jobling (1990), *Can. J. Zool.*, **68**, 2185–91, and Jørgensen and Jobling (1993), *Aquaculture*, **116**, 233–46.

Regime	Atlantic salmon		Arctic charr	
(BL s^{-1})	Food intake	Growth rate	Food intake	Growth rate
0	29.6	31.2	64	137
1.0	22.6	23.3	51	46
1.5	25.6	24.5	50	55
2.0	22.8	29.2	39	37

controls. Thus, the increased growth observed in fish held in flowing water may, at least in part, result from improvements in social environment and a weakening of feeding hierarchies.

11.5 HORMONAL MANIPULATIONS AND GROWTH

Since the swimming muscles make up 40–60% of the body weight of most fish species, any changes in the rate or efficiency of muscle growth will have large consequences for overall rates of body weight gain. A number of endocrine factors are known to influence muscle protein synthesis and growth, and some of the possible hormonal actions and interactions are shown in Fig. 11.3. Growth hormone (somatotropin), thyroid hormones and insulin may all have, or may interact to produce, stimulatory effects on muscle protein synthesis, turnover and deposition. Consequently, a number of studies have been carried out in order to examine the effects of treatment with exogenous hormones, either in isolation or in various combinations, on the growth of a number of fish species. The anabolic effects of growth and thyroid hormones have been investigated in several studies, but of all the hormones implicated in the regulation of feeding, metabolism and growth it is the sex steroids and their analogues that have attracted the most attention. Interest in these steroids arose because

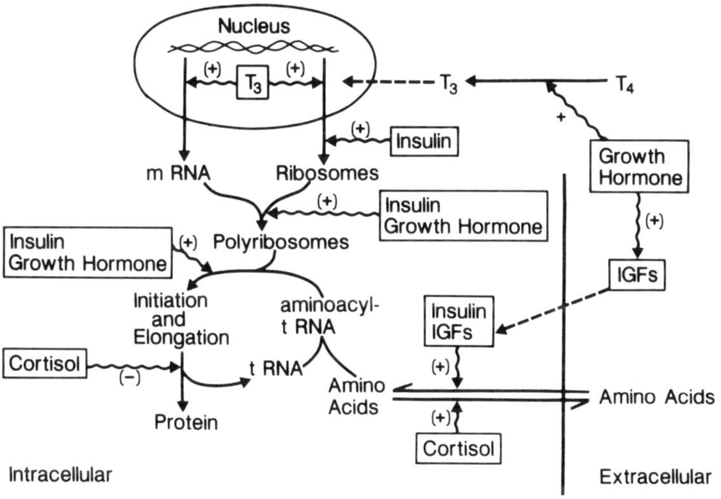

Fig. 11.3 Hormonal control and regulation of protein synthesis in muscle tissue. IGF, insulin-like growth factors (also known as somatomedins); T_3, triiodothyronine; T_4, thyroxine.

of the situation in the livestock industry, where it was clearly demonstrated that hormonal manipulations using exogenous sex steroids could lead to both growth promotion and improvements in food utilization efficiency.

Growth-promoting effects in fish have been reported for all of the types of hormones (growth and thyroid hormones; insulin; sex steroids) listed above, but the doses giving maximum effects upon growth differ between species, and some anomalous results have been reported. For example, hormonal treatments that have been reported to result in increased growth of a range of species have either been ineffective, or have led to growth retardation, in channel catfish.

Growth hormone

Growth hormone (somatotropin), which is produced by the alpha-cells of the proximal pars distalis (anterior lobe) of the pituitary gland (hypophysis), plays an important role in the regulation of metabolism and growth in fish. Growth is either severely retarded or ceases in hypophysectomized individuals. The negative effects of hypophysectomy can, at least in part, be alleviated either by the administration of pituitary extracts containing growth hormone, or by injections of growth hormone alone.

Growth hormones are proteinaceous hormones with somewhat similar chemical structures and biological functions across a range of vertebrate groups. There are, however, significant differences in the amino acid sequences and molecular weights between the growth hormones of the different vertebrate classes. Some differences in growth hormone structure may also exist within smaller systematic groups such as the mammals or the teleosts. These differences in structure may result in native and non-native growth hormone preparations having different biological potencies when tested in any given species.

Much of the work with fish species has been conducted using growth hormone preparations of mammalian origin (bovine, ovine, porcine), but with the isolation and characterization of growth hormone from a number of fish species the use of native preparations has become increasingly common. At the same time, advances in gene technology have provided a means for the production of recombinant growth hormones in relatively large quantities. These developments have provided increased opportunities for the study of the metabolic and anabolic effects of growth hormone using a hormone (recombinant) preparation that should be equipotent to that from the native source.

Treatment of fish with exogenous growth hormone has generally been found to result in an increase in growth rate, but the stimulatory effects of growth hormone have been reported to differ widely depending upon species and dosages used. The efficacy of growth hormone treatment may also be age, or size, dependent, with the growth-promoting effects being relatively greater for older and larger fish. In other words, administration of exogenous growth hormone does not disturb the general pattern of a decreasing specific growth rate with increasing size and age of fish, but the stimulation of growth due to hormone injections is dramatically more pronounced in older fish.

The increase in growth observed following the application of growth hormone is the net result of the influence of the hormone on a range of processes. The administration of growth hormone has been reported to result in higher rates of food consumption in a range of salmonid and ictalurid species, and improved food conversion has also been observed in a number of studies. The reports of increased food consumption following administration of growth hormone tend to be anecdotal in nature, and in studies where food consumption has been closely monitored, there is little evidence that there is any increase following growth hormone treatment. Improved food conversion is, however, a general finding in studies in which fish have been treated with exogenous growth hormone.

The improvements in food conversion observed in fish treated with exogenous growth hormone are probably largely the result of changes in patterns of protein and lipid metabolism, but growth hormone may also lead to a stimulation of the intestinal absorption of nutrients, particularly amino acids.

The metabolic effects of growth hormone may include a stimulation of protein synthesis, and increases in RNA:DNA ratios have been recorded in hybrid striped bass, female *Morone saxatilis* × male *M. chrysops*, treated with exogenous growth hormone. These changes may lead to increased protein deposition, but this does not necessarily imply that there should be changes in the relative compositions of the body tissues with respect to percentages of water, protein, lipid and ash.

Application of growth hormone may also affect carbohydrate metabolism leading to increased insulin synthesis and release, and there may also be an increase in lipid metabolism leading to the mobilization of lipid reserves and a reduction in the lipid content of body tissues. These changes are the result of both direct effects of growth hormone on metabolism and indirect effects resulting from either interactions with, or stimulation of, other hormonal and neuronal mechanisms.

Because growth hormones are proteinaceous in nature it has generally been believed that if these hormones were administered orally they would be attacked and denatured by the enzymes of the digestive tract. Denaturation of the hormones would result in them losing their biological potency. Consequently, injection has been the most widely employed method for the administration of the growth hormone preparation. Injected doses of hormone found to produce anabolic effects range between 0.25 and 100 μg per g body weight per week, with doses of 25–50 $\mu g.g^{-1}$ producing growth stimulation in tilapias and carps. The most effective dose for salmonids appears to be approximately 10–20 $\mu g.g^{-1}$ when the hormone is injected, but when the hormone is delivered using hormonal implant and constant infusion methods, the effective dose may be substantially reduced. Neither injection nor implantation can be used for the treatment of large numbers of fish, and alternative routes of administration have been investigated.

Following oral administration of growth hormone, there appears to be incomplete digestion of the growth hormone molecule, and some of the fragments absorbed into the bloodstream retain growth-promoting activity. Thus, oral delivery of growth hormone may result in accelerated growth of fish compared with the growth of untreated controls. The oral doses required for growth promotion are, however, greater than those found in studies employing either injection or infusion methods. The reduced efficacy of oral treatment is usually interpreted as indicating that there has been partial digestion and denaturation of the administered hormone. Immersion of fish in solutions of growth hormone may also give rise to growth promotion, but again the doses required are much higher than those usually administered using other methods.

As an alternative to the direct use of growth hormone, some attempts have been made to examine the possible growth-promoting effects of compounds thought to be potent stimulators of growth hormone release. For example, a synthetic peptide (enkaphalin analogue) which is known to stimulate growth hormone secretion from the rat pituitary has been shown to promote increased growth and improved food conversion in tilapia, *Tilapia nilotica* × *T. aurea*, following immersion treatment. In addition, oral administration of clonidine, a drug that is known to stimulate secretion of growth hormone in rats and Man, has been reported to have an anabolic effect in coho salmon fry (but not yearlings).

Thyroid hormones

Thyroid hormones appear to be essential for normal growth and development, and these hormones may have a central role to play in governing metamorphosis in some fish species. For example, immersion of larvae in solutions of thyroid hormones has been found to lead to accelerated development and improved survival in the tilapia species, *Oreochromis mossambicus* and *O. niloticus*, and to enhance the growth rates of larval carp, *Cyprinus carpio*, milkfish, *Chanos chanos*, and eel, *Anguilla anguilla*. Administration of exogenous thyroid hormones has also been shown to promote initiation of metamorphosis and give rise to increased rates of development in Japanese flounder, *Paralichthys olivaceus*.

When thyroid hormones are administered to larger fish at near-physiological doses, they appear, in general, to have an anabolic effect. Administration of high doses of these hormones may, however, have catabolic effects and lead to growth anomalies.

It is uncertain to what extent the anabolic effects of thyroid hormones represent a direct action of these hormones on growth. The growth-promoting effects may be the result of stimulation of, and synergism with, other hormones, accompanied by more generalized influences on overall rates of metabolism. Thyroid hormones have been implicated in the stimulation of the release of growth hormone from the hypophysial cells, in the promotion of lipolysis and in the stimulation of food intake.

Many investigators have attempted to increase fish growth by the administration of thyroid hormones, but the results are not consistent even when studies have been carried out on a single species. The factors that appear to be of most importance in governing the results obtained seem to be the dose and form (thyroxine (T_4) or triiodothyronine (T_3)) of the hormone and the method of administration (injection, immersion or oral).

Thyroxine may be effective when added to the water or given by injection, but oral administration of T_4 has been found to be relatively ineffective. It is possible that thyroxine is poorly absorbed from the digestive tract owing to irreversible binding to luminal proteins or feed ingredients. An alternative explanation for the lack of an anabolic effect is that thyroxine is absorbed from the gut but is sequestered and metabolized by the liver, thereby reducing potency.

Oral administration of triiodothyronine, on the other hand, is effective in producing enhancement of growth and improvement of food conversion. Inclusion levels of 5–100 mg T_3 per kg diet have produced anabolic effects in some species of tilapia, eels and salmonids, but similar doses were ineffective in stimulating growth of carp. Thus, there are obviously some interspecific differences in the responses shown to the administration of exogenous triiodothyronine.

The influence of thyroid stimulating hormone (TSH) on growth of fish has been examined, but the anabolic effects have been small. When TSH was administered to hypophysectomized fish there was no evidence of a growth-promoting effect. This suggests that the thyroid hormones do not act alone in influencing fish growth, and that some pituitary hormones are required before thyroid hormones can exert an anabolic effect.

Insulin

Insulin, a polypeptide produced and secreted by the pancreas, has an important role in governing carbohydrate metabolism in mammals, but, in fish, the role of insulin appears to be more important in protein metabolism. Insulin secretion appears to be stimulated by amino acids, and insulin has been shown to promote the incorporation of amino acids into tissue proteins. This points to the hormone having important influences on protein synthesis. Little is, however, known about the possible growth-promoting effects of long-term treatment with insulin, although there are suggestions that administration of insulin can result in increased weight gain and improved food conversion in some salmonid species.

Anabolic steroids

Anabolic steroids, which include the androgens (male sex hormones) and oestrogens (female sex hormones), have been used as dietary supplements in order to enhance the growth of various species of livestock, and experimental studies have been conducted on a range of fish species. Since the steroid hormones are resistant to attack by digestive enzymes they have the advantage over proteinaceous hormones that they can be administered in the food without loss of biological activity. Thus, it is possible to administer the hormone without having to expose the fish to the trauma of handling and injection, and large groups of fish can be treated relatively easily.

Generally, the androgens have been found to be more effective than oestrogens in promoting growth in fish, and the naturally occurring androgens testosterone and 11-ketotestosterone have been shown to have an anabolic effect when fed to fish. It is, however, the growth-promoting effects of the synthetic androgen analogues, and particularly methyltestosterone, that have been most widely investigated. These androgen analogues differ slightly in molecular structure from the natural androgens. In the production of these analogues the aim has been to synthesize steroids that have enhanced anabolic properties compared with the natural androgens, whilst at the same time having reduced androgenic effects. Whilst some of the compounds used (e.g. dimethazine, norethandrolone) seem to promote growth without having marked androgenic effects, the most commonly used anabolic steroid, methyltestosterone, is potently androgenic. Thus, whilst this steroid has been shown to enhance growth in a wide range of fish species (including salmonids, eels, tilapias, common carp, Asiatic catfish, *Heteropneustes fossilis*, and the snake-head, *Channa striatus*), administration of the hormone may also cause androgenic effects such as changes in the gonads and the development of secondary sexual characteristics.

When hormonal treatment is given to larval and juvenile fish, before gonadal differentiation has occurred, complete sex reversal may be induced. The immersion of eggs in solutions of steroid hormones at various times during the course of incubation, and the administration of steroids to young fish via the diet has been shown to induce sex reversal in at least 15 gonochoristic fish species (i.e. non-hermaphroditic species, in which individuals have either testes or ovaries). This gives the possibility of using steroid treatment to obtain fish of one sex, and this has been exploited commercially in the production of single-sex populations. There are advantages in producing and rearing single-sex populations in situations where one sex either grows much faster, or grows to a larger size, than the other.

The effectiveness of the steroid in inducing sex reversal depends upon the timing of the treatment, the dosage given and the test species. The effective oral dose of the androgen methyltestosterone ranges between 3 and 25 mg per kg dry diet, and administration is usually continued for a period of 10–90 days depending upon the age of the fish, the species and rearing conditions (e.g. water temperature). The best example of the use of sex-reversal procedures is in the production of all-female stocks of rainbow trout. This has been so successful that many commercial trout farms now produce only female fish for the market.

The first step in the process is to produce a population of fish in which all are phenotypically male. From the time that the fry first begin to take exogenous food they are fed a diet containing methyltestosterone at a concentration of 3 mg per kg dry diet. This diet is fed for a period of 750 degree-days (e.g. for 75 days at 10 °C), and is then replaced by untreated food which is fed until the time that the fish mature. This regime ensures that all the fish will develop into phenotypic males, but some of the fish (approximately half) will be genotypic females. These masculinized (sex-reversed) female fish will produce milt (sperm) containing only the female sex chromosome. The sex-reversed females can be distinguished from normal males by the fact that they do not develop a sperm duct and are, therefore, unable to release sperm freely. Normal mature males can be 'stripped', i.e. when the abdomens of mature normal males are gently pressed the fish will exude milt, whereas similar treatment will not lead to a release of sperm by sex-reversed female fish. Sperm must be obtained from the sex-reversed females by dissecting the fish and then removing and macerating the testes. When the eggs taken from normal females are fertilized with sperm from the sex-reversed females, all the offspring produced will be female fish (Fig. 11.4).

Whilst this method is effective in producing single-sex populations and can be carried out relatively easily on fish farms, the method does have the disadvantage that it involves the slaughter of fish. Thus, the sex-reversed female fish can only be used as broodstock on one occasion. The problem arises because the sex-reversal treatment, involving the feeding of relatively high doses of methyltestosterone, leads to abnormal develop-

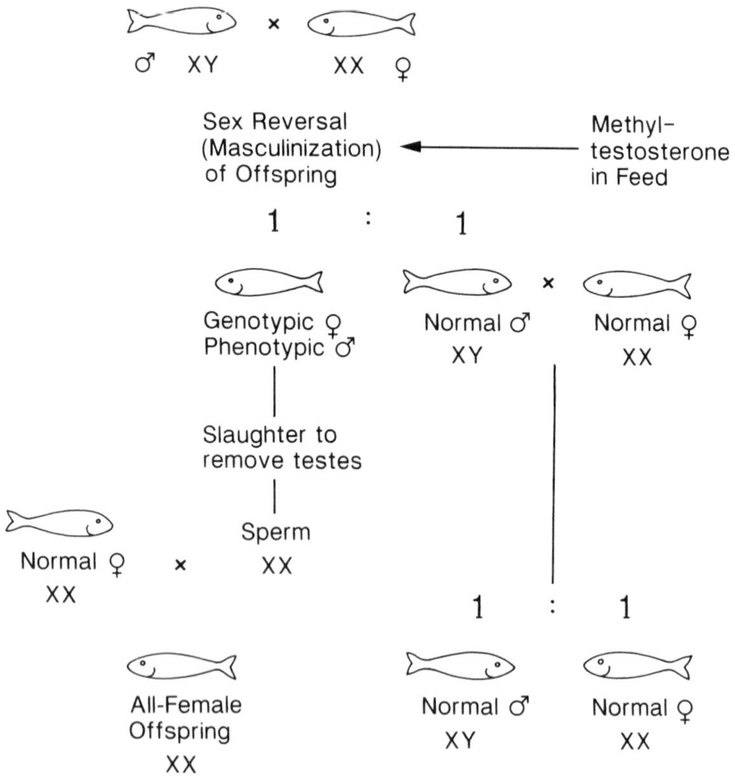

Fig. 11.4 Production of all-female fish populations. The procedure involves the sex-reversal of genotypic female fish induced by the feeding of methyltestosterone, an androgenic steroid.

ment. In other words, the sperm ducts of the sex-reversed fish are non-functional. This problem can be overcome by feeding lower doses (0.5 mg methyltestosterone per kg diet) of the sex steroid. When low doses of the steroid are fed, many of the sex-reversed females become fully functional males that can be stripped to obtain sperm for the fertilization of eggs. If these functional male–sex-reversed females are to be used as broodstock, the farmer or breeder must be certain that these fish are genetic females and not normal males.

In order to ensure that this can be achieved some extra steps must be introduced into the sex-reversal process. One way in which this can be done is to feed the offspring resulting from the mating of normal and sex-reversed females (Fig. 11.4) with food containing low doses of methyl-testosterone. These fish, which are all known to be genetic females, will then be sex-reversed and will develop into fully-functional males that can be stripped in the same fashion as normal males. These sex-reversed females can then be used as broodstock for several seasons.

In an alternative production procedure, the initial batch of eggs intended for the production of sex-reversed females is fertilized using sperm that have been irradiated with ultraviolet light in order to destroy the genetic material. These sperm can activate egg development but there will be no contribution made by the genetic material provided by the male. In order to restore the full genetic complement, and ensure normal development of the eggs and embryos, the eggs must be given a heat or pressure shock shortly after fertilization. Some eggs will not survive this treatment. Mortalities at the egg stage are not, however, a serious problem because very little economic investment will have been made up until this stage. Because the genetic material in the sperm had been destroyed prior to fertilization of the eggs, it follows that all the fish that hatch contain only genetic material contributed by the female fish. All the fish are, therefore, genetically female. Whilst these fish could be grown on for the market, this would not be a particularly sound policy owing to the investment of time and costs that would be involved in producing very large numbers of the gynogenetic fish. A much better strategy is to sex-reverse these fish by feeding methyltestosterone and then use them as broodstock. By feeding low doses of methyltestosterone to these fish they will develop as functional males (phenotypic males–genotypic females) that can be stripped of milt in the same fashion as normal males (Fig. 11.5). Consequently, these fish can be used as broodstock for a number of seasons, and the numbers of fish required to be produced are much smaller than when using destructive methods.

Because single-sex populations of females produced by fertilizing the eggs of normal females with sperm from sex-reversed females comprise fish that are one generation away from those that were treated with hormone, they do not contain steroid residues in their flesh. These female fish can, therefore, be safely marketed without there being any risk of negative publicity.

Repartitioning agents

A number of phenylethanolamines with β-adrenergic receptor activity have been evaluated as repartitioning agents, i.e. agents for increasing the proportion of lean body tissue, in domestic livestock. Cimaterol, an orally active pharmaceutic analogue of the catecholamines, has for example been reported to increase the protein:fat ratio in the carcasses of domestic animals such as pigs, lambs, poultry and cattle. Similarly, treatment with clenbuterol leads to an increase in protein accretion in the skeletal muscles, and β-agonists may, therefore, be effective repartitioning agents.

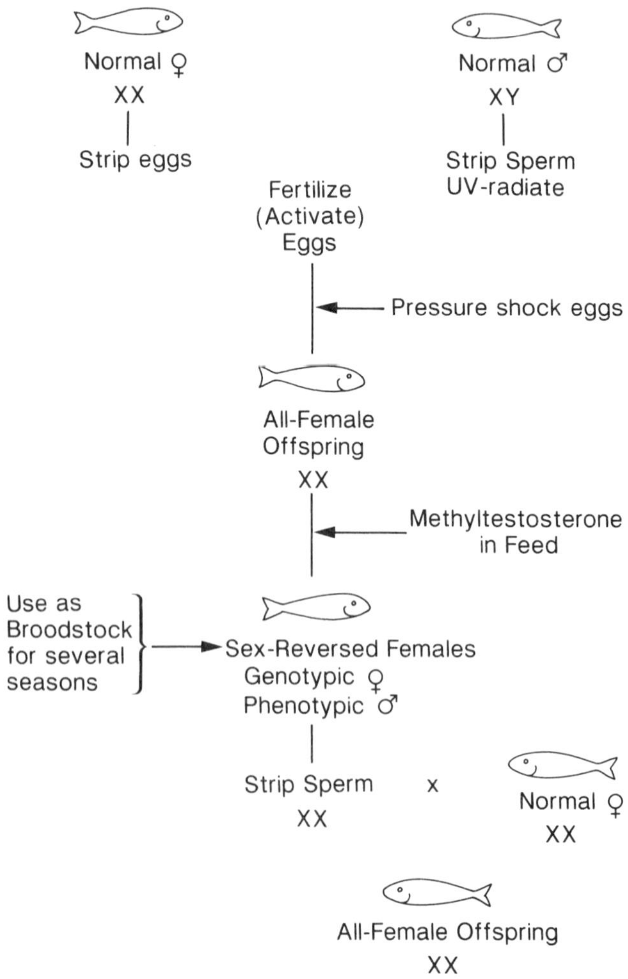

Fig. 11.5 Production of all-female fish populations. Gynogenetic females are produced by thermal shocking of eggs fertilized with inactivated sperm. Functional male fish are produced by sex-reversing the females by feeding with low doses of methyltestosterone, an androgenic steroid.

Whilst treatment of animals with agents of this type may result in there being increases in protein synthesis in the skeletal muscles, it appears that the main mechanism via which the β-adrenergic agonists lead to a modulation of protein accretion is related to reductions in rates of protein degradation.

Ractopamine is another phenylethanolamine which has been shown to influence nutrient deposition in livestock, and dietary supplementation at levels of 10–30 $mg.kg^{-1}$ has led to increased growth performance, improved nitrogen retention and reduced carcass fat in cattle, pigs and turkeys. Channel catfish appear to respond to dietary inclusion of ractopamine in a similar manner to mammals and birds, with fish fed diets containing 20 mg ractopamine per kg showing increased rates of growth and reductions in the percentages of muscle and mesenteric fat compared with untreated controls.

11.6 SELECTION PROGRAMMES FOR GROWTH

It would, perhaps, be easy to believe that individuals capable of growing rapidly should have a greater capacity to synthesize new body tissue than individuals that grow more slowly. The relationship between anabolic capacity, assessed as rate of protein synthesis, and growth potential has been investigated for a number of species of laboratory and domestic animals. There has been found to be a clear relationship between rate of protein synthesis and growth rate for individuals of a given genotype, but when different genetic strains are compared it is rare to find that the rate of protein synthesis is greatest in the most rapidly growing strain. Rates of protein synthesis displayed by individuals of rapidly growing strains can be greater than, the same as or less than rates recorded for individuals from slow-growing strains.

An explanation for these observations can be found by examining growth in the broader context of anabolic–catabolic relationships. Protein growth (deposition) is dependent not only on the rate of synthesis, but also on protein degradation rates (breakdown). Therefore, it is the relationship synthesis–breakdown, and not the rate of protein synthesis, that is of overriding importance in the determination of growth potential. For example, dwarf chickens show higher rates of both protein synthesis and breakdown than do broilers, but growth rates of broilers are higher. In practice, it appears that strains which display rapid growth have low rates of protein breakdown, rather than high rates of synthesis. It is metabolically advantageous that high growth rates are linked to low rates of degradation, rather than high rates of synthesis, because low rates of breakdown imply that protein turnover is inherently low. A low rate of protein turnover implies, in turn, that the metabolic costs of maintaining the status quo will be lower in rapidly growing, than in slow-growing, strains or genotypes. A further consequence of this is that strains having low rates of turnover would also be expected to make effective use of food resources and grow efficiently. Thus, strains that are known to have inherently low rates of protein breakdown and turnover are also often those which are found to grow most rapidly and most efficiently.

A breeding programme based upon the selection of genotypes or strains showing high rates of synthesis is unlikely to result in the development of fast-growing strains. Tissue turnover appears to be a much better physiological correlate for growth than does rate of synthesis, so selection of strains displaying low rates of tissue turnover (or low metabolic rates) is more likely to be successful in producing strains which grow rapidly and make efficient use of the food supplied.

11.7 SOMATIC (P_S) VERSUS REPRODUCTIVE (P_R) GROWTH

Compared with the male sperm, the eggs produced by female fish represent a large energetic investment, and this is reflected in the relative sizes of the male and female gonads in the period immediately prior to spawning. For example, in Atlantic salmon the gonadosomatic index (GSI = (Gonad weight/Somatic weight) × 100) of the female may be 20% or more, whereas the testes are relatively small and the GSI of the males will usually be within the range 3–5%. This should not, however, be taken to imply that the overall reproductive costs are greater in the female than in the male. In addition to the costs of gamete development, there may be additional energetic costs associated with migration from feeding to breeding areas, defence of breeding sites and the preparation of a nest, courtship and the guarding of mates, the act of mating and shedding of gametes, and parental care.

It is usually the males that engage most actively in courtship and breeding-territory defence, and in species that show egg guarding and parental care it is often the male that performs this duty. Thus, although the energetic investment in reproductive growth (P_R) may be expected to be greater for females than for males, the energetic costs of engaging in reproductive behaviour may be markedly higher for the males. Consequently, the overall energetic investment in reproduction may not differ to any great degree between the sexes.

For example, during the course of the autumn–winter spawning season, female Atlantic salmon suffered a body weight loss of 30%, whilst males decreased in weight by about 25%. There was, however, a greater decrease in the lipid content of the tissues of the males than the females, and it was estimated that fish of both sexes had suffered a 50% decrease in body energy content during the course of the spawning season. The nature of these losses differed markedly between the sexes. In the female, approximately half of the loss could be accounted for by the shedding of gametes, with the remainder being derived from metabolism of somatic tissue reserves. In the male, on the other hand, over 90% of the energy loss was due to metabolism of somatic tissue reserves, and very little of the decline in body energy content was due to the shedding of the gametes. Thus, although the overall energetic costs of reproduction were similar for both sexes, the greatest costs for the female appeared to be related to gamete production, whereas the greatest costs for the male appeared to be those related to engaging in various reproductive behaviours.

Thus, within the context of the energy budget (Equation 6.2), a large proportion of the male reproductive costs can be considered to form part of the metabolic (M), rather than the production (P) component. It is obviously extremely difficult to make an accurate assessment of the metabolic costs involved in the performance of the wide range of behavioural activities that contribute to the successful completion of reproduction, and few attempts have been made to determine these costs under controlled conditions. In view of these difficulties, most studies relating to the energetic costs of reproduction have been concentrated upon the examination of reproductive growth (P_R) in the female by means of studies on the time course of egg development and patterns of energetic deposition in the ovaries.

The egg will comprise 20–40% dry matter (mostly protein and lipid) and the energy content will usually be within the range 5–8 $kJ.g^{-1}$. In fish the yolk incorporated into the oocyte during vitellogenesis is usually either the sole, or the major, source of nutrients for the developing embryo and post-hatch larva prior to the time of exogenous feeding. There may also be some reliance on nutrients from the yolk for some time after the young fish has begun to take exogenous food. Thus, the level of investment made by the female in yolk production is expected to be of critical importance for the survival of the developing young.

There is little synthesis of yolk in the ovaries and oocytes. The liver is the major site of synthesis, and vitellogenin is transported from the liver to the ovaries in the bloodstream. The vitellogenin is then incorporated into the developing oocytes. Vitellogenesis is quite often a protracted process, requiring several days or weeks for completion. During this period, the oocyte increases markedly in size due to the incorporation of the yolk. In salmonids, for example, the diameter of the oocyte may increase from 50 μm to approximately 5 mm during the course of vitellogenesis, and this represents a million-fold increase in volume. This increase in oocyte volume will lead to a marked increase in the size of the ovaries, and immediately prior to spawning, the ovaries of some fish species may represent 20–30% of the total body weight. This increase in size of the ovaries is usually most dramatic in species, such as salmonids and cyprinids, that have a relatively short spawning season. In many of these species there is simultaneous development and ovulation of large numbers of oocytes. In species which have a longer spawning season the oocytes may develop in small batches and spawning events will occur several times during the course of the year. In the batch-spawning species the ovaries will seldom exceed 5–10% of the total body weight, but a measurement of ovarian size will not give a true reflection of the total annual investment in reproductive growth.

In the context of the energy budget, production can be divided into somatic (P_S) and reproductive (P_R) growth such that:

$$P_S + P_R = R - (F + U + M) \qquad (11.5)$$

Provided that the other components of the budget remain unchanged during the period of reproductive growth, P_S and P_R can be considered as being directly competitive processes for finite resources. There is, however, evidence that changes in components other than somatic growth may contribute to partially offset the costs of reproduction. In some cyprinid species, for example, levels of activity may be lower during the period prior to spawning than at other times of the year. It has been suggested that the energetic savings made by reducing activity levels result in a greater proportion of the available resources being channelled towards reproductive growth.

Nevertheless, somatic and reproductive growth can be viewed as being competitors for limited resources, and the drop in somatic growth rate that occurs when fish mature is often taken as being indicative of this competition. Production of gametes may be given priority over somatic growth in mature fish, such that the annual increment of body weight increase in mature individuals is often small. Thus, the investment in somatic growth may fall from approximately 25% of the annual energy budget in young, immature fish to 0–5% in mature individuals. At the same time reproductive growth increases from 0% of the budget in immatures to 20% or so in mature fish.

Several species breed only once before dying, and in these semelparous fish, the energetic investment made in gametes and spawning activities is considered to compromise survival. The majority of fish species are, however, iteroparous, with mature individuals being active for several breeding seasons. However, even amongst the iteroparous species, the costs of reproduction may be considerable. In the period immediately following spawning, the fish may be in poor condition and weak. Several weeks or months may be required for recovery from the energetic drain imposed by gamete production and the rigours of spawning activity. Amongst the iteroparous species, there is often a temporal alternation between the repletion of the somatic reserves and the development of the gonads. This leads to the fish showing distinct patterns of energy storage and depletion linked to cycles of reproductive growth. In this way, the fish may build up sufficient reserves to enable reproductive growth to continue during periods when food supplies are limited and rates of ingestion are insufficient to support gonadal growth and development.

Thus, somatic energy reserves are repleted during times of plentiful food supply, whilst the energetic demands of reproductive growth will lead to a depletion of these energy depots at a later stage during the cycle. It is, for example, often observed that the condition (K = (Weight/Length^3) × 100) of the fish changes with season, and the condition factor (K) usually reaches its nadir in the immediate post-spawning period. At the same time, the water content of the muscle may increase and the relative proportion of other constituents decrease. In extreme cases, such as is observed for some species of flatfish, the swimming musculature may contain approximately 96% water (3% protein and 1% lipid) at the end of the spawning season. It is, first and foremost, the white, anaerobic muscle that shows these changes in composition, with the red muscle composition being little affected.

In mature individuals, reproductive growth will occur to the detriment of somatic growth, but when food supplies are restricted there will also be some limitation imposed upon reproductive growth. The effects of food restriction on the reproductive growth of fish has been studied for a small number of species under laboratory conditions. One effect of food supply is that it may have a major influence upon the size and/or age at which the fish initially becomes mature. Furthermore, iteroparous species of fish held under conditions of very low food availability may not reproduce each year, but may take one or more years to build up the energy reserves required to support gamete development and spawning activities. Thus, the poorer the food supply, the lower the proportion of the population that proceed with gonadal development and come into spawning condition. Relationships of this kind between the food supply and the proportion of spawning adult fish within populations appear to occur in a range of fish species, including rainbow trout, winter flounder, *Pseudopleuronectes americanus*, and the haddock, *Melanogrammus aeglefinus*.

When conditions are adequate to enable the fish to spawn each year, food supply may influence reproductive growth either through effects upon fecundity (egg number/body weight) or by influences upon egg size and quality. Restricted food supply can result in reduced fecundity either by bringing about a reduction in the numbers of oocytes that begin development (reduced oocyte recruitment) or by influencing the numbers of recruited oocytes that complete development. Reductions in oocyte recruitment under conditions of restricted food supply have been reported to occur in brown trout and winter flounder, whereas follicular atresia and oocyte resorption may be prevalent in rainbow trout, brook trout, *Salvelinus fontinalis*, and the guppy. Food restriction can also lead to females producing smaller eggs containing reduced quantities of yolk, as has been seen in the haddock.

For any given species, a limitation in food supply can lead to a simultaneous reduction in egg size and female fecundity, with severe food restriction usually leading to a reduction in the proportion of the female population that spawns. To what extent the different effects of food limitation will be observed depends both upon the degree of food restriction and the species under study, and there is the possibility that several of the above-mentioned effects can be seen simultaneously in fish populations subjected to food restriction.

For example, the effects of different levels of food restriction on a range of reproductive parameters have been examined in rainbow trout. Hatchery populations of fish were held in large groups and were fed on dry pellet feed at different levels ranging from 0.4% to 1.5% of fish body weight per day for a period of one year before their first spawning as two-year-old fish. Fewer fish spawned on the lowest rations (0.4% and 0.75%) than on the other feeding regimes, and there was also a slight, but insignificant, trend for the egg size to be larger in fish held on higher rations. The greatest effects of restricted feeding were, however, on the numbers of eggs produced by the different groups of females. Fecundity was significantly reduced in the females held on the lowest rations, even after due allowance was made for the effects of differences of fish size on egg production. The reductions in fecundity observed in the females fed at the lowest levels may have been determined early in the cycle of oocyte development through influences upon the rates of recruitment of previtellogenic oocytes into vitellogenesis. The final number of eggs produced could also be influenced by there being effects of feeding level on the proportions of maturing oocytes that undergo atresia during the course of development. In rainbow trout, atresia of developing oocytes may be seen at all stages of development from the time that the oocytes begin to sequester exogenous yolk until final maturation and ovulation. Under conditions of adequate food supply, fewer than 10% of the developing oocytes may become atretic, restricted feeding may lead to 20% or more of the vitellogenic oocytes becoming atretic, and a prolonged period of food deprivation may lead to 100% atresia with the fish 'missing' a spawning.

11.8 COMPENSATORY OR 'CATCH-UP' GROWTH

Following spawning, the energy reserves of fish may be seriously depleted and the fish may be thin and in poor condition. During the post-spawning period the reserves will be replenished, and, provided that food supplies are adequate, rates of weight gain may be very rapid. Fish also experience a depletion of energy reserves and a decline in condition during periods of poor food supply, such as during the winter months when prey availability may be reduced. Thus, under the conditions prevalent in the wild, there may be pronounced cycles of depletion and repletion of energy reserves, and growth will often be limited by food supply.

Fish, and other animals, appear to be able to adapt to these feast-and-famine conditions by showing marked growth spurts when food supplies are increased following a period of starvation or food restriction. It is usually the fish that are in the poorest condition that show the greatest response and display the most rapid rates of weight gain when adequate feeding conditions are restored (Fig. 11.6).

This type of growth response is usually referred to as compensatory, 'catch-up' or recovery growth, and the response can be generally described in the following terms. Under conditions of adequate food supply the growth curve of an animal will follow a predetermined 'ideal' course, but a period of starvation or food restriction will cause the pattern of growth to deviate from the ideal. On a return to conditions of adequate feeding a series of mechanisms will be initiated. This leads to a type of self-correction as the animal gains weight rapidly in an apparent attempt

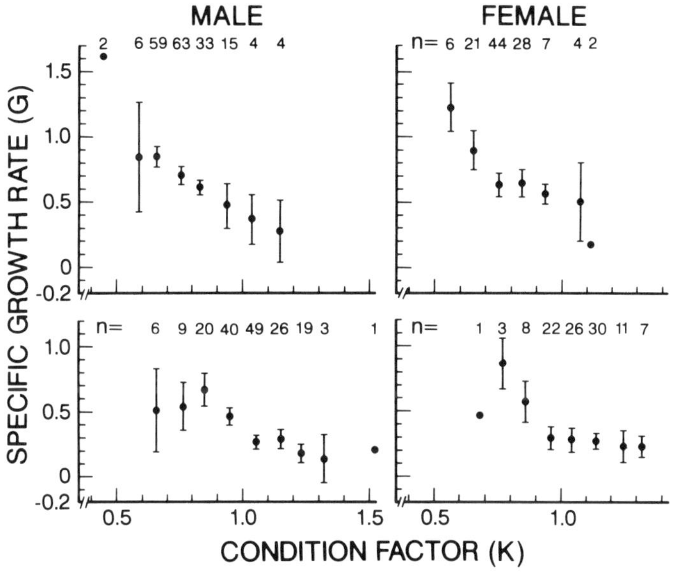

Fig. 11.6 Relationships between condition factor ($K = (W/L^3) \times 100$) and specific growth rates (*G*) of large, mature male and female Atlantic cod, *Gadus morhua*, in the period following capture from the wild, and during post-spawning recovery. Reproduced with permission from Pedersen and Jobling (1989), *Aquaculture*, **81**, 161–8.

to restore the growth pattern to the ideal (Fig. 11.7). Thus, when food supplies are increased following a period of undernutrition, deprived animals are almost invariably observed to show more rapid rates of weight gain than conspecifics that have been fed adequately throughout their lives.

In some cases, recovery from starvation or food restriction has been reported to be complete, i.e. following transfer to unlimited food supplies, the previously food-restricted animals achieve the same body size as animals that had been allowed continuous access to adequate food resources. Complete recovery has not, however, always been reported, and partial recovery has been observed in many studies of the growth responses of restricted–refed animals. There may be species differences in the responses shown to restricted–refeeding regimes, and there may also be differences between both developmental stages and the sexes in the ability to recover from periods of undernutrition. In addition, both the intensity and duration of the nutritional restriction also appear to influence growth during the recovery period.

Recovery growth following periods of undernutrition has been well documented for a number of fish species, e.g. salmonids, cyprinids and gadoids, and instances of both complete and partial recovery have been reported. Whilst recovery, or compensatory, growth appears to be almost universal amongst animals, the physiological basis of the phenomenon is incompletely understood.

In many cases, restricted–refed animals have been reported to show higher rates of food intake than size-matched conspecifics allowed continuous access to adequate food supplies. In other words, a restoration of

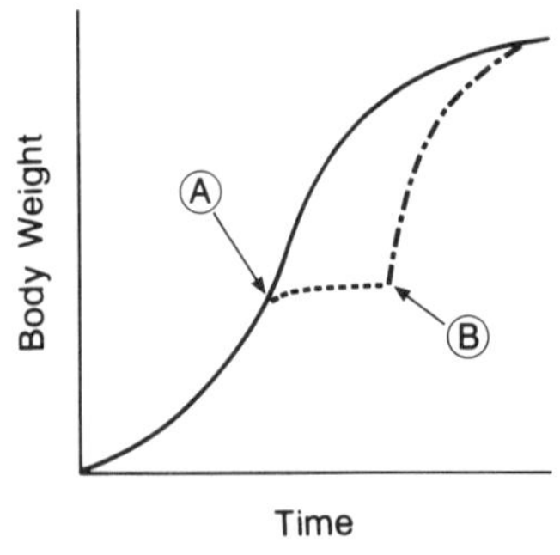

Fig. 11.7 Growth and development of normal animals (solid curve), and deviations from the normal growth curve during periods of restricted feeding and compensatory growth. (A) Time point at which feeding restriction was imposed; (B) return to unlimited food supply. Note that growth rate is rapid during the recovery period. It is assumed that there is complete recovery, in that the food-deprived individuals 'catch-up' the animals following the normal growth curve.

adequate feeding following a period of undernutrition results in animals becoming hyperphagic, and this overeating would be expected to promote rapid weight gain. The extent to which fish that have been deprived of food display hyperphagia on a return to adequate feeding conditions may be related to the degree to which the storage depots were depleted during the course of food restriction. Thus, juvenile Atlantic salmon have been shown to display increased appetite following a period of food deprivation, and the propensity of the fish to feed appeared to be related to the nutritional status of the fish, assessed in terms of lipid content. Similarly, an inverse relationship has been found between food intake and the size of the lipid depots in juvenile Arctic charr held under conditions that led to differences in the extent to which body energy reserves were depleted (Fig. 11.8).

The hyperphagia and rapid weight gain observed during recovery growth are often accompanied by improved food conversion. In some cases, however, restricted–refed animals have been reported to show improved food conversion efficiencies without becoming hyperphagic.

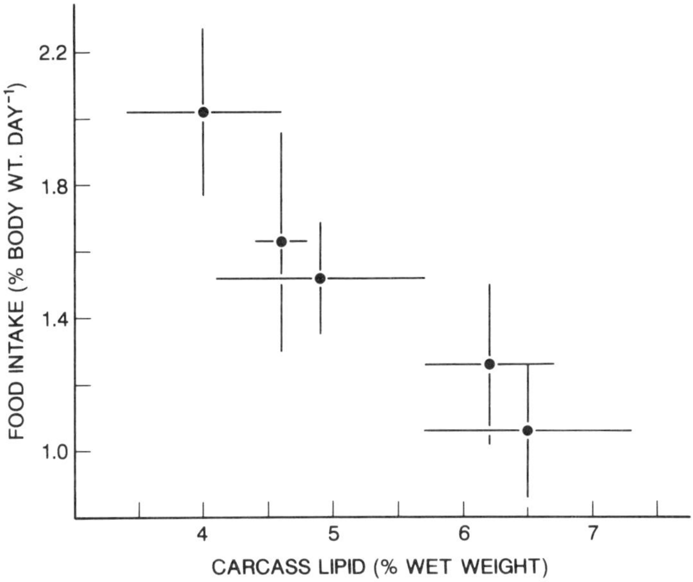

Fig. 11.8 Plot showing food intake by juvenile Arctic charr, *Salvelinus alpinus*, on a return to adequate feeding conditions following periods of rearing that led to differences in the extent to which carcass lipid depots were depleted. Values shown are means (lines depict 95% confidence intervals) of food intakes for individual (n = 8 or 9) fish monitored close to the times at which fish were sampled for body composition analysis. Reproduced with permission from Jobling and Miglavs (1993), *J. Fish Biol.*, **43**, 487–9.

Because food conversion is usually reported in terms of weight gain per unit weight of food consumed, it is possible that the improvements in conversion observed in restricted–refed animals could be related to differences in the composition of the tissues deposited during recovery and normal growth. The differences in food conversion could also be an indication that there are differences in metabolic expenditure between restricted–refed and continuously fed animals.

In a number of studies of recovery growth conducted on domestic animals, the composition of the weight gain has been found to differ between restricted–refed and continuously fed animals, but results are inconsistent. Relatively few studies of this nature have been carried out on fish. In the cod the tissues of the swimming muscle are the first to be repleted when adequate feeding conditions are restored following a period of starvation. Repletion involves a rapid increase in the weight of the muscle, a relative increase in muscle lipid and glycogen content and a corresponding decrease in the percentage of muscle water. The relative muscle water content must drop below about 82% before the lipid stores of the liver begin to show any marked increase. Thus, during the initial stages of recovery from a period of food deprivation the cod give priority to the deposition of muscle tissue, which has a low energy density (approximately 6–8 $kJ.g^{-1}$). Only later do the fish begin to deposit energy-rich lipid reserves (approximately 38 $kJ.g^{-1}$) in the liver. For fish that have had continuous access to adequate food supplies, weight gain will reflect both muscle growth and the deposition of energy storage reserves. In well-fed cod, a major portion of the body energy gain may be deposited as lipid reserves in the liver.

Thus, differences in gross weight gain and food conversion between restricted–refed and continuously fed fish need not give a realistic reflection of energetic changes. Consequently, rapid growth and an improvement in food conversion during a period of recovery growth does not necessarily imply that the physiological responses of the fish to cycles of food deprivation and refeeding include an improvement in energetic efficiency.

In some studies, the composition of the tissue gain of restricted–refed animals has been reported as being identical to that of animals displaying normal patterns of growth. Results of this type suggest the possibility that improved conversion may be linked to differences in metabolic expenditure and energetic efficiency between restricted–refed and continuously fed individuals. Animals adapt to food deprivation by reducing energy expenditure (Chapter 8), and restrictions in food intake generally result in lower rates of weight loss than would be expected if metabolic rates were maintained at pre-deprivation levels. These observations have been used as the basis of a hypothesis developed in an attempt to explain the rapid growth and improved food conversion recorded during recovery from periods of food deprivation.

It has been suggested that, in the period following transfer from a restricted to a liberal feeding regime, the metabolic rate may not immediately return to the same level as that of a continuously fed animal. In other words, there may be some time lag in the readaptation of the metabolic rate to the fully fed state, such that low rates of energy expenditure may be maintained for a short time even though the animals are no longer food-restricted. The consequences of animals showing metabolic responses of this kind would be as follows. Low rates of metabolic expenditure occurring concurrently with either normal or high rates of food intake would result in large amounts of food energy being available for growth. Thus, rates of weight gain should be rapid during the initial phase of recovery from a period of undernutrition. Maintenance of low rates of metabolic expenditure during the period of recovery would be expected to lead to a large weight gain per unit food intake, i.e. efficient food conversion. Whilst the hypothesis of a depressed metabolic rate in restricted–refed animals is an attractive one, the expectations of the hypothesis are not always borne out by experimental results, in that animals displaying recovery growth do not invariably display lower rates of energy expenditure than conspecifics that have been adequately fed. Consequently, alternative explanations must be sought.

Part of the decline in metabolic rate seen during food deprivation appears to be the result of a reduction in the rate of turnover of cellular biochemical components, particularly protein (Chapter 8). Rates of both the synthesis and breakdown of tissue components appear to decline markedly with the imposition of feed restriction. Growth is a reflection of the difference between the magnitude of synthesis and that of breakdown. Negative growth, or weight loss, obviously means that breakdown exceeds synthesis, whereas positive growth, or weight gain, is a reflection that rates of synthesis are greater than degradation rates. The decreases in rates of synthesis and breakdown that accompany food deprivation have the consequence that there is a reduction in both the turnover of the tissue components and in energy expenditure, with both reaching a nadir after a prolonged period of starvation.

Transfer from a restricted to a liberal feeding regime is expected to result in an increase in synthetic activity. Since the energetic costs of tissue synthesis are generally considered to be quite high (Chapter 8), any increase in synthetic activity would also be expected to result in marked increases in energy expenditure and metabolic rate. Thus, the high growth rate, and efficient food conversion, observed during recovery from food deprivation are unlikely to be the result of energy expenditures being markedly depressed during this period.

Growth reflects the difference between synthesis and breakdown of tissue components, and, whilst an increase in the rate of synthesis is usually accompanied by an increase in the rate of breakdown, there is little evidence to suggest that there is a very tight coupling between the two processes. In other words, it is possible that there is some uncoupling of rates of synthesis and breakdown during the recovery from undernutrition. Following the restoration of a liberal feeding regime, it is hypothesized that rates of synthesis and energy expenditure may increase quite rapidly whereas rates of breakdown may take longer to return to the levels observed in continuously fed animals. According to this scenario, animals that are recovering from a period of food deprivation would display rapid rates of growth and efficient food conversion despite having metabolic rates that were similar to those of continuously fed conspecifics.

The bulk of the experimental data point to the fact that rates of weight gain increase, and food conversion is improved, in fish recovering from periods of food deprivation. There are undoubtedly several mechanisms involved in producing these responses. These mechanisms may include the development of hyperphagia, changes in energetic expenditure and improvements in the efficiency of energy utilization. There may also be differences in the composition of the tissues deposited during periods of recovery as opposed to normal growth. The relative contributions of the different mechanisms to the compensatory growth response remain, however, largely unknown. Nevertheless, some attempts have been made to examine the feasibility of exploiting the mechanisms induced by restricted–refeeding regimes in the commercial production of both livestock and fish.

Trials have been conducted with a number of types of livestock (broilers, turkey poults, weanling pigs, cattle, sheep) in order to investigate the influences of various feeding regimes, incorporating different levels of restriction or fast, on commercially important parameters such as meat yield, lean versus body fat ratios, production times and production efficiencies. There have also been small numbers of studies conducted with commercially exploited fish species in order to examine the effects of restricted–refeeding regimes on rates of growth and food utilization efficiencies.

Results from trials carried out with rainbow trout suggest that, under some conditions, the fish may show complete recovery following a period of food restriction or deprivation, and that food utilization efficiencies of restricted–refed fish are better than those of fish allowed continuous access to adequate food supplies. Similar results have been obtained in a pilot-scale commercial study conducted with cod. Thus, the implication is that the imposition of a restricted–refeeding regime may lead to an improvement in the utilization of the food supplied without there being any loss in weight gain, or any increase in the length of the production cycle.

Results obtained in some experiments conducted with rainbow trout, and in trials carried out with other salmonids, have led to different conclusions being drawn. In these experiments, the imposition of a restricted–refeeding regime led to the fish showing improved food conversion during the recovery growth phase, but the overall time taken to reach market size was longer on the restricted–refeeding than on the normal production feeding cycle. Thus, in these cases, the restricted–refeeding regimes did not result in any overall gains that could be exploited commercially.

Whether or not any net benefits accrue from the imposition of a restricted–refeeding regime may depend upon factors such as the severity and duration of the restriction, the size and age of the fish and the rearing environment. Thus, the effects of these factors need to be elucidated before the potential benefits of incorporating restricted–refeeding regimes into production cycles can be adequately assessed.

Chapter twelve

Bioenergetics – concluding summary

Considerable research effort has been directed towards the study of fish bioenergetics and growth. Measurements have been made of the different components of the energy budget, and bioenergetics budgets have been used in fisheries research for a variety of purposes. Common applications of bioenergetics studies have been in the estimation of growth or production of fish populations and in the prediction of food consumption rates to evaluate the impact of predatory fish species on their prey. Bioenergetics can also have useful applications within aquaculture, since the construction of energy budgets provides an ideal means for the evaluation of the relative importance of the various environmental factors which influence growth.

The majority of the available bioenergetics information has been obtained from laboratory studies conducted either on species that are easy to keep in captivity, or on those that are commercially important. Most experimental work has been carried out on young, actively growing fish. It has been possible to construct energy budgets for the young of a number of carnivorous fish species, growing actively under conditions of plentiful food supply. In the majority of cases, it has been found that 20–35% of the ingested energy is deposited as growth. These bioenergetics studies give a reasonably good picture of the energy partitioning in fish held under laboratory or fish farming conditions.

There are, however, many unsolved problems and areas where information is inadequate with regard to the energy partitioning and growth in wild fish populations. The application of new techniques to the study of fish bioenergetics can assist in finding solutions to some of these problems. For example, energy expenditure of fish in the wild has rarely been quantified, and until recently it was not possible to collect such information. Developments in telemetry have enabled researchers to monitor activity patterns of wild fish, and correlations between the energy expenditure on activity and some physiological parameters have also been sought. Information of this type has contributed to the quantification of field metabolic rates.

A major problem within fish bioenergetics has been the accurate assessment of food ingestion, but radiographic methods now allow the food intake of individual fish to be monitored, enabling in-depth studies of feeding and growth relationships to be conducted. Radiographic methods do not contribute to solving the problem of the estimation of food intake by wild fish, but do have important applications within the field of aquaculture. In traditional growth trials the relationships between food supply and growth are investigated, but the information that can be obtained from these types of study has severe limitations. Because data are restricted to groups, and not individuals within the group, results must be reported in terms of the gross changes occurring within the group as a whole. Thus, although differences between groups may be observed and reported, the traditional growth trial does not permit any detailed analysis of the causal factors underlying these differences. The monitoring of food intake and growth of individuals within the group permits such an analysis to be made, and this can provide valuable additional information about the relative importance of the different biotic and abiotic factors influencing growth.

Other developments within studies of fish bioenergetics and growth include the application of techniques that enable studies of anabolic and catabolic processes to be made at the tissue level. Investigation of the hormonal control and integration of metabolic and growth processes is also becoming increasingly possible. These studies have commercial implications in that they can contribute to the development of methods for increasing rates of fish production and improving food conversion in the aquaculture industry.

The results of fish bioenergetics and growth studies are being increasingly used as a tool in fisheries science, and it is now common practice to use some form of bioenergetics modelling in both aquaculture production planning and fisheries management. The accuracy and reliability of such modelling will be determined both by the methods used in estimating, and by the precision of, the input terms. If the penalty for incorrect prediction is high, then increased effort should be channelled into obtaining precise estimates of the critical parameters in the model. Sensitivity analysis is a useful, cheap and rapid way of evaluating how uncertainty in the input parameters affects the model predictions. This type of analysis also reveals the relative importance of the different components of the energy budget, enabling identification of the input parameters that are required to be estimated with the greatest precision.

Thus, the study of fish bioenergetics enables the investigation of problems related to fisheries management and production, and also has a central place within aquacultural research. In the foregoing chapters the basic principles of bioenergetics have been discussed and some of the methods used for obtaining estimates of the components of the energy budget have been described. Attempts have also been made to indicate potential problems and pitfalls involved in carrying out experimental studies, and areas of uncertainty requiring further research effort have been indicated.

PART TWO: FURTHER READING

Books

Bagenal, T. (ed.)(1978) *Methods for Assessment of Fish Production in Fresh Waters: IBP Handbook 3*, Blackwell Scientific, Oxford.

Blaxter, K. (1989) *Energy Metabolism in Animals and Man*, Cambridge University Press, Cambridge.

Brafield, A.E. and Llewellyn, M.J. (1982) *Animal Energetics*, Blackie, Glasgow.

Brown, M.E. (ed.)(1957) *The Physiology of Fishes, I. Metabolism; II. Behavior*, Academic Press, New York.

Cossins, A.R. and Bowler, K. (1987) *Temperature Biology of Animals*, Chapman & Hall, London.

Gerking, S.D. (ed.)(1967) *The Biological Basis of Freshwater Fish Production*, Blackwell Scientific, Oxford.

Gerking, S.D. (ed.)(1978) *Ecology of Freshwater Fish Production*, Blackwell Scientific, Oxford.

Grodzinski, W., Klekowski, R.Z. and Duncan, A. (eds)(1975) *Methods for Ecological Bioenergetics: IBP Handbook 24*, Blackwell Scientific, Oxford.

Hoar, W.S., Randall, D.J. and Brett, J.R. (eds)(1979) *Fish Physiology*, Vol. VIII, Academic Press, London.

Kamler, E. (1992) *Early Life History of Fish: An Energetics Approach*, Chapman & Hall, London.

Rankin, J.C. and Jensen, F.B. (eds)(1993) *Fish Ecophysiology*, Chapman & Hall, London.

Schreck, C.B. and Moyle, P.B. (eds)(1990) *Methods for Fish Biology*, American Fisheries Society, Bethesda, MD.

Thorpe, J.E. and Huntingford, F.A. (eds)(1992) *The Importance of Feeding Behavior for the Efficient Culture of Salmonid Fishes*, World Aquaculture Society, Baton Rouge, LA.

Tytler, P. and Calow, P. (eds)(1985) *Fish Energetics – New Perspectives*, Croom Helm, London.

Weatherley, A.H. and Gill, H.S. (1987) *The Biology of Fish Growth*, Academic Press, London.

Wootton, R.J. (1990) *Ecology of Teleost Fishes*, Chapman & Hall, London.

Review articles and papers

Beamish, F.W.H. (1978) Swimming capacity, In *Fish Physiology*, Vol. VII (eds W.S. Hoar and D.J. Randall) Academic Press, London, pp. 101–87.

Beitz, D.C. (1985) Physiological and metabolic systems important to animal growth – An overview. *J. Anim. Sci.*, **61**, 1–20.

Boujard, T. and Leatherland, J.F. (1992) Circadian rhythms and feeding time in fishes. *Env. Biol. Fishes*, **35**, 109–31.

Brett, J.R. (1972) The metabolic demand for oxygen in fish, particularly salmonids, and a comparison with other vertebrates. *Resp. Physiol.*, **14**, 151–70.

Bromage, N., Jones, J., Randall, C., Thrush, M., Davies, B., Springate, J., Duston, J. and Barker, G. (1992) Broodstock management, fecundity, egg quality and the timing of egg production in the rainbow trout (*Oncorhynchus mykiss*). *Aquaculture*, **100**, 141–66.

Davison, W. (1989) Training and its effects on teleost fish. *Comp. Biochem. Physiol.*, **94A**, 1–10.

Forster, R.P. and Goldstein, L. (1969) Formation of excretory products, In *Fish Physiology*, Vol. I (eds W.S. Hoar and D.J. Randall), Academic Press, London, pp. 313–50.

Goolish, E.M. (1991) Anaerobic swimming metabolism of fish: sit-and-wait versus active forager. *Physiol. Zool.*, **64**, 485–501.

Goolish, E.M. (1991) Aerobic and anaerobic scaling in fish. *Biol. Rev.*, **66**, 33–56.

Hawkins, A.J.S. (1991) Protein turnover: a functional approach. *Functional Ecol.*, **5**, 222–33.

Houlihan, D.F. (1991) Protein turnover in ectotherms and its relationships to energetics. *Adv. Comp. Environ. Physiol.*, **7**, 1–43.

Jobling, M. (1981) The influences of feeding on the metabolic rate of fishes: A short review. *J. Fish Biol.*, **18**, 385–400.

Jobling, M. (1981) Temperature tolerance and the final preferendum – rapid methods for the assessment of optimum growth temperatures. *J. Fish Biol.*, **19**, 439–55.

Jobling, M., Baardvik, B.M., Christiansen, J.S. and Jørgensen, E.H. (1993) The effects of prolonged exercise training on growth performance and production parameters in fish. *Aquaculture International*,

Jobling, M., Christiansen, J.S., Jørgensen, E.H. and Arnesen, A.M. (1993) The application of X-radiography in feeding and growth studies with fish: A summary of experiments conducted on Arctic charr. *CRC Rev. Fish. Sci.*, **1**, 223–37.

Kelly, J.M. and McBride, B.W. (1990) The sodium pump and other mechanisms of thermogenesis in selected tissues. *Proc. Nutr. Soc.*, **49**, 185–202.

Lobley, G.E. (1986) The physiological bases of nutrient responses; growth and fattening. *Proc. Nutr. Soc.*, **45**, 203–14.

Millward, D.J. (1989) The nutritional regulation of muscle growth and protein turnover. *Aquaculture*, **79**, 1–28.

Mommsen, T.P. and Walsh, P.J. (1992) Biochemical and environmental perspectives on nitrogen metabolism in fishes. *Experientia*, **48**, 583–93.

Randall, D.J. and Wright, P.A. (1987) Ammonia distribution and excretion in fish. *Fish Physiol. Biochem.*, **3**, 107–20.

Reeds, P.J. (1987) Metabolic control and future opportunities for growth regulation. *Anim. Prod.*, **45**, 149–69.

Part Three

Environmental Tolerances and Preferences

Chapter thirteen

Environmental tolerances – general introduction

When fish, and other organisms, experience environmental disturbances which lie outside the 'normal' range the effects may be dramatic. The environmental disturbances will interfere with a range of biochemical and physiological systems essential for the sustenance of life. Long-term exposure to such disturbances may result in death. Short-term exposure may result in changes that, whilst not being lethal, impair the ability of the fish to function normally with respect to activities such as feeding, growth and reproduction.

Exactly what constitutes the normal environmental range will vary from animal to animal, depending upon the evolutionary history of the species. For example, polar species of fish are tolerant of cold water temperatures and die at temperatures that would be considered normal for tropical species. Whilst there are clear interspecific differences in environmental tolerances and preferences, the experiences of an animal during its lifetime may also have marked influences upon the ability of the individual to withstand environmental disturbances. Thus, the tolerance limits of a species are rarely rigidly fixed, but are subject to change depending upon the experiences of individuals within the species. The processes resulting in these changes are usually termed *resistance acclimation* when the effects of a single variable are discussed, or *resistance acclimatization* when there are interactions between several variables, as would be the case in the natural environment of the fish.

Whether or not exposure to an environmental disturbance proves to be lethal will be dependent upon the length of time the fish is subjected to the disturbance, so it is not possible to give single values for the 'lethal limits' of fish species. Thus, when lethal limits are reported it is essential that information also be given about the measurement conditions and exposure period. For example, a fish may tolerate exposure to a high

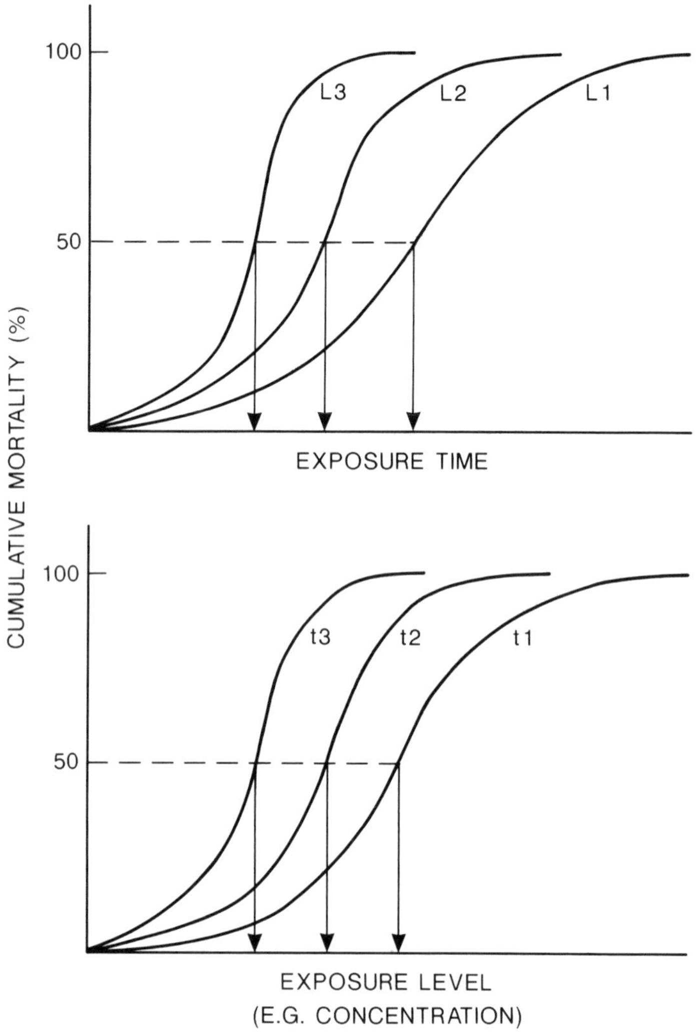

Fig. 13.1 Graphs illustrating the measurement of tolerance limits by time mortality (upper figure) and dosage mortality (lower figure). The upper figure shows the cumulative mortality that occurs with time following exposure to increasing levels (dosages) ($L1 \rightarrow L2 \rightarrow L3$) of a given environmental factor. The times to 50% mortality (LT_{50}) at each level of exposure are indicated by the vertical arrows. The lower figure shows the cumulative mortality that occurs following exposure to given levels or dosages (e.g. concentrations) of a given factor for increasing fixed exposure times ($t1 \rightarrow t2 \rightarrow t3$). Levels of a given factor required to give 50% mortality (LD_{50}) at increasing exposure times ($t1 \rightarrow t2 \rightarrow t3$) are indicated by the vertical arrows.

concentration of a given chemical for a few minutes, but the same concentration will prove to be lethal with longer-term exposure.

The methods used for the determination of lethal limits involve the investigation of the effects of different combinations of the factor under test (chemical, temperature, pH etc.) and exposure time on groups of fish. Results are usually reported as the combinations which produce 50% mortality. In practice, there are two possible approaches that can be adopted in making these determinations. Firstly, the fish can be exposed to the test factor at a given level and the mortality of the animals can be monitored over time with increasing length of the exposure period (time mortality) (Fig. 13.1). Alternatively, the mortality of animals exposed to different levels of the test factor for a given time period can be studied (dosage mortality) (Fig. 13.1). In each case the results are plotted graphically and the conditions giving 50% mortality are estimated. These types of study give values of either median lethal doses (LD_{50}) for different exposure times or of median lethal times (LT_{50}) for different levels of

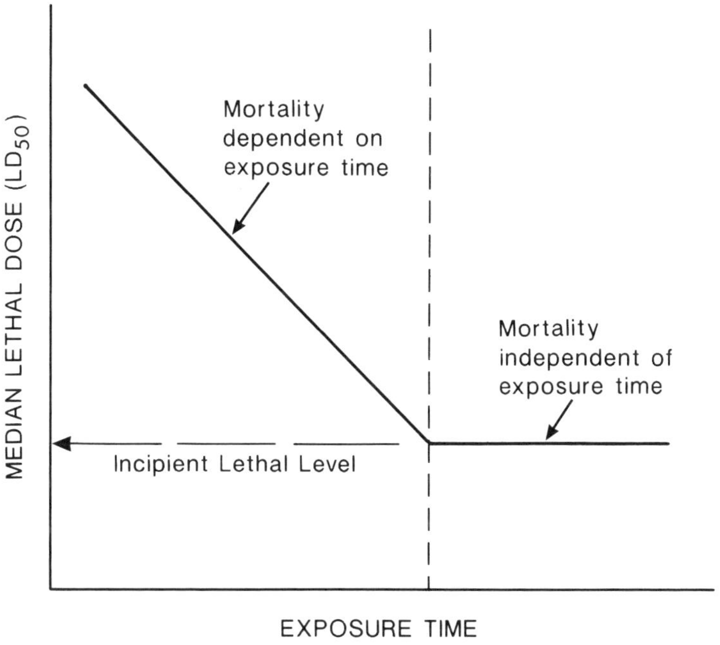

Fig. 13.2 The effect of increasing the length of the exposure time upon the median lethal level or dosage (LD_{50}) of a given environmental factor as established from dosage–mortality curves. Note that the LD_{50} decreases as exposure time increases until it finally reaches a plateau. The minimum LD_{50} represented by the plateau indicates the 'incipient lethal level' which separates the 'zone of resistance' from the 'zone of tolerance' for the particular factor.

exposure. The practical value of these results may be limited, both because they apply only to a specific set of conditions and because direct comparisons between different studies are not possible. Attempts have, therefore, been made to develop standard procedures for the estimation of lethal limits, so that results obtained in different test laboratories and for different fish species can be compared directly.

With increasing exposure periods, the dosage–mortality curves are gradually shifted to the left and this results in a progressive reduction in LD_{50} with longer exposure times. At very long exposure times the LD_{50} will approach a constant value. This is illustrated in Fig. 13.2, which shows that a plot of LD_{50} values against exposure time will give a relationship with a distinct break point. Thus, the LD_{50} decreases as exposure time increases and plateaus at the break point. This minimal LD_{50} represents the *incipient lethal level* and is an estimate of the dose which, theoretically, 50% of the population could tolerate for an indefinitely long period of time. The incipient lethal level is, therefore, estimated by the determination of dosage mortality after a long exposure time. In toxicological studies, for example, it has been common practice to determine incipient lethal levels based upon the results of tests employing exposure times of either 96 h or 10 000 min. Whilst a number of attempts have been made to introduce internationally recognized standard procedures for use in toxicological studies, there are still differences between countries in recommended methodology.

Chapter fourteen

Temperature

14.1 THERMAL RELATIONS OF FISHES

The thermal responses of fish can be divided into tolerance, resistance and preference, with tolerance limits being determined using the methods described in Chapter 13. Thus, experiments are conducted in order to determine the *incipient lethal temperatures* (upper and lower) by exposing groups of fish to a range of temperatures for an extended time period (usually 96 h). The limits of temperature tolerance will, however, be modified by the previous thermal history of the fish such that the experimentally determined value for the incipient lethal temperature will depend upon the conditions to which the fish had been exposed prior to the start of the test. Thus, fish held for an extended period under cold conditions are more cold tolerant and more heat sensitive than fish held in warm water. In other words, the incipient lethal temperatures will differ for cold- and warm-acclimated individuals. Consequently, in order to obtain a complete picture of the temperature tolerance of the fish, it is necessary to measure the upper and lower incipient lethal temperatures (UILT and LILT) for fish that have been acclimated to a wide range of constant temperatures. The upper and lower incipient lethal temperatures will increase as acclimation temperature increases, but this cannot continue indefinitely. At some point the UILT becomes equal to the acclimation temperature, and this ultimate upper incipient lethal temperature (UUILT) represents the highest temperature to which the fish can be acclimated. For some species of fish the ultimate lower incipient lethal temperature (ULILT) coincides with the freezing point of water, but for other species the value of the ULILT is several degrees above freezing.

The plotting of the UILT and LILT values as a function of acclimation temperature gives a graphical representation of the complete thermal tolerance of the fish and, because of its shape, this form of representation is known as a temperature tolerance polygon (Fig. 14.1). The area enclosed by the polygon defines the *thermal tolerance zone* of the fish and

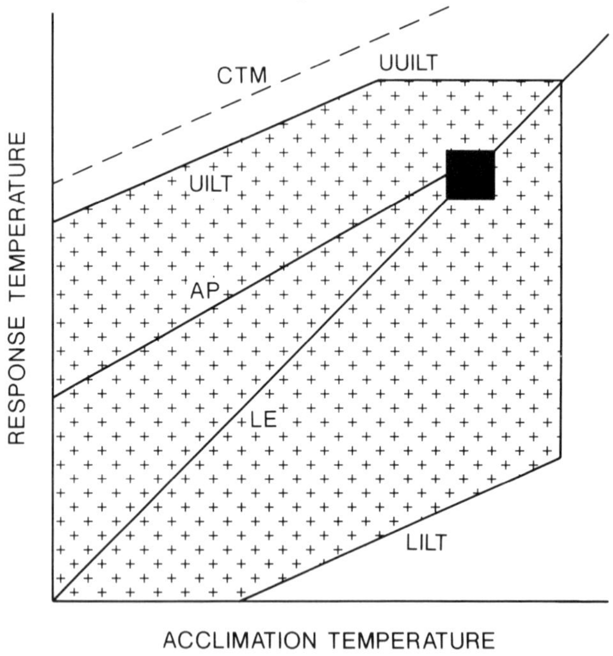

Fig. 14.1 Graphical representation of the thermal relations of fish. AP, acute thermal preferendum; CTM, critical thermal maximum: survival time above CTM is virtually zero; LE, line of equality, where the response temperature is equal to the acclimation temperature; LILT, lower incipient lethal temperature; UILT, upper incipient lethal temperature; UUILT, ultimate upper incipient lethal temperature. UUILT is the maximum temperature to which the fish can be acclimated. The incipient lethal limits mark the boundaries between the zone of tolerance and the zone of resistance, and also define the limits of the temperature tolerance polygon. The tolerance zone is marked with crosses, and the black square is the zone of final preferendum (see text).

the area can be quantified in terms of degrees squared $[(°C)^2]$. Thus, eurythermal fish, which are tolerant of a wide range of temperatures, will have tolerance polygons enclosing large areas (e.g. goldfish, *Carassius auratus*, and brown bullhead, *Ictalurus nebulosus*, have tolerance polygons that enclose in excess of 1100 degrees squared), whereas stenothermal species tolerate a much more restricted temperature range (e.g. the tolerance polygons of salmonids are usually 650 degrees squared or less).

The temperature tolerance polygon described above refers to lethal limits, i.e. the combination of acclimation and lethal temperatures that 50% of the fish population can withstand for an indefinite period. It is possible to expand the tolerance polygon concept to include corresponding graphs that designate the thermal limits for processes such as growth and reproduction. The thermal limits for growth are narrower than those for survival, so that the growth polygon will lie within the boundaries delimiting the lethal tolerance temperatures. Eurythermal fish, which have large lethal limits polygons, will, also, generally be capable of displaying positive rates of growth over a broader range of temperatures than steno-thermal species, but the temperature range over which successful reproduction is possible may be quite limited.

Hitherto, in the discussion of thermal limits and tolerance polygons, reference has been made to species tolerances. The impression may have been gained that the UUILT and ULILTs are genetically fixed limits of temperature resistance. This is not the case because these limits, and the area covered by the tolerance polygon, can be influenced, to some extent, by factors other than temperature. Both abiotic (e.g. season, photoperiod, salinity) and biotic (e.g. ontogenetic stage, nutritional status, disease) factors have been shown to interact with temperature in governing thermal limits and the area of the tolerance polygon. For example, the heat resistance of the mummichog, *Fundulus heteroclitus*, has been found to be greater during summer than during winter, even when fish have been exposed to identical temperature conditions in the laboratory prior to testing.

Thermal tolerances may also change during the course of growth and development, with there being marked differences in tolerance limits for fish at different stages of the life cycle. For example, newly-hatched cod, *Gadus morhua*, larvae suffer heavy mortality when temperatures exceed 16–17 °C, but larger cod may tolerate temperatures up to 24–25 °C. On the other hand, the juveniles (underyearlings and yearlings) of several temperate zone freshwater species (e.g. cyprinids, percids and centrarchids) appear to tolerate a wider range of temperatures than do adults of the same species. During the course of the summer months the small fish are usually found in shallow, inshore waters where daytime temperatures can be quite high, and where temperatures may fluctuate markedly between day and night. The adult fish, on the other hand, tend to inhabit cooler, deeper water where temperature conditions are more stable. Thus, seasonal and ontogenetic changes in temperature tolerances may lead to an extension of the 'ecologically relevant' tolerances beyond those represented by the conventional tolerance polygon constructed on the basis of laboratory experiments conducted on fish of a given age under one specific set of rigidly controlled conditions.

Outside of the area enclosed by the tolerance polygon lies the *zone of resistance*. The upper boundary of the resistance zone is delimited by the temperature at which the fish becomes debilitated, loses equilibrium and is no longer able to escape the lethal conditions. The major problems experienced in trying to determine the limits of the resistance zone are that outside the zone of temperature tolerance there are strong interactions between the lethal temperature and exposure time (cf. Fig. 13.2), or between the rates at which water temperatures are raised and the temperature at which the fish lose equilibrium. Thus, the higher the rate of heating, the higher the temperature at which there is loss of equilibrium. In a number of studies the upper limit to the resistance zone has been determined as the *critical thermal maximum* (CTM), which is a measure of thermal resistance obtained by increasing the water temperature at a rate of 1 °C min^{-1} until the fish lose equilibrium. Survival times above CTM are virtually zero.

Whilst fish may be able to tolerate exposure to a wide range of temperatures, they will, if given a choice, elect to spend most time within waters of a certain temperature. The temperatures selected (*acute thermal preferenda*) are dependent upon the previous thermal experiences of the fish. Fish acclimated to high temperatures will usually have acute preferenda that are higher than those of cold-acclimated fish (Fig. 14.1).

Acute thermal preferenda are usually determined by placing the fish in a long trough which has warm water flowing in at one end and cold water at the other. This enables temperature gradients to be generated within the trough and, by observing the position of the fish at regular intervals for a period of a few hours, the acute thermal preferendum can be estimated.

If the fish are left exposed to the temperature gradient for a longer period the temperature preferendum will be observed to change. The preferenda of individuals with initially different thermal backgrounds will gradually gravitate towards each other. For example, brown bullheads acclimated to 7 °C were found to have an acute preferendum of 16 °C, whereas fish previously held at 24 °C for an extended period of time preferred 26 °C. With extended exposure to the temperature gradient there was a gradual upward drift in the temperatures selected by the different groups of fish, and after 24 h the fish selected a temperature of 29–30 °C, irrespective of their previous thermal history. This temperature is known as the *final temperature preferendum.*

The final temperature preferendum is, therefore, usually defined as 'the temperature around which the fish will ultimately congregate irrespective of their previous thermal experience', but has also been defined as 'that temperature at which the preferred temperature is equal to the acclimation temperature'. There are, therefore, two methods by which the final preferendum can be estimated. The first method is the one described above, in which the fish are held in a temperature gradient for an extended period until they gravitate to the final preferendum. The alternative method consists of determining the acute thermal preferenda of groups of fish acclimated to different temperatures. The acute preferenda are then plotted against acclimation temperatures and a line drawn through the points. Where this line intersects with the line of equality (a line drawn at an angle of 45° to the x-axis) the acclimation temperature is equal to the acute temperature preferendum, and this represents the estimate of the final preferendum (Fig. 14.1). Both methods of estimation have been widely used for the estimation of final temperature preferenda of fish species.

There are a series of practical problems involved in the generation of temperature gradients, and the maintenance of adequate temperature control usually requires that the experimental troughs be relatively small and contain a shallow depth of water. This means that experiments conducted in troughs must be carried out using small fish, either on species that have a small adult body size, or on the juveniles of larger fish species.

An alternative method, which enables the temperature preferences of slightly larger fish to be studied, involves the use of a shuttlebox system. The fish swims to and fro between the chambers of the shuttlebox and thereby controls the temperature to which it is exposed. The chambers will be fitted with heating and cooling units which are switched either on or off when the fish breaks a light beam when moving from one chamber to the other (Fig. 14.2). When the fish enters a chamber, the heater will be switched on and the water temperature will gradually increase. The fish can avoid being exposed to a high temperature by moving to the other chamber, and as the fish leaves the chamber the cooling unit will be switched on, leading to a fall in temperature.

Thus, by swimming from one chamber to the other the fish can maintain the temperature within quite narrow limits. The fish will not usually swim between chambers so rapidly that the water temperature is rigidly controlled at one specific temperature, and it is more likely that the temperature will be maintained within a range spanning a few degrees. Consequently, it is more realistic to consider the final preferendum as a zone rather than a single fixed temperature (Fig. 14.1).

The thermal preferenda displayed by fish of a given species do not appear to be genetically fixed and, just as was observed to be the case for thermal tolerances, the preferenda may be influenced by a variety of abiotic and biotic factors. For example, temperature preferenda may be influenced by levels of dissolved oxygen, with fish tending to select lower temperatures in hypoxic than in normoxic waters. Thus, fish tend to display behavioural hypothermia when exposed to hypoxic conditions, with the temperatures selected being a few degrees lower than those preferred when the fish are in normoxic waters.

The onset of behavioural hypothermia appears to occur when dissolved oxygen falls below some critical threshold value, and in both cod and goldfish the response seems to be triggered when the oxygen tension falls to about 50 mm Hg. As hypoxia becomes more pronounced, the fish

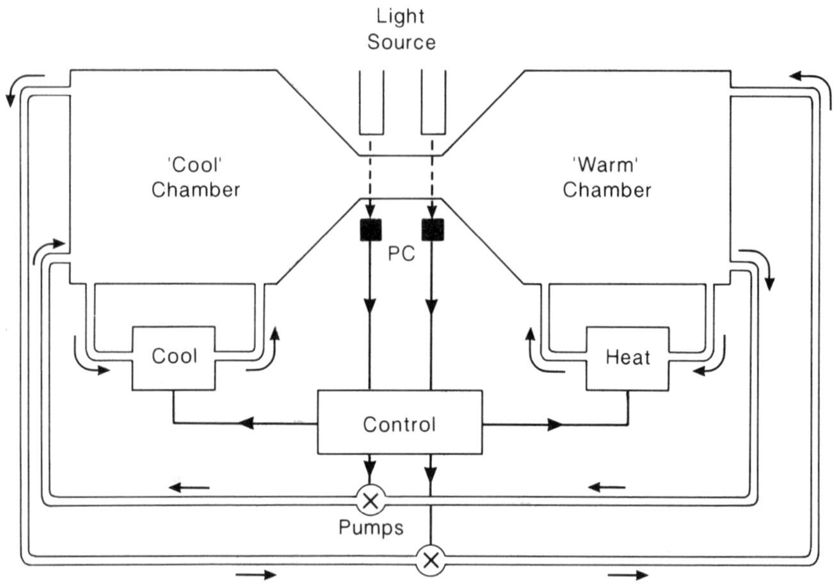

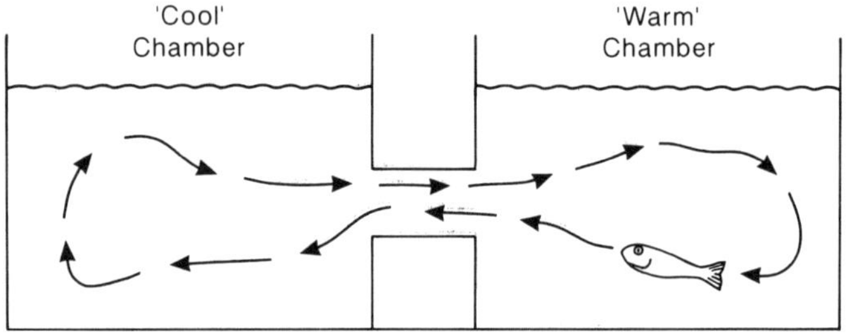

Fig. 14.2 A shuttlebox system for use in temperature preference studies with fish. The upper figure shows the shuttlebox viewed from above, and the lower figure shows a side view. The fish can regulate the temperature by moving from one chamber to the other and breaking the light beam, thereby initiating either heating or cooling. If the fish swims from the 'warm' to the 'cool' chamber, the heating is turned off and the cooling system on, and vice versa. Passage through the tube connecting the 'warm' and 'cool' chambers is registered by means of photocells (PC), and the heating and cooling units, and the pumps are controlled by computer. By circulating water between the 'warm' and 'cool' chambers by means of pumps, a constant temperature difference can be maintained between the two chambers.

select progressively lower temperatures, the cod, for example, selecting a temperature of about 13.5–14 °C under normoxic conditions, but only 9 °C when the oxygen tension falls below 30 mm Hg. Similarly, when exposed to anoxic or highly hypoxic water the goldfish may select temperatures 10–12 °C lower than those selected under normoxic conditions.

Behavioural hypothermia may be of considerable benefit and survival value to animals exposed to hypoxic conditions. Firstly, selection of a lower temperature will lead to a fall in metabolic rate in an ectothermic animal such as a fish, thereby reducing overall oxygen demand. Secondly, since oxygen is more soluble in cold than in warm water, the amounts of oxygen available under a given set of conditions will be greatest for fish that opt to maintain themselves within waters having the lowest temperature. Finally, temperature is known to influence the ability of haemoglobin to bind to oxygen, oxygen affinity decreasing as temperature is increased. Thus, by selecting lower temperatures under hypoxic conditions the fish might enable the maintenance of high levels of oxygen loading of the haemoglobin at the gills, and thereby ensure the provision of an adequate supply of oxygen to the tissues despite conditions of environmental hypoxia.

Temperature preferenda may also be influenced by the developmental stage of the fish, and ontogenetic changes in both temperature tolerances and preferences have been observed in a number of species. For example, the final preferenda of rainbow trout may be dependent upon age, with younger fish having preferenda that are a few degrees higher than those of adults.

Information about temperature tolerances and preferences may prove to be of value in the classification of fish species, or life history stages, into 'thermal guilds' according to their temperature requirements. This may be of especial importance when viewed in the light of the fact that the final preferendum zone is often found to coincide closely with the temperatures where a range of physiological processes (swimming speed, metabolic scope, growth rates) are either maximized or optimized.

14.2 THERMAL ACCLIMATION AND COMPENSATORY RESPONSES

In the preceding section it has been indicated that the thermal tolerances and preferences of fish are dependent upon previous thermal history. In addition, the thermal characteristics will change with time when a fish is moved from one temperature to another. In other words, when a fish is moved from a higher to a lower temperature, there will be a gradual loss in tolerance to high temperature and an improvement in resistance to low temperatures, and the opposite set of changes will be observed in fish moved from cold to warm water. Thus, the fish will gradually become acclimated to the new thermal conditions, but the rate at which acclimation occurs will be dependent upon the direction of the temperature change. Fish will generally acclimate more rapidly to the new conditions following an increase in temperature than they do when moved to cooler water. This difference is due to the fact that the changes resulting in acclimation are themselves dependent upon a series of metabolic and other cellular processes, the rates of which are also temperature dependent. Thus, whilst acclimation to an increase in temperature may occur within the course of a few days, acclimation to low temperatures may require periods of up to several weeks.

An abrupt alteration in temperature will lead to short-term (acute) changes in the rates of various physiological processes, but if the temperature is then held constant for an extended period of time, there will usually be a gradual change in the physiological rates as acclimation occurs. Several patterns of compensatory changes have been observed, and these are classified by comparing the rate of the physiological process measured directly after transfer to a new temperature with the rate measured following the acclimation period (Chapter 8).

For example, if fish are moved from cool (5 °C) to warm water (20 °C) there will be an abrupt increase in the measured rates of the various physiological functions (acute response). If after a period at 20 °C the rates remain unchanged, then no adaptation to the new conditions has occurred. If, however, the physiological rates are seen to have changed with time and differ from those measured as the initial acute response, then there has been some adjustment to the new thermal conditions. Compensation will have been *complete* if the rates measured at 20 °C are identical to those originally measured at 5 °C. *Partial* compensation will have occurred if the rates measured at 20 °C are lower than those measured at the same temperature immediately after transfer, but are still above those rates originally observed at 5 °C. Other forms of response (termed *overcompensation* and *inverse* (*paradoxical*)) have been observed following acclimation, but partial compensation is probably the response most usually reported.

A wide range of physiological activities are subject to change during acclimation, some of which (e.g. temperature tolerance and preference, metabolic rate or oxygen consumption, swimming performance) are studied by examining the responses of the whole animal to the temperature change. Changes occurring at the tissue, cellular and subcellular level have, however, also received attention. These latter changes may involve:

1. alterations in tissue morphology and ultrastructure;
2. changes in the biochemical pathways of cellular metabolism;
3. enzymatic adaptations involving changes in both enzyme concentrations and the types of isoforms present within the cell.

Tissue and cellular restructuring

Following transfer of fish from a warm to a cooler environment, there may be changes in both the gross morphology and the ultrastructure of certain body tissues. For example, following cold acclimation there is an increase in the proportion of red fibres in the muscles of a number of cyprinid and non-cyprinid fish species, and there are also ultrastructural changes in both the red and white fibres.

Thermal acclimation may, for example, lead to modifications to some of the myofibrillar contractile proteins. In addition, the volume of the muscle fibre occupied by mitochondria is higher in goldfish acclimated to 5 °C than in fish acclimated to 25 °C, and mitochondrial proliferation is greater in the red than in the white muscle fibres. Increases in mitochondrial volume percentage with cold acclimation are also seen in other species. These increases are the result of the mitochondria increasing in numbers, not in size. The proportion of mitochondria located between the muscle myofibrils tends to increase during cold acclimation, with a concomitant reduction in the proportion of mitochondria at the periphery of the fibre. Cold acclimation may also give rise to an increase in the surface area of the sarcoplasmic reticulum, leading to a closer relationship between the sarcoplasmic reticulum and the myofibrils.

The net result of these ultrastructural changes is to reduce the diffusional path lengths for metabolites and ions between the sarcoplasmic and mitochondrial compartments. Thus, the ultrastructural alterations that occur during cold acclimation are considered to partially offset the effects of temperature in limiting the rates of multi-enzyme pathways that are dependent upon the diffusion of a range of intermediate compounds between cellular compartments and organelles.

All of the structural modifications contribute to improved muscular function following acclimation to low temperature, and the results of these changes may be observed as differences in muscle recruitment patterns between acutely transferred and acclimated fish swimming in cold water.

With increasing swimming speed there is a sequential recruitment of muscle fibres in the order red (slow, aerobic), pink (intermediate) and white (fast, glycolytic), with only the slow, aerobic red fibres being active at very low swimming speeds.

The effects of temperature on muscle fibre activity and recruitment have been extensively studied in the carp, *Cyprinus carpio*, and it was found that when 15 °C acclimated fish were acutely transferred to either 20 °C or 10 °C, the swimming speeds at which the different fibre types were recruited differed. At 20 °C the fast white fibres were recruited at a swimming speed of approximately 2.5 $BL.s^{-1}$, but in fish acutely transferred to 10 °C recruitment of the same fibres occurred in fish swimming at a speed of about 1.5 $BL.s^{-1}$. Thus, in the fish acutely transferred to the lower temperature, the effect of temperature in reducing the power output of the slow, red fibres was counteracted by more rapid recruitment of the faster-contracting fibre types.

Following acclimation to low temperature, the maximum cruising speed was found to increase relative to that of fish acutely transferred to low temperature. The increase in maximum cruising speed was correlated with a rise in the swimming speed at which the faster-contracting fibre types were first recruited. These changes were the net result of both an increase in the relative proportion of slow, red fibres in the muscle, and changes in the contractile properties of the muscle fibres with acclimation to low temperature.

A separate adaptive strategy to cold acclimation involves changes in cellular membrane lipids, and it has often been observed that the degree of unsaturation (i.e. number of double bonds in the fatty acid chains) of membrane lipids is dependent upon the ambient temperature. In order to function properly the membranes must be in a fluid state, and the state of fluidity is largely governed by the fatty acid composition of the membrane phospholipids. The different phospholipids (e.g. phosphatidylserine (PS), phosphatidylinosine (PI), phosphatidylethanolamine (PE), phosphatidylcholine (PC), sphingomyelin, cardiolipins) are present in the cell membranes in different proportions, with the exact phospholipid composition being dependent upon the tissue examined. The membranes of fish tissues, like those of other animals, contain PC as the major phospholipid, followed by PE, with the other phospholipids being present in lesser proportions.

The lipids of animal cells contain fatty acids that usually have 16–22 carbon atoms, and these fatty acids may have up to six double bonds (the greater the number of double bonds between successive carbon atoms, the more unsaturated the fatty acid). During temperature acclimation there may be changes in patterns of lipid synthesis leading both to changes in fatty acid composition and to changes in the relative proportions of the different phospholipids. Thus, by restructuring the composition of the membranes in response to changes in environmental temperature (homeoviscous adaptation), the fluidity of the membranes can be maintained.

During cold acclimation the total amount of phospholipid in the cell membranes does not alter, but homeoviscous adaptation may result from changes in:

1. the relative proportions of the different phospholipids;
2. the fatty acid composition of the phospholipids;
3. the distribution of the fatty acids within the phospholipid molecule.

When rainbow trout are transferred from 20 °C to 5 °C, the membranes of several tissues (gill, liver, kidney) show increases in the proportion of PE and a corresponding decrease in PC. Similar relative increases in the PE content have been seen in the intestinal membranes of goldfish, liver and muscle membranes of carp and in the liver membranes of green sunfish, *Lepomis cyanellus*, following a period of cold acclimation. In contrast to the two main phospholipids, PC and PE, the relative proportions of the other phospholipids appear to be much less influenced by changes in temperature. On the basis of the above, a change in the ratio of the two major phospholipids, PC and PE, would appear to provide a sensitive measure indicative of changes in environmental temperature.

The changes that occur in the fatty acid compositions of the phospholipids with cold acclimation are mainly related to an increase in the proportions of unsaturated fatty acids. This is accompanied by a corresponding decrease in the relative proportions of saturated fatty acids. There is also some tendency for the chain length of the fatty acids to increase in cold-acclimated fish. In general, the increase in unsaturation is brought about by replacement of saturated with polyunsaturated fatty acids, with proportions of monounsaturated fatty acids remaining relatively unaffected by a temperature change. The extent to which there are cold-induced changes in fatty acid composition varies with the phospholipid class and the type of membrane, but there is invariably seen to be an increase in unsaturation of the membrane lipids following cold acclimation. For example, following exposure to cold conditions, there is an increased incorporation of the desaturation products of both the (n−3) and (n−6) fatty acid series into the phospholipids of the rainbow trout kidney. In the phospholipids of the liver, there is a more marked increase in the proportions of unsaturated fatty acids (PUFAs and HUFAs) of the (n−3) series at low temperatures. There is a large increase in 22:6 (n−3) in PC in response to low temperature, whereas in PI there tends to be a more marked increase in 20:5 (n−3) than 22:6 (n−3).

These cold-induced changes in the fatty acid composition of the phospholipids are brought about by several mechanisms acting in concert. Following a drop in temperature, there is an increase in the production of unsaturated fatty acids due to an increased activity of the fatty acid desaturase enzymes. In addition, in cold-acclimated fish the enzyme systems involved in the elongation and desaturation of fatty acids appear to act preferentially on fatty acids of the (n−3) series, which results in increased production of the two fatty acids 20:5 (n−3) and 22:6 (n−3).

These long-chained, unsaturated fatty acids are also preferentially incorporated into the phospholipids, which leads to a rapid increase in the relative proportions of these fatty acids in the phospholipids.

Metabolic reorganization

Rates of oxygen consumption displayed by both whole animals and different tissue types change during thermal acclimation, indicating that the intensity of the cellular metabolism alters as a result of the acclimatory processes (Chapter 8). In addition to changes in intensity, there may be alterations in the importance of the various metabolic pathways that lead to the generation of ATP, such that the relative rates at which respiratory substrates (lipids, proteins, carbohydrates) are utilized may differ between warm- and cold-acclimated animals.

The possible influences of changes in temperature on metabolic patterns can be studied either by examining the activities of various metabolic enzymes in the tissues of cold- and warm-acclimated animals, or by examining the metabolic fates of a range of compounds that may act as respiratory substrates.

For example, the livers of both warm-acclimated rainbow trout and warm-acclimated mummichog contain greater levels of triacylglycerols and lower levels of glycogen than do the livers of cold-acclimated conspecifics, which suggests that there may be substantial temperature-dependent differences in liver tissue metabolism. In striped bass, *Morone saxatilis*, on the other hand, there tends to be increased storage of triacylglycerols in the livers of cold-acclimated fish, and carbohydrate appears to be used preferentially as the respiratory substrate at low temperatures. Thus, there may be interspecific differences in the temperature-dependent metabolic responses shown by a given tissue type.

The situation is further complicated by the fact that different tissues may display dissimilar changes in metabolic pathways following temperature acclimation. For example, whilst the liver of the striped bass stores lipid and metabolizes carbohydrate in the cold, the response of the muscle tissue appears to be different. In muscle tissue, lipid oxidation is much higher and carbohydrate oxidation lower following a period of cold acclimation, which is the reverse of the changes seen in the liver of this species. Thus, the situation regarding shifts in substrate utilization and the modification of metabolic pathways in response to thermal acclimation is a complex one.

The changes observed may also be dependent upon factors other than temperature *per se*. Nutritional factors, such as dietary composition and the overall size of the storage reserves, may be expected to have important influences upon the patterns of metabolic reorganization observed during thermal acclimation. In addition, the nature and tissue distributions of the major storage products (e.g. lean fish tend to store either glycogen or triacylglycerols in the liver, whereas fatty fish have most of their reserves in the form of triacylglycerols distributed in the muscle and peritoneal tissues), may be expected to have an influence, as might reproductive status and the ontogenetic stage of development.

Enzymatic adaptations

The changes that occur during thermal acclimation must involve a series of adaptations at the enzymatic level, and there have been a number of studies carried out in order to examine the effects of temperature changes on rates of enzymatic activity in a range of fish tissues. One way in which this can be done is to grow isolated cells in culture at different temperatures and then study the changes in enzymatic activity that occur with time.

Liver cells (hepatocytes) have been most often used in this sort of study, and it has been found that the relative activities of enzymes such as citrate synthase and cytochrome *c* oxidase tend to increase in low-temperature cultures, and decrease in high-temperature cultures. In other words, acclimation to low temperature results in compensatory changes at the enzyme activity level. Whilst tissue cell culture has been used for the examination of the effects of temperature acclimation on enzyme activities, alternative methods are more frequently employed.

In these types of studies it has become common practice to acclimate separate groups of fish to low and high temperatures and then examine tissue enzyme activities of the cold- and warm-acclimated fish at some intermediate temperature. For example, when groups of fish have been acclimated to 5 °C and 25 °C it has been usual to make measurements of enzyme activities at 15 °C, and the results are presented in the form of a ratio. This ratio expresses the relative activities of the enzyme at the intermediate temperature and allows comparisons to be made between cold- and warm-acclimated fish. A ratio of 1 would mean that the enzyme activities of the cold- and warm-acclimated fish were the same, whereas an enzyme activity ratio over 1 indicates that activities are higher in cold- than in warm-acclimated fish. An enzyme activity ratio over 1 is, therefore, taken as providing evidence that there have been temperature-dependent changes in activity levels during the course of an acclimation period.

Fish muscle tissue has been much studied with respect to the effects of thermal acclimation on enzyme activity levels. The activities of enzymes such as cytochrome *c* oxidase, succinate dehydrogenase and citrate synthase have been found to be elevated (i.e. enzyme activity ratio greater

Table 14.1 Influence of thermal acclimation on the activities of some mitochondrial and glycolytic enzymes in fish species. Fish were acclimated to either 'warm' (25 °C) or 'cold' (5 °C) conditions, and enzyme activities were measured at an intermediate temperature (15 °C). Entries are activity ratios, as defined in text

	Common carp, *Cyprinus carpio*	Green sunfish, *Lepomis cyanellus*	Striped bass, *Morone saxatilis*
Mitochondrial enzymes			
Cytochrome *c* oxidase	1.58	1.94	1.97
Citrate synthase	2.25		1.57
Glycolytic enzymes			
Pyruvate kinase	1.00	0.74	0.66
Lactate dehydrogenase	0.99	0.90	0.92

than 1 when tested at intermediate temperature) in cold-acclimated fish of several species (e.g. cyprinids, centrarchids, threespine stickleback, *Gasterosteus aculeatus*, striped bass). On the other hand, many of the glycolytic enzymes (e.g. phosphofructokinase, pyruvate kinase, lactate dehydrogenase) appear to be little altered during thermal acclimation, and the enzyme activities of cold- and warm-acclimated fish are similar when measured at an intermediate temperature (Table 14.1). For example, whilst carp muscle shows temperature-dependent changes in activities of citrate synthase, cytochrome *c* oxidase and Ca^{2+}-activated ATPase, thermal acclimation does not appear to result in compensatory changes in the activities of the glycolytic enzymes pyruvate kinase and lactate dehydrogenase. Thus, it appears that the enzymes that have functions in aerobic energy liberation and ion transport show the greatest temperature-dependent compensatory changes, whereas there may be either no or only minor changes in glycolytic enzymes.

There is, therefore, convincing evidence that thermal acclimation leads to increased activities of some enzymes in cold-acclimated fish. This could be achieved in a number of ways. For example, the same end result could be obtained by there being either changes in the cellular concentrations of the enzyme and/or changes in the enzyme isoforms present. Both of these mechanisms may be involved in bringing about the observed changes in enzyme activities following a period of thermal acclimation and, in some cases, both may be acting simultaneously. It has, for example, been found that the cellular concentration of cytochrome *c* oxidase increases in the muscle tissues of both green sunfish and goldfish following acclimation to cold conditions.

Because a change in temperature may have marked influences on the catalytic properties of enzymes, an increase in enzyme concentration may not always have the effect of offsetting the reductions in catalytic activity caused by low temperatures. Under such circumstances, levels of enzyme activity could be maintained by there being alterations in the enzyme isoforms in the cell. In other words, a change in temperature would result in shifts in the relative proportions of two or more enzyme isoforms such that overall enzyme activity remained closely matched to the prevailing temperature conditions. This could be achieved either by the synthesis of specific enzyme isoforms within particular relatively narrow temperature ranges, or by the maintenance of a complex set of isoforms that vary in relative proportions with changes in temperature. It is known that several enzymes are polymorphic (i.e. exist in several isoforms) but to what extent enzyme polymorphism contributes to the activity changes observed during thermal acclimation has been relatively little studied.

Heat-shock proteins

Organisms of many kinds respond to a rapid elevation in temperature by synthesizing a group of proteins known as *heat-shock proteins.* Induction of the synthesis of heat-shock proteins occurs rapidly at temperatures 5–15 °C above the normal environmental temperature, whereas the synthesis of other cellular proteins appears to be suppressed under similar conditions. The synthesis of heat-shock proteins increases the thermotolerance of the animal to subsequent exposure to elevated temperatures. Thus, the heat-shock proteins are thought to protect the cells against the pathological effects of heat.

Some of the heat-shock proteins may have regulatory transport functions within the cell, whereas others may play a structural role, because they tend to accumulate around specific organelles and the cytoskeletal network. The precise functions of the heat-shock proteins, and their modes of action, are not known, but there is an increasing body of evidence to suggest that they may be synthesized following exposure to a range of stressors (e.g. heavy metals and other chemical agents, radiation, viral infections) other than temperature. Thus, the induction of heat-shock protein synthesis may represent a general cellular response to exposure to adverse environmental conditions, rather than being indicative of heat-shock *per se.*

Avoidance of freezing

Damage to cells and tissues as a result of freezing can be caused in several ways, and freeze damage may be averted by a range of protective mechanisms. Freezing may cause damage by the formation of ice leading to the disruption of cytoplasm or cell membranes, and the freezing of intracellular water is inevitably lethal. Freeze injury may also arise due to cellular dehydration. When tissues begin to freeze, ice crystals form first in the extracellular fluid, and the remaining extracellular fluid becomes hypertonic with respect to the intracellular fluid. Thus, cells may begin to lose water osmotically, and become dehydrated. The resulting changes in ionic strength may then lead to disruption of cell functions and death.

Freeze avoidance relies upon the prevention of ice formation and this can be achieved by either colligative (i.e. depending upon the concentration, not the nature, of the solute) or non-colligative means. Seawater is hypertonic with respect to the plasma of teleost fish. The concentration of solutes in seawater, principally sodium chloride, is approximately three times that in the plasma of the majority of teleosts. The colligative effects of these solutes results in a lowering of the freezing point relative to that of pure water. The colligative effects of the solutes are greater in seawater than in the blood plasma of teleost fish and, consequently, the body fluids of fish freeze at temperatures a degree or so above the freezing point of full-strength seawater (−1.86 °C). As a result of this, marine fish species that inhabit high- and temperate-zone-latitude ocean environments may risk being exposed to seawater at temperatures below the freezing point of their body fluids. One means by which these fish could avoid freezing would be to increase the solute concentration of the body fluids and thereby depress the freezing point by a colligative mechanism.

Some temperate zone marine fish, such as the winter flounder, *Pseudopleuronectes americanus*, increase plasma levels of sodium chloride following cold exposure, whereas other species, such as the mummichog, increase the concentration of plasma glucose and other solutes during the winter months. The increases in plasma solute concentrations shown by these species are, however, only sufficient to decrease the freezing point of the body fluids by about 0.2–0.3 °C. When exposed to cold-water conditions a small number of teleost fish species, including the rainbow smelt, *Osmerus mordax*, increase serum osmolalities to levels significantly higher than those of typical teleosts. The body fluids of the fish may become almost isosmotic with full-strength seawater, with the major osmolyte being the sugar alcohol, glycerol. The appearance and accumulation of glycerol in the body fluids is strongly temperature dependent. Low levels of glycerol occur at temperatures of 0 °C and above, whereas high levels are found in the body fluids of rainbow smelt exposed to temperatures close to the freezing point of seawater. Serum concentrations of glycerol can reach levels that are sufficiently high to cause a freezing point depression of 0.75–0.8 °C, with other osmolytes, such as electrolytes, urea, free amino acids and glucose, further contributing to the depression of the freezing point due to colligative properties.

A number of fish species have developed antifreezes that prevent freezing via non-colligative mechanisms, and the rainbow smelt has both colligative and non-colligative antifreezes. A non-colligative antifreeze depresses the freezing point of a fluid to a lower temperature than would be predicted on the grounds of colligative properties alone. The freezing point of the body fluids may be depressed 200–500 times more than expected on the basis of the colligative reactions, and the non-colligative antifreezes result in a lowering of the freezing temperature of the plasma without causing osmotic imbalance. The melting point of the plasma is, on the other hand, unaffected by these antifreezes, because this depends solely on the colligative properties of the fluid. The separation of the freezing and melting points of fluids is termed *thermal hysteresis*, and a number of antifreeze molecules have been shown to possess this unusual property. These antifreezes act by either interfering with or preventing the formation of ice crystals, possibly by binding physically to a newly formed ice crystal and thereby inhibiting further growth. The non-colligative antifreezes are large molecules, being either antifreeze glycoproteins (AFGP) or antifreeze peptides (AFP). Five major classes of non-colligative antifreeze compounds have been identified.

Antarctic notothenid fish possess glycopeptides composed of repeating units of a tripeptide (alanyl-alanyl-threonine) bound to a disaccharide (galactose + N-acetylgalactosamine). These glycopeptides (AFGP) occur in at least eight different size ranges (molecular weight 2600–34 000), and in some of the tripeptide sequences the amino acid proline may be substituted for alanine. The same glycopeptides have been found in the plasma of the rock cod, *Gadus ogac*, a Northern Hemisphere species which occurs off the coast of Labrador. Glycopeptides are also employed as antifreezes by the Atlantic cod, *Gadus morhua.*

In some northern temperate and subarctic marine fishes, polypeptides (AFP), rather than glycopeptides (AFGP), serve as the antifreeze compounds. The AFPs of the winter flounder and short-horned sculpin, *Myoxocephalus scorpius*, are in the form of helical coils and contain alanine as the major amino acid (over 60%). The AFP of the sea raven, *Hemitripterus americanus*, on the other hand, has a relatively low alanine content (14%), but is comparatively rich in cystine (6–7%), and the AFP of the ocean pout, *Macrozoarces americanus*, is neither rich in alanine nor rich in cystine. The Antarctic zoarcids appear to have a peptide antifreeze that differs in molecular structure from that of their relatives in the Northern Hemisphere.

The antifreezes (AFGPs and AFPs) are synthesized in the liver and secreted into the blood and extracellular fluids. They are present all year round in Antarctic fish species that are continuously exposed to cold conditions, but production of the antifreezes is seasonal in the northern temperate species. In some species, such as the winter flounder and tomcod, *Microgadus tomcod*, antifreeze production commences during autumn in advance of the onset of winter conditions, and before the fish have been exposed to low temperatures. In these species, the synthesis of the antifreezes appears to be influenced by photoperiod, and production is suppressed by long daylengths (15 h or more of daylight). On the other hand, antifreeze glycopeptides do not begin to appear in the blood of large, adult Atlantic cod before the fish have been exposed to water temperatures below zero for a period of several days. In juvenile cod, however, plasma concentrations of antifreeze start to rise when the water temperature falls to about 2 °C, so that the young fish develop high plasma concentrations of antifreeze earlier in the winter than do larger, older individuals. Thus, there appear to be distinct ontogenetic changes in the patterns of synthesis and turnover of the antifreeze glycoproteins in the cod, with the juveniles both being capable of withstanding lower temperatures in the presence of ice than the adults, and being physiologically prepared to meet a low-temperature challenge should one occur early in the winter. The antifreeze activity is absent from the body fluids during the spring and summer, and there is a loss of blood antifreeze within the space of a few weeks following exposure to temperatures of 10 °C or over.

The rainbow smelt possesses both colligative (glycerol) and non-colligative (AFP) antifreezes, with the major portion of the freezing point depression of the body fluids being due to the high concentrations of glycerol that accumulate on exposure to cold-water conditions. The AFP may be restricted to the plasma, being excluded from the cells due to its large size. Thus, the AFP may provide limited protection to the cells against freezing. Glycerol, on the other hand, passes freely through cell membranes and can provide protection to the cells. The production and maintenance of high concentrations of glycerol may, however, be energetically expensive, and glycerol appears in the body fluids only when minimum temperatures are encountered. Thus, the AFP may be the first line of defence against freezing, whilst glycerol serves as a back-up, the production of glycerol being initiated under extreme conditions.

Chapter fifteen

Dissolved gases

15.1 INTRODUCTION

Air at sea level contains about 20.9% oxygen, and under standard conditions atmospheric pressure (AP) is equivalent to 760 mm Hg. The partial pressure (P_{O_2}) of oxygen will, under these circumstances, be about 159 mm Hg (760 × 0.209). In practice, the partial pressure under different conditions can be calculated as:

$$P_{O_2} = (AP - VP) \times \%O_2 \qquad (15.1)$$

where VP is the water vapour pressure in mm Hg and $\%O_2$ is the percentage of oxygen in the atmosphere. Oxygen partial pressures in the air at sea level will generally fall within the range 154–158 mm Hg.

Gases are soluble in water, and if water is brought into contact with a gas atmosphere, some of the gas molecules will enter the water and go into solution, i.e. the gas will become dissolved in the water. Eventually an equilibrium will be reached when an equal number of gas molecules enter and leave the solution per unit time, and the amount of gas that is then dissolved depends upon the pressure of the gas in the gas phase. The dissolved gas exerts no measurable pressure and it is, therefore, somewhat misleading to use the term *partial pressure* when referring to dissolved gases. The term *tension* is usually preferred, with the tension of a gas in solution being defined as the pressure of the gas in the gas mixture with which the particular solution is in equilibrium. In practice, the terms partial pressure and tension are often used interchangeably. Thus, water in contact, and equilibrated, with atmospheric air will have an oxygen tension of dissolved oxygen equal to that of the air. The equilibrium point is reached, however, when the amount of dissolved oxygen in the water is quite low. For example, a sample of oxygen-saturated fresh water at 10 °C at sea level will contain 7.93 ml $O_2.l^{-1}$ and the oxygen tension will be 157.3 mm Hg. If the sample were only 50% saturated with oxygen, it would contain 3.97 ml $O_2.l^{-1}$ (7.93/2) and the oxygen tension would be

78.7 mm Hg (157.3/2). Thus, under the conditions prevailing in aquatic environments the water will seldom be expected to contain more than 10 ml $O_2.l^{-1}$ (about 15 $mg.l^{-1}$) (for oxygen at STP 1 ml = 1.428 mg), which is substantially less than the 209 $ml.l^{-1}$ found in the air.

In addition to the pressure of the gas in the gas phase, factors such as the nature of the gas itself, the temperature and the presence of other solutes in the water will also have a marked influence on solubility.

Two coefficients have been employed to express the results of solubility measurements with gases. The first is known as the *absorption coefficient*, and this may be defined as the volume of gas (in millilitres STPD) dissolved in 1 l of water when the pressure of the gas is 1 atm (760 mm Hg). The amount of gas is expressed as the volume this gas would occupy if the dry gas were at 0 °C and 1 atm pressure; this is designated standard temperature and pressure dry (STPD). Thus, if v_0 is the volume of gas dissolved reduced to STPD, V is the volume of the solvent, and P_G is the partial pressure of the gas in atm, then the absorption coefficient is given by $v_0/(V \times P_G)$. By re-arranging this expression to:

$$v_0 = \text{absorption coefficient} \times (V \times P_G) \qquad (15.2)$$

it can be seen that the amount of gas dissolved in a given volume of solvent will depend upon the partial pressure of the gas. If the gas partial pressure is doubled, then twice as much gas will be dissolved. In other words, a gas will dissolve in any given solvent according to its partial pressure in the gas phase, but there will be marked differences in gas concentration between the two media.

The second coefficient used in describing the solubility of gases is the *coefficient of solubility*. This may be defined as the volume of the gas taken up by a unit volume of the solvent under a particular set of temperature and pressure conditions. In other words, if v is the volume of the dissolved gas, measured at temperature T and a partial pressure P_G atm, then:

$$v = (T \times v_0)/(T_0 \times P_G) \qquad (15.3)$$

where T_0 is 0 °C, i.e. 273 K. According to this definition, a re-arrangement of the equation would give:

$$v_0 = (v \times P_G \times T_0)/T \qquad (15.4)$$

which shows that the solubility of a specific gas will decrease as temperature increases. For example, under equilibrium conditions between atmospheric oxygen (P_{O_2} within the range 154–158 mm Hg) and fresh water the solubility of O_2 decreases from about 8 $ml.l^{-1}$ at 10 °C to 5.6 $ml.l^{-1}$ at 30 °C. The solubility of gases will also be influenced by the presence of solutes in the water, and an increase in dissolved solids will reduce the solubility of dissolved gases. Dissolved gases will not, however, affect the solubilities of other gases, and any gas will dissolve according to its own partial pressure in the gas phase, independent of the presence of other gases. The fact that dissolved solids influence gas solubility means that the solubilities of atmospheric gases are lower in seawater than in fresh water, the difference in oxygen solubility between the two media being approximately 20%.

In addition it must be realized that the solubilities of different gases are markedly different. For example, if the solvent is water, and temperature is held at 15 °C, the solubility of oxygen will be 34 ml O_2 per l water when the gas pressure is 1 atm. Corresponding solubilities for CO_2 and N_2 are about 1020 and 17 ml per l water, respectively. Thus, CO_2 is 30 times more soluble in water than O_2, whereas N_2 is only half as soluble as O_2.

Finally, it is important to distinguish between both the coefficient of solubility and absorption coefficient, and the *capacitance coefficient.* The capacitance coefficient may be defined as the increase in concentration that results from each unit increase in partial pressure of a particular gas. It must be realized that the capacitance coefficient includes not only the gas that is physically dissolved in the solvent, but also chemically bound gas. In water, for example, where there are only very small amounts of chemically bound molecular oxygen, the coefficient of solubility and the capacitance coefficient for O_2 will be almost identical. In the same water, however, carbon dioxide can be present as dissolved CO_2, as the bicarbonate ion (HCO_3^-), and as carbonate (CO_3^{2-}), so the capacitance coefficient is higher than the coefficient of solubility. In body fluids, where both carbon dioxide and oxygen may be chemically bound in large quantities, the capacitance coefficients are very much higher than the coefficients of solubility. Thus, when a solvent contains large amounts of a gas in chemically bound form, it is possible for the gas to diffuse *down* its partial pressure gradient, and, at the same time, *up* a concentration gradient.

15.2 DISSOLVED OXYGEN

The gradient of oxygen tension (difference in oxygen tension between the fish blood and the surrounding water) is the driving force governing oxygen uptake, but the oxygen concentration in the water indicates how much water must be pumped over the gills in order for the fish to obtain a given amount of oxygen. For example, as temperature increases, the oxygen content of fully saturated water will decrease, but oxygen tension is little affected by temperature. Thus, because of the reduced oxygen content of the inspired water, a fish respiring in warm water must pump more water over the gills than a fish in cold water in order to deliver a given amount of oxygen per unit time. This is necessary despite the fact that the gradient in oxygen tension which is driving uptake is little changed at the higher temperature.

The oxygen tension may not always remain close to that recorded under normoxic conditions (when the water is saturated with dissolved oxygen). Biological activities can result in levels of dissolved oxygen above saturation or well below it. Thus, in contrast with the atmosphere, in which P_{O_2} and oxygen concentration are relatively constant, the P_{O_2} and oxygen concentration in natural water bodies can vary dramatically. The variations arise from a number of sources, and the variations in P_{O_2} and oxygen concentrations in natural waters may be either temporal or spatial. Some of the most dramatic fluctuations may be seen in shallow eutrophic lakes where there can be marked diurnal changes in the balance between photosynthetic oxygen production and the oxygen demand for respiration and chemical processes. Thus, oxygen tension may rise to over 200 mm Hg during the day, and fall dramatically to 20–30 mm Hg during the night.

Variations in P_{O_2} and oxygen concentrations also occur on a seasonal basis, and are pronounced in situations where there is a stratification of the water body. Such effects are particularly pronounced in temperate freshwater lakes where the formation of a warm surface layer during the summer effectively seals off the underlying cool, dense water from contact with the atmosphere. Oxygen demand may exceed supply in these deeper waters, such that levels of dissolved oxygen decline as the summer progresses. These changes can result in 'summer kills' of fish and other aquatic organisms. During the winter months, contact between the water surface and the atmosphere may be reduced by ice formation. Oxygen may be consumed through the decay of organic matter, but adequate replenishment may be prevented by the ice cover. A gradual decline in levels of dissolved oxygen can eventually lead to 'winter kills' of fish.

Any factor that either results in a reduction in the surface of contact between the water and the atmosphere, or leads to reduced rates of mixing, will generally have negative effects upon levels of dissolved oxygen. Whilst the photosynthetic activity of algae and higher plants results in a production of oxygen, floating vegetation can act as a physical barrier to oxygen replenishment of water bodies, and mats of vegetation may reduce wind-blown mixing of the surface water layer. In addition, the decomposition of rotting vegetation consumes considerable quantities of oxygen, so aquatic animals that inhabit highly eutrophic water bodies containing large amounts of decaying plant material may be frequently subjected to hypoxic or anoxic conditions. Tropical swamps, streams polluted with high levels of organic matter, or tidal pools containing rotting seaweed are all habitats in which fish are likely to experience hypoxic conditions.

When fish are exposed to hypoxic conditions they may show one of several responses to the oxygen deficiency. Firstly, they may seek to avoid waters subject to low levels of dissolved oxygen by moving to alternative habitats. This presupposes that the fish are able to accurately monitor the oxygen content of the water and, thereafter, select waters having relatively high levels of dissolved oxygen.

Fish of several species are known to show avoidance reactions to waters deficient in oxygen, and this avoidance response often appears to occur before the onset of respiratory distress. Thus, the fish seem to possess the sensory mechanisms required for rapid detection of hypoxic conditions. In laboratory preference trials, fish frequently select locations with high levels of dissolved oxygen, and only seldom make excursions into oxygen-deficient waters. For example, when held in a gradient apparatus, young brook trout, *Salvelinus fontinalis*, showed a rapid avoidance response to severely oxygen-deficient water (1–2 $mg.l^{-1}$, 9–17% saturation) and moved to water having an oxygen concentration of 5 $mg.l^{-1}$ (45% saturation) or over. With longer-term exposure to the dissolved oxygen gradient the fish avoided the hypoxic water, and they were able to distinguish between water having dissolved oxygen concentrations of 4–5 $mg.l^{-1}$ (40% saturation) and 5–6 $mg.l^{-1}$ (50% saturation).

Thus, it seems clear that the fish are able to detect the levels of dissolved oxygen in the water and to distinguish between hypoxic and normoxic conditions. There have been several studies carried out in order to locate the 'oxygen receptors' in fish, and the bulk of the evidence suggests that there are at least two different locations for such receptors. One set of receptors is located superficially, in contact with the inspired water that flows over the gills. These externally facing receptors appear to be largely associated with the pseudobranch or the first gill arch, but it is also possible that the receptors have a more diffuse distribution in the buccal cavity, pharynx and on the gill arches. Since these receptors are in direct contact with the inspired water it is thought that they provide the information that leads to the rapid behavioural (avoidance) responses of the fish on encountering hypoxic conditions. Information from these receptors also appears to be involved in the induction of the increased rate of gill ventilation which is seen as an immediate response to hypoxia, and may be involved in the instigation of the bradycardia (reduced heart rate) seen in fish during longer-term exposure to hypoxia. The second set of 'oxygen receptors' appear to detect the P_{O_2} in the arterial blood leaving the gills. These receptors appear to be located in, or just down-stream from, the gills, and these 'arterial blood oxygen receptors' seem to provide the information that leads to the changes in gill ventilation that accompany more prolonged exposure to hypoxic conditions.

If alternative habitats, with adequately oxygenated water, are not available to the fish, they must resort to methods other than avoidance in order to survive the hypoxic conditions. Some species breathe air, a few species are capable of surviving on anaerobiosis, but many small fish species perform aquatic surface respiration (ASR) in which they use the higher levels of dissolved oxygen at the air–water interface.

15.3 AQUATIC SURFACE RESPIRATION (ASR)

Many species of fish, particularly small, tropical teleosts, such as cyprinodontids and poeciliids, respond to decreasing levels of dissolved oxygen by showing aquatic surface respiration. This involves the fish rising to the surface and using the thin oxygenated layer at the air–water interface for aquatic respiration, i.e. the fish continue to breathe water and do not migrate to the surface in order to gulp air. ASR can also be performed in the proximity of air pockets or bubbles trapped under surface ice. The behavioural response is widespread, being observed for both temperate and tropical species that inhabit potentially hypoxic habitats. In such habitats, 80% or more of the fish species may display this form of behaviour when exposed to oxygen-deficient water. ASR is not, however, a behaviour that is induced as a last resort in the attempt to survive hypoxic conditions. ASR may occur in waters where the levels of dissolved oxygen would not compromise survival of the fish, but by adopting this behaviour, the fish appear to be able to maintain rates of feeding, growth and activity at higher levels than would otherwise be possible.

The fish appear to initiate ASR at a distinct threshold level of dissolved oxygen, and then the amount of time engaged in this behaviour increases rapidly as oxygen deficiency becomes more pronounced. By the time that P_{O_2} has declined to 10–25 mm Hg, fish of the majority of species will be spending 50% of their time engaged in ASR. The P_{O_2} at which the fish engage in ASR is, however, also influenced by other factors, such as temperature. For example, in the fathead minnow, *Pimephales promelas*, the 50% threshold for ASR was approximately 10 mm Hg (0.7 mg $O_2.l^{-1}$) at 10 °C, 17 mm Hg (1.0 mg $O_2.l^{-1}$) at 20 °C and 28 mm Hg (1.4 mg $O_2.l^{-1}$) at 30 °C. These differences are probably related to the fact that rates of oxygen demand increase with increasing temperature, whilst, at the same time, the oxygen content ($mg.l^{-1}$) of fully saturated water decreases as temperature increases.

When fish adopt ASR they will increase their vulnerability to predators, such as birds, that attack from above. It would, therefore, be predicted

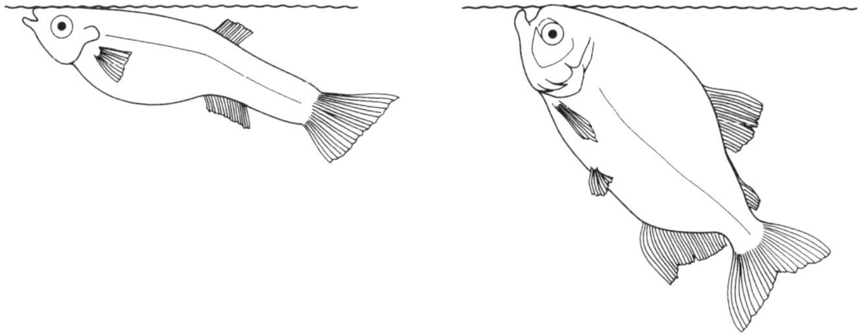

Fig. 15.1 Aquatic surface respiration (ASR) displayed as a response to exposure to hypoxic conditions. Note that the guppy, *Poecilia reticulata,* subtends only a small angle relative to the water surface when engaging in ASR. The pacu, *Piaractus mesopotamicus,* which has a high-domed head and terminal mouth, needs to subtend a much steeper angle than the guppy in order to contact the air–water interface.

that the fish would require a lower P_{O_2} threshold before adopting ASR and/or would spend less time at the surface in the presence of a predator than under conditions of reduced predation risk. Reductions in ASR behaviour have been observed in a number of experimental studies where fish have been exposed to predation risk.

When the fish engage in ASR, they will be breathing by skimming the air–water interface, and the efficiency with which this can be done will depend upon the shape of the head and the relative position of the mouth (Fig. 15.1). Fish that are relatively flat dorsally and that have pointed heads and supraterminal mouths will be required to subtend only a small angle (10–20°) relative to the water surface when engaging in ASR. In addition, these fish will be able to skim the air–water interface whilst swimming forwards in a position in which the head is pointing only slightly upwards. Fish with blunter, more rounded heads and terminal mouths will need to subtend steeper angles (30–75°), and species with subterminal mouth positions may have to hang almost vertically in the water column in order to come into contact with the air–water interface. Whilst a subterminal or ventrally placed mouth does not preclude a species from engaging in ASR, these fish would be expected to be at a some disadvantage compared with species having supraterminal mouths, since they will be unable to skim the air–water interface.

In some characids and serrasalmids, inhabiting the Amazon Basin and other Neotropical areas, the use of the air–water interface during ASR is accompanied by the temporary formation of dermal swellings on the lower jaw. The majority of these species have a terminal or subterminal mouth position under normoxic conditions, but the dermal swelling that occurs when the fish are exposed to hypoxia leads to a temporary transformation to a more supraterminal position. In the pacu, *Piaractus mesopotamicus*, a serrasalmid species, the dermal swellings begin to appear about 1 h after the onset of ASR, and the swellings gradually increase in size if oxygen-deficient conditions persist. On a return to normoxic conditions the swellings regress and the lower jaw regains its normal size and shape. The fish has a high dome to its head and the function of the morphological modification of the lower jaw appears to be to facilitate ASR. Thus, the dermal extension may help to keep the lower jaw in contact with the air–water interface when the fish swims forward at an angle of 40–50° to the surface with only the very top of its head exposed.

Thus, the behavioural response of ASR, which may be accompanied by hypoxia-induced morphological adaptations in some species, enables fish to survive in the oxygen-poor waters of tropical swamps and floodplains. Similar behaviour is shown by a range of temperate-zone species that may be subjected to oxygen-deficient waters at different seasons of the year.

15.4 AIR BREATHING – BIMODALLY BREATHING FISH

Since air contains much more oxygen than water, and is a much less dense and viscous medium, aerial respiration should be much less costly from an energetic point of view than aquatic respiration. The advantage of aerial respiration becomes increasingly great as the level of dissolved oxygen declines. Nevertheless, even in habitats in which there may be frequent periods of hypoxia, the majority of fish species are water breathers that are incapable of aerial respiration. Furthermore, many air-breathing fish species are bimodal, and rely on aquatic respiration, either exclusively or predominantly, until dissolved oxygen levels fall below some threshold. Aerial respiration has, however, evolved in most groups of bony fish, and there is a great diversity of air-breathing strategies and structures amongst the different species (Table 15.1). In addition to developing new structures for air breathing, the species that breathe bimodally also show modifications to the nervous and circulatory systems to meet the demands of coordinating the activities of two different gas-exchanging organs which are operative in two different media. For example, unimodal water breathing is generally characterized by relatively regular and rhythmic ventilatory and cardiovascular patterns, and levels of gases in the blood are usually maintained within relatively narrow limits. In bimodally breathing fish, on the other hand, the branchial ventilation pattern may be highly variable, aerial respiration may occur at irregular intervals, and levels of blood gases may fluctuate more widely than those in unimodal species. Thus, aerial respiration may be largely an 'on-demand phenom-

Table 15.1 Accessory air-breathing organs (ABOs) used by a range of bimodally breathing fish species

Organ used for aerial respiration	Fish	Comment
Gills/skin	*Hypopomus, Anguilla, Periophthalmus*	Facultative air breather
Mouth/opercular cavity		
Pharyngeal epithelium	*Electrophorus*	Obligatory
Diverticula	*Clarias, Anabas*	Facultative
	Ophiocephalus	Obligatory
Gastrointestinal tract		
Stomach	*Eremophilus, Ancistrus*	Facultative
Intestine	*Hoplosternum*	Obligatory
	Misgurnus	Facultative
Gas bladder	*Arapaima, Lepidosiren, Protopterus*	Obligatory
	Amia, Lepisosteus, Neoceratodus	Facultative

enon, with bimodally breathing fish often showing great flexibility in their use of air breathing.

Hypoxia and aerial respiration

Hypoxia appears to be the most important factor influencing the rate of air breathing in bimodally breathing species, with some *facultative* air breathers only initiating aerial respiration once the level of dissolved oxygen falls below a very low critical threshold. 'Continuous', *obligate* air breathers, which use aerial respiration at all dissolved oxygen levels, will tend to increase the proportion of oxygen they obtain via air breathing as levels of dissolved oxygen decline. Aquatic hypoxia may, however, have only a minor effect on the rate of air breathing shown by those obligate air-breathing species that have either extremely reduced or intermittently ventilated gills.

The gill ventilatory and cardiovascular responses to oxygen-deficient water vary considerably amongst the air-breathing species, with the response shown appearing to be linked to the efficiency with which the fish can extract oxygen from the air. Most air-breathing fish respond rapidly to a fall in dissolved oxygen levels, probably as a result of prompt detection of the oxygen-deficient conditions by externally directed oxygen receptors. The initial response to hypoxia displayed by many species is an increase in gill ventilation, but a further fall in oxygen tension will result in a depression of gill ventilation. A reduction in ventilation of the gills during severe hypoxia may serve to reduce the risk of oxygen loss from

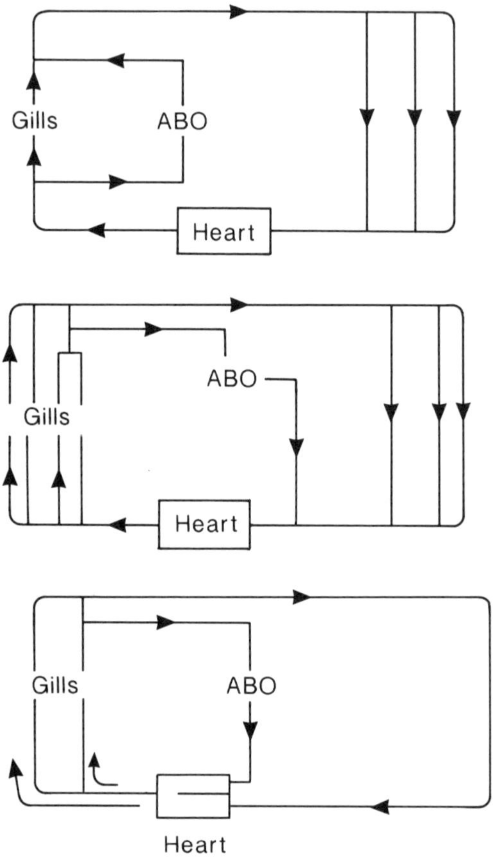

Fig. 15.2 Simplified schematic diagrams of modifications to the circulatory system displayed by bimodally breathing fish using different types of air-breathing organ (ABO). In species in which the ABO consists of modified gills, the buccal mucosa or chambers extending from the opercular cavity, the ABOs are in parallel with the gills so that blood enters directly into the systemic circulation (upper figure). In the majority of bimodally breathing species, the blood that has passed through the ABO must also flow through the gills before entering the systemic circulation. There are modifications to the circulatory system when the gas bladder functions as an ABO, as in some teleosts (middle figure), and in the lungfish (Dipneusti) (lower figure). Note the partial separation of the blood flow in the lungfish.

blood that has been oxygenated in the 'accessory' air-breathing organ (ABO) and must pass through the gills before being distributed to the systemic circulation (Fig. 15.2). At first sight this might appear to represent a severe 'design deficiency', because this type of circulatory pattern both provides the opportunity for oxygen to diffuse from the blood to the oxygen-deficient water as it passes over the gills, and means that the systemic circulation and body tissues will be supplied with mixed blood in which the oxygen-carrying capacity is not fully utilized. Nevertheless, a pathway of blood flow from the ABO via the heart and gills to the systemic circulation is seen in the majority of the air-breathing fish species, almost irrespective of the type of structure used as an ABO. Some species may, however, be able to shunt the blood flow through the gills in such a fashion that there is a reduced flow through the secondary lamellae, and this would reduce the risk of losing oxygen from the blood to the surrounding water. Other modifications to gill structure, such as an overall reduction in surface area and a thickening of the secondary lamella epithelium, in comparison with the situation in water-breathing species, may also serve to minimize the diffusion of oxygen from the blood to the water when air-breathing fish experience hypoxic conditions.

Air-breathing organs (ABOs)

A few species, e.g. the knifefish, *Hypopomus brevirostris,* the common eel, *Anguilla anguilla,* and some mudskippers, *Periophthalmus* spp., are able to breathe using their gills when exposed to air. The gills may be modified to prevent collapse when in contact with air, the secondary lamellae being generally blunt, relatively stiff and broad. In these species, exposure to the air may also lead to increased gas exchange over the skin, and this would assist in the maintenance of rates of oxygen uptake.

In only a small number of species are the air-breathing organs arranged in parallel with the gills, so that the blood flowing through the ABO can enter the systemic circulation directly (Fig. 15.2). This arrangement ensures that the systemic arterial blood, which comes from either the gills or the ABO, is highly oxygenated. In fish species (e.g. *Clarias* spp., *Saccobranchus* spp.) having the gills and ABO arranged in parallel, the efferent vessels from the ABO will usually fuse with the efferent branchial circulation before joining the systemic circulation. In these species the ABO consists of either modified gills, highly vascularized regions of the buccal mucosa, or chambers extending from the opercular cavity.

Other species which use structures in the buccal and pharyngeal region as ABOs do not, however, have this type of circulatory arrangement. Thus, in the majority of species, blood that has passed through the ABO must also flow through the gills before entering the systemic circulation. Some air-breathing species use a combination of the gills and a heavily vascularized buccal cavity to extract oxygen from air held within the buccal–branchial cavity. The swamp eel, *Synbranchus marmoratus,* for example, has an ABO that consists of minute papillae spread over much of the bucco-pharyngeal mucosa. In other species the buccal epithelium may be highly folded, with complex papillations, forming a large surface for aerial gas exchange. The electric eel, *Electrophus electricus*, has an ABO of this type, and the foldings of the ABO account for approximately 15% of the total body surface area.

Air chambers, extending from the buccal or pharyngeal cavities, are found in a range of species inhabiting tropical marshes of Asia and Africa. In the snake-head, *Ophiocephalus striatus*, the air chambers are located dorsally and are partially enclosed within the skull, whereas in the swamp-eel, *Amphipnous cuchia*, the highly vascularized air chambers are more open and are often referred to as 'pharyngeal lungs'.

Air-breathing organs may also be derived from the tissues of the opercular cavity, either through the tissue becoming highly vascularized and thin, or via the development of specialized diverticula. Several gouramies, members of the suborder Anabantoideiidae (e.g. *Anabas* spp., *Macropodus* spp., *Trichogaster* spp.) possess paired labyrinth organs that take the form of elaborate plates or tufts lying within the opercular cavity. In these species the ABOs develop as outgrowths from the first branchial arch, to become complex highly vascularized labyrithine organs having a large surface area.

In some species of fish showing aerial respiration, part of the gastrointestinal tract has been modified to serve as an ABO, and oxygen can be absorbed from air that is swallowed. The air is then voided either through the anus or via gulping up the air and expelling it through the mouth. In the Columbian catfish, *Eremophilus mutisii*, the central portion of the stomach is highly vascularized and is used for aerial respiration. This species is a frequent, but not obligatory, air breather, and the frequency of air breathing is affected by the level of dissolved oxygen. Air ventilation takes place during rapid surfacing, with expiration of the old air preceding inspiration. Several other species of siluriform catfish are also known to use the modified gastrointestinal tract as an ABO.

The gas bladder as an ABO

The gas bladder is derived from the same embryonic tissues as the gastrointestinal tract, and the gas bladder may originally have developed as an accessory respiratory organ. The gas bladder has a respiratory function in the lungfish (Dipneusti), reedfish and bichirs (Brachiopterygii), and in some species of the ray-finned fish (Actinopterygii). Amongst the ray-finned fish a respiratory gas bladder is found in the gars, *Lepisosteus* spp., the bowfin, *Amia calva*, and in a small number of teleost species. The gas bladders of air-breathing fish differ greatly in their structural complexity. In some species, such as the bowfin, the gas bladder remains essentially sac-like, with the major difference from the swim bladder found in water-breathing species being an increase in the vascularization of the inner wall of the gas bladder. Several species, however, have a gas bladder in which the inner wall has become a complex septate structure. This occurs to the greatest degree in the lungfish, where internal septa, ridges and pillars effectively partition the gas bladder into a series of small compartments which open to a central cavity. Even within the lungfish, however, there is wide variation, with the single 'lung' of the Australian lungfish, *Neoceratodus forsteri*, being less divided than the paired 'lungs' of the African, *Protopterus* spp., and South American, *Lepidosiren paradoxa*, lungfish.

In most species that use the gas bladder for aerial respiration, surfacing and protrusion of the mouth into the atmosphere is followed immediately by expiration. Emptying of the gas bladder, caused largely by recoil of the ribs and compression of the body wall, is followed by an active expulsion of the gas due to an increase in pressure within the buccal cavity. This is then followed by an active inspiration phase. The gas bladder is inflated with a single, or small number of, breaths and then the fish resubmerges. Ventilation and filling of the gas-bladder is achieved using a similar buccal, muscular cycle to that used for ventilating the gills. During inhalation the floor of the buccal cavity drops, there is an increase in volume and air is drawn into the mouth. The mouth is then closed and an elevation of the buccal floor compresses the air and drives it through the pneumatic duct into the gas bladder.

In a small number of species, such as the banded knife-fish, *Gymnotus carapo*, and the trahira, *Hoplerythrinus unitaeniatus*, the air-breathing cycle is reversed. The cycle is peculiar in that it starts with the inhalation of fresh air at the water surface before the stale air has been expelled from the gas bladder. Air is actively inspired, and inspiration is due to the reduced pressure in the bucco-pharyngeal cavity brought about by an increase in volume. Inspired air is then forced from the bucco-pharyngeal region into the oesophagus, which expands to act as a short-term reservoir. Muscular activity in the oesophagus then drives the air into the gas bladder via the pneumatic duct, which joins the oesophagus very close to its junction with the stomach. The muscles of the bucco-pharyngeal region play only a very minor role during this phase of the cycle. The gas bladder is divided into two distinct chambers, and the newly inspired air is initially forced into the anterior chamber to prevent mixing with the stale air which is still in the posterior, respiratory chamber of the gas bladder. Once the anterior chamber has been filled with fresh air the stale air is expelled from the posterior chamber and enters the oesophagus via the pneumatic duct. Muscular contractions in the oesophagus then force this air into the bucco-pharyngeal cavity, and compression of the buccal cavity causes expulsion of gas bubbles from the mouth. At the same time gas flows from the anterior to the posterior chamber of the gas bladder, completing the ventilatory cycle.

The role of receptor systems in air breathing

Irrespective of the type of structure employed as an ABO, aquatic hypoxia appears to stimulate a suite of responses that serve to initiate, or increase, aerial oxygen uptake and minimize branchial oxygen loss. The switch from aquatic to aerial respiratory modes is probably mediated by externally orientated oxygen receptors.

Oxygen receptors

The oxygen receptors that initiate aerial respiration in the facultative air-breathing South American catfishes *Ancistrus chagresi* and *Hypostomus plecostomus* have been suggested to have an external location, and there is good evidence for externally orientated receptors in lungfish, the bowfin and gar, *Lepisosteus osseus*, which all use the gas bladder as an ABO. In the bowfin, stimulation of these receptors leads to reflex bradycardia, and also initiates air-breathing responses. Similarly, in the gar, stimulation of these receptors leads to a switch in the emphasis from aquatic to aerial respira-tion. The externally orientated receptors in the gar may exert some control over the depression of gill ventilation seen on the switch to air breathing, and stimulation of these receptors also leads to an increase in air-breathing frequency.

In the lungfish, on the other hand, the initial effect of chemoreceptor stimulation is an increase in gill ventilation, followed by an increase in air-breathing frequency. Thus, several air-breathing fish species appear to be capable of monitoring dissolved oxygen levels with the help of externally orientated receptors, but the behavioural changes and ventilatory responses observed following stimulation of these receptors may vary between species.

Whilst there is good evidence for the existence of externally directed oxygen receptors in air-breathing fish species, experimental manipulations that lead to reductions in the supply of oxygen to the body tissues have been shown to result in the initiation of aerial respiration. Thus, air-breathing species must also be capable of monitoring either internal oxygen levels and/or tissue oxygen demand.

Most bimodally breathing fish species increase air-breathing frequency and gill ventilation when the oxygen tension (P_{O_2}) of the arterial, or mixed venous, blood falls. This suggests that there are internally orientated (intravascular) chemoreceptors that can induce changes in air-breathing behaviour and ventilation patterns of the gills and ABO. In the bowfin, for example, there is convincing evidence that the gill ventilatory responses to hypoxia are, in part, under the control of internally orientated O_2 receptors.

The detection of low internal oxygen tensions may, in addition to resulting in changes in ventilation patterns, also result in there being changes in perfusion of the gills and ABO with blood. For example, in the lungfish, a decrease in internal P_{O_2} appears to result in an increased perfusion of the vessels of the 'lung' (gas bladder) in the period immediately following the inspiration of air at the water surface.

There appear to be at least two sets of oxygen receptors, externally and internally orientated, that have an influence upon aerial respiration and ventilation patterns of air-breathing fish. Questions arise as to how the signals from these receptors are integrated to provide a set of coordinated responses to hypoxic conditions. It seems likely that the initial responses to hypoxia are brought about as a result of signals received from externally orientated receptors. Should hypoxia persist and compromise blood oxygen tension, signals from the internally directed receptors may lead to either initiation of a new, or a strengthening of the existing, suite of responses. In the gar, for example, both gill ventilation and the frequency of air breathing are influenced by both sets of receptors. Rates of air breathing increase as internal P_{O_2} falls, but for any given internal P_{O_2}, the frequency of air breathing is higher the more oxygen-deficient the water. Thus, signals from both the internal and external receptors are integrated to determine the rate of air breathing. Gill ventilation may also change in response to both internal and external stimuli, but aquatic hypoxia has been shown to inhibit gill ventilation irrespective of whether or not the blood levels of dissolved oxygen are low. This powerful inhibitory effect of external hypoxia suggests that the externally orientated receptors exert an overriding control over gill ventilation, and a similar pattern of control may be expected in other bimodally breathing fish that possess fully functional gills.

Other chemoreceptors

Other chemoreceptors that may contribute to respiratory control in air-breathing fish are carbon dioxide- and pH-sensitive receptors. In water-breathing species the internal P_{CO_2} is usually considerably lower than that observed for terrestrial species, and a control of breathing based on the monitoring of metabolically produced CO_2 seems to be of little consequence. In some habitats, such as tropical swamps, however, dissolved CO_2 can build up to such high levels that the binding properties of the haemoglobin are adversely affected, and blood acid–base balance may also be disturbed.

In bimodally breathing species, air breathing inevitably leads to increased retention of CO_2 and an increase in internal P_{CO_2}. Thus, the external and internal CO_2 levels may be expected to have some influence on the control of breathing in tropical air-breathing species. Exposure to increased external P_{CO_2} may lead to a general increase in respiration, both aquatic and aerial, whereas at higher CO_2 levels aquatic respiration may be suppressed whilst the rate of air breathing is increased. The receptors that elicit these responses are probably directed towards the external medium, but there could also be internal receptors located in, or directly downstream from, the gills.

Evidence for an external or branchial location for a set of carbon dioxide receptors is provided by the results of studies on the swamp eel. Water breathing is strongly inhibited when the fish are exposed to CO_2-rich water, and the inhibition persists as internal P_{CO_2} increases. Transfer to CO_2-free water results in a rapid restoration of aquatic respiration, suggesting that the fish possess a sensory mechanism that is capable of differentiating between external and internal levels of carbon dioxide. Similarly, in the African lungfish, exposure to high levels of dissolved carbon dioxide in the water leads to an inhibition of aquatic respiration before there are any marked changes in internal P_{CO_2}, which lends support to the idea that there are externally orientated CO_2-sensors.

In contrast to the response shown to elevated aquatic CO_2 levels, the filling of the ABO with CO_2-enriched air may evoke little change in respiratory response, as has been found in the Australian lungfish. Although the filling of the gas bladder with a CO_2-rich gas mixture led to a marked elevation in the carbon dioxide levels of blood in the efferent vessels leaving the gas bladder, there was little effect on the P_{CO_2} of the blood reaching the systemic circulation. It must be remembered that the blood flows from the ABO to the heart and through the gills before passing to the systemic circulation, and it is clear that the CO_2 absorbed into the blood from the air in the gas bladder must have been excreted as the blood passed through the gills. Thus, during the performance of aerial respiration in bimodally breathing fish, the primary function of the gills is the excretion of carbon dioxide, and this does not require the maintenance of high levels of gill ventilation.

Mechanoreceptors

Whilst chemoreceptors, primarily the oxygen receptors, appear to exert the dominant control over respiration in bimodally breathing fish, there is convincing evidence that mechanoreceptors located within the ABO also influence the breathing pattern. For example, in the bowfin a deflation of the gas bladder results in the fish ascending to the water surface in order to inspire. Furthermore, if the gas bladders of fish that are ascending to breathe are artificially inflated by injections of gas, the fish abort the ascent and return to the bottom. This is a response to mechanical distension of the gas bladder, because the same response is induced irrespective of whether the gas used to inflate the gas bladder is air, oxygen or nitrogen. Thus, this stimulation of breathing by deflation, and inhibition by inflation, of the gas bladder appears to be the result of the stimulation of mechanoreceptors initiating a reflex response (Hering–Breuer reflex).

Similarly, inflation of the gas bladder of the African lungfish, *Protopterus annectens*, has been shown to result in a prolongation in the interval between surfacing and air-breaths. Inter-breath interval was not, however, solely under the control of feedback signals from the ABO mechanoreceptors, because the composition of the gas mixture used to fill the gas bladder also had an influence upon inter-breath interval. For any given level of gas bladder inflation pressure, reductions in oxygen content of the gas introduced into the ABO shortened the inter-breath interval, implying that oxygen receptors (possibly located in the walls of the blood vessels draining the gas bladder) also exert some control over the frequency of air-breathing.

The gar appears to differ from the bowfin with respect to the influences the ABO mechanoreceptors have on air breathing. In the gar, deflation of the gas bladder has little effect on the respiratory response when the fish are exposed to normoxic water, but does stimulate air breathing when fish are held in oxygen-deficient water. Thus, whether or not the gar shows a response to gas-bladder deflation is dependent upon ambient dissolved oxygen levels in the surrounding water, so it seems likely that it is an interaction between mechano- and chemoreceptor stimulation that governs the response. Similarly, the inhibitory effect of gas-bladder inflation on air breathing may arise due to the integration of signals received from the mechanoreceptors and from the receptors that respond to changes in internal P_{O_2}.

Irrespective of the types of stimuli and receptors that are involved, it is known that inflation of the ABO and the onset of spontaneous air breathing are closely linked to cardiovascular changes, such as increased heart rate, an increase in cardiac output and an elevation in the blood flow to the ABO. Thus, the mechanoreceptors may have an important role to play in the coordination of ventilation and perfusion of the ABO.

Coordination of aerial and aquatic respiration

There is considerable variability in both the structures used, and in the respiratory patterns displayed, by bimodally breathing fish. Despite this variability, the overall control and coordination of aerial and aquatic respiration appears to depend upon input signals from a limited number of sets of peripheral receptors. The oxygen receptors appear to exert the dominant control over the periodic breathing patterns observed in the bimodally breathing species, and whilst the internally directed receptors that monitor blood P_{O_2} may control the overall level of ventilation, the externally orientated receptors seem to act to cause the switch from aquatic to aerial respiration under hypoxic conditions. Receptors that monitor ambient levels of carbon dioxide may have a modifying effect upon these ventilatory patterns under conditions where levels of dissolved CO_2 are high. The mechanoreceptors associated with the ABO may control the onset and termination of air breathing in some species (e.g. the bowfin) but the major role of these receptors appears to be in the coordination of ventilation and perfusion.

The respiratory priorities set by the chemo- and mechanoreceptors can, however, be overridden in cases where the fish are exposed to potentially harmful water-borne agents, e.g. toxic chemicals, or other adverse environmental conditions. For example, exposure to water containing strong acids, cyanide or noxious compounds results in an inhibition of gill ventilation, irrespective of dissolved oxygen levels, and may induce some bimodally breathing species to switch from aquatic to aerial respiration. Thus, the stimulation of a variety of chemoreceptors, associated with the olfactory and gustatory systems, may initiate a series of emergency defence responses that protect the fish during times of exposure to potentially lethal conditions.

15.5 PHYSIOLOGICAL RESPONSES TO HYPOXIA IN UNIMODAL FISH

In the unimodal water-breathing fish, there are marked changes in cardiovascular and gill ventilatory patterns in response to changes in dissolved oxygen levels. As with the bimodally breathing species, there appear to be both internally and externally orientated oxygen receptors that serve to coordinate the responses shown by the fish when they experience hypoxic conditions.

Oxygen receptors in unimodal fish

In the Atlantic cod, there is a rapid response to mild hypoxia that involves a marked increase in gill ventilation frequency, and this response is probably mediated via signals emanating from externally directed oxygen receptors located on the first gill arch. Increased ventilation frequency following the onset of hypoxia also results in an increased gill ventilation volume (frequency × stroke volume). If hypoxia becomes more severe, and the blood becomes hypoxemic, gill ventilation volume is increasingly influenced by changes in ventilatory stroke volume. It is possible that the changes in stroke volume are brought about as a result of the detection of falling blood P_{O_2} levels by internally directed oxygen receptors, but this has not been unequivocally demonstrated.

Thus, stimulation of the oxygen receptors initiates a series of reflex responses that results in increased gill ventilation volume, and severe hypoxia can result in the rate of water flow over the gills being increased between 5- and 20-fold depending upon fish species. To what extent the increase in gill ventilation is brought about by increases in ventilatory frequency or stroke volume varies, both with the severity of the hypoxia and with species. For example, in the common carp the 16-fold increase in the gill ventilatory volume that accompanies hypoxia appears to be largely due to a more than 12-fold increase in ventilatory frequency. For carp held at 25 °C, ventilatory stroke volume appears to remain more or less constant until external P_{O_2} falls to approximately 40 mm Hg, and even under severely hypoxic conditions the increase in stroke volume may be relatively modest. In other species, such as the Atlantic cod and rainbow trout, increases in ventilatory stroke volume make a more substantial contribution to the increase in water flow over the gills that occurs under conditions of oxygen deficiency.

Stimulation of the externally directed oxygen receptors also leads to changes in the performance of the cardiovascular system. The most noticeable response to oxygen-deficient conditions is a rapid onset of reflex bradycardia (reduction in heart rate), and under hypoxic conditions the heart rate may decrease to just over half of that recorded under normoxia. The reduction in heart rate does not, however, result in a marked decrease in cardiac output (heart rate × stroke volume), because the hypoxic bradycardia is also accompanied by an increase in stroke volume. Thus, under hypoxic conditions, cardiac output is often found to remain virtually unchanged, with the increase in stroke volume compensating for the reduction in heart rate.

These ventilatory and cardiovascular changes will usually lead to the maintenance of rates of oxygen uptake even under conditions of a relatively severe hypoxic challenge, but they may lead to conditions under which the efficiency of gas transfer over the gills could be severely compromised.

Gill ventilation, perfusion and gas exchange

Gas exchange at the gills is optimized when:

$$(V_W \times \beta_W)/(V_B \times \beta_B) = 1 \tag{15.5}$$

where V_W (ventilation volume) is the volume of water pumped over the gills per minute, V_B is the blood flow through the gills (usually expressed as cardiac output), and β_W and β_B are the capacitance coefficients of the water and blood, respectively. The presence of haemoglobin in the blood leads to the capacitance coefficient of the blood being much higher than that of water, and consequently optimum conditions for gas exchange occur when gill ventilation volume is about 10–20 times the cardiac output.

Under normoxic conditions the ratio of blood flow to ventilation volume is usually within the range 1:15 to 1:20, so that the conditions for gas exchange are close to optimal. A 15-fold increase in ventilation volume, in the absence of any change in cardiac output, as occurs under hypoxic conditions, could lead to the ratio of blood flow to ventilation volume falling to 1:150 or more, a ratio which is far from the ideal.

Under normoxic conditions, the majority of fish species remove 50–80% of the oxygen from the water as it flows over the gills, but the marked decrease in the blood:water flow ratio that occurs with hypoxia would be expected to result in a reduction in oxygen extraction efficiency. As expected, fish are unable to maintain a high extraction efficiency in oxygen-deficient waters, especially when hypoxia is severe. For example, common carp held under normoxic conditions at 25 °C remove about 75% of the oxygen from the water flowing over the gills, but the percentage removed gradually falls with increasing hypoxia, and oxygen extraction efficiency is less than 40% at an external P_{O_2} of about 40 mm Hg.

Thus, the decrease in blood:water flow ratio that accompanies hypoxia creates conditions that become progressively less favourable for efficient oxygen extraction. Some of the cardiovascular responses observed under hypoxia do, however, tend to offset the effects of the unfavourable blood:water flow ratio, and reduce the negative effects to a minimum. For example, in a number of species, such as the rainbow trout and the spotted dogfish, *Scyliorhinus canicula*, hypoxia appears to evoke a cardiac reflex known as *cardio–respiratory synchrony* in which the pattern of the heart beating becomes coupled to the respiratory cycle. The flows of water over, and blood through, the gills are *unidirectional and continuous*, but the rates of flow are *pulsatile and not constant*. With cardio–respiratory synchrony the heart tends to beat as the mouth opens, and the function of cardio–respiratory synchrony may be to couple the pulsatile flows of the two media, so that the rapid flow of blood caused by ventricular contraction coincides with the rapid flow of water that occurs during the expiratory phase of the respiratory (ventilatory) cycle.

Hypoxic bradycardia results in an increase in the length of the cardiac cycle, and cardiac output is maintained due to increases in stroke volume. One consequence of this is that there is an increase in pulse pressure, and in some species exposure to hypoxic conditions has been found to lead to a doubling of ventral aortic blood pressure. This increase in blood pressure may cause a change in the distribution of the blood flow through the gills, with the increase in lamellar recruitment being the most notable effect. Under normoxic conditions, fish at rest may have only 50–60% of the gill secondary lamellae perfused with blood, but with the onset of hypoxia, secondary lamellar recruitment may lead to over 80% of the lamellae being perfused. This increased perfusion of the secondary lamellae obvi-ously leads to an increase in the surface area for gaseous exchange.

In addition to lamellar recruitment, the effects of increased blood pressure may include changes in blood flow patterns from perfusion of the basal channels of the secondary lamellae to an increased perfusion of the central and peripheral channels. This, in turn, can have the dual effect of increasing the effective surface area for gas diffusion, and may result in a 'stretching' of the secondary lamellar epithelium, which would reduce diffusion distance (Fig. 15.3).

The consequences of these changes can be seen by a consideration of the Fick diffusion equation:

$$D = (\mathrm{P_W} - \mathrm{P_B}) \times A \times \mathrm{K}/d \tag{15.6}$$

which demonstrates that the rate of gas transfer (D) is dependent upon the gas diffusion gradient between the water and blood [$(\mathrm{P_W} - \mathrm{P_B})$], the surface area for diffusion (A), the diffusion characteristics of the tissue (K is Krogh's diffusion constant) and the diffusion distance (d). Under normoxic conditions the external $\mathrm{P_{O_2}}$ ($\mathrm{P_W}$) will be approximately 155 mm Hg, and the $\mathrm{P_{O_2}}$ in the venous blood flowing to the gills ($\mathrm{P_B}$) will usually be within the range 30–50 mm Hg, giving a large gradient to drive gas diffusion. Under hypoxic conditions, $\mathrm{P_W}$ is reduced and, in the initial stages, $\mathrm{P_B}$ may be only slightly reduced, so the onset of hypoxia results in a marked decrease in the gas diffusion gradient. An increase in the surface area for diffusion (A) and a decrease in the diffusion distance (d) will serve to offset the negative effects of the reduced diffusion gradient, and will enable rates of gas transfer (D) to be maintained.

Thus, under mild hypoxia the $\mathrm{P_{O_2}}$ in the arterial blood leaving the gills to flow to the systemic circulation may be maintained within the range 80–110 mm Hg. With increasing levels of hypoxia the arterial $\mathrm{P_{O_2}}$ cannot be maintained, and it gradually falls. A decline in arterial $\mathrm{P_{O_2}}$, or fall in blood oxygen content, leads to the liberation of catecholamines from the chromaffin tissue of the head kidney (pronephros). The increase in circulating levels of catecholamines evokes a further suite of responses which further offset the potentially negative effects of oxygen deficiency.

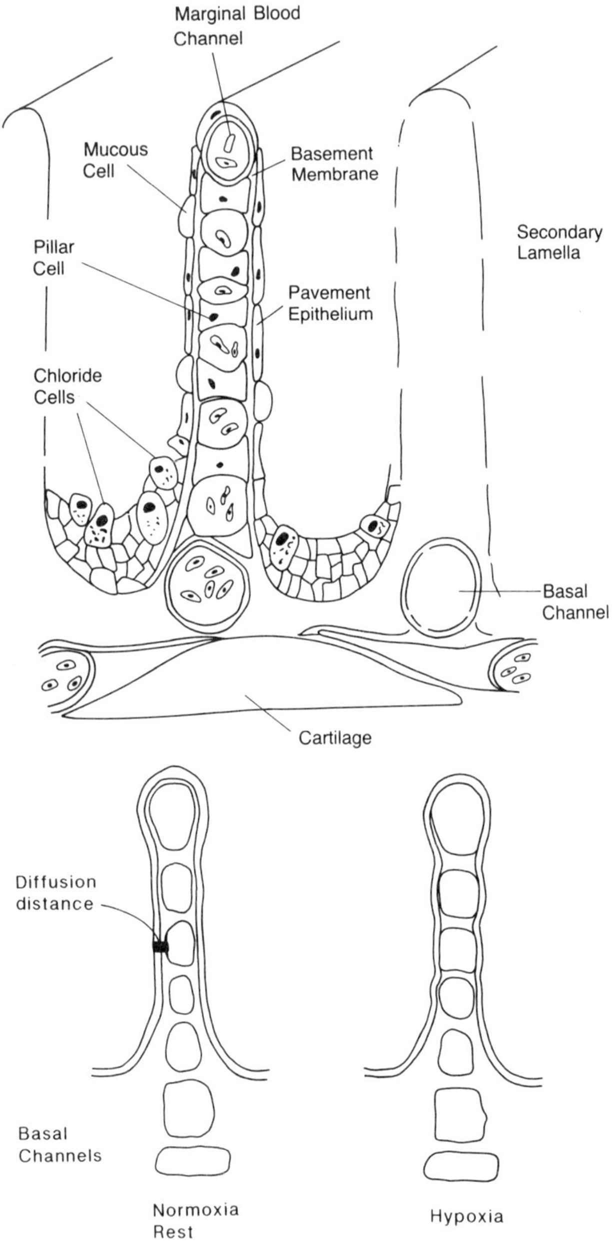

Fig. 15.3 Schematic diagram of a section through the secondary lamellae of a fish gill showing the general structure, and the possible effects of increased perfusion of the lamellae with blood. A redirection of blood flow from the basal channels leads to increased perfusion of the secondary lamellae and a reduction in diffusion distance.

Catecholamines and responses to hypoxia

During hypoxia, nervous control mechanisms may be instrumental in causing a redistribution of blood flow between the visceral and somatic parts of the systemic circulation. Thus, when fish are exposed to oxygen-deficient conditions there may be a reduction in the blood flow to the gastrointestinal tract. Whilst this response appears to be, largely, under nervous control, the release of catecholamines that occurs as hypoxia increases in severity may act to further reduce the flow of blood to the gut.

Catecholamine release may also serve to reinforce the changes in patterns of blood flow through the gills, leading to a further increase in lamellar recruitment, and an increased distribution of blood from the basal to the central and peripheral channels of the secondary lamellae. It is also possible that catecholamines increase the permeability of the gill epithelium, which would lead to an enhancement of diffusion of oxygen from the water to the blood.

Catecholamines have little influence upon heart rate, so do not reinforce the hypoxic bradycardia. Increased levels of circulating catecholamines do, however, contribute to the maintenance of cardiac output via an enhancement of cardiac stroke volume.

The bradycardia that occurs with the onset of hypoxia, prior to any fall in blood P_{O_2}, leads to an increased residence time of the blood in the heart between contractions. This leads to an increase in the time available for the blood to perfuse the cardiac tissues, possibly leading to improved oxygen diffusion from the blood into the cardiac muscle.

With increasing hypoxia the P_{O_2} of both the arterial and venous blood may be reduced. A decrease in the P_{O_2} of the blood reaching the body tissues may lead to the onset of anaerobiosis, with a concomitant fall in blood pH, and a decrease in venous blood P_{O_2} will reduce the gradient for gaseous diffusion between the blood and cardiac muscle. The primary actions of the catecholamines appear to be directed towards the maintenance of muscular contraction in the face of the dual challenge imposed by declining blood P_{O_2} and falling plasma pH.

During the initial stages of hypoxia the increase in the rate of water flow over the gills results in there being an enhancement of carbon dioxide excretion as the blood passes through the secondary lamellae. One consequence of this is that there is an initial increase in blood pH, by 0.15–0.20 pH units. This, in turn, may lead to an increase in the affinity of the haemoglobin to bind oxygen, and thereby enhance oxygen uptake at the gills. With more pronounced hypoxia, and a decline in P_{O_2} in the systemic blood supplying the tissues, there is increased likelihood of the onset of anaerobiosis. Anaerobiosis, and the release of lactate, would eventually lead to a metabolic acidosis. Metabolic acidosis, and the concomitant fall in blood pH, would be expected to lead to changes in the affinity of the haemoglobin for oxygen, and thereby compromise the ability of the haemoglobin to bind to and transport oxygen. Thus, if the oxygen-transporting capabilities of the haemoglobin are to be maintained in the face of the challenge of falling blood pH, it is essential that there exist mechanisms that ensure that there is a restoration of intracellular pH in the red blood cells. The initiation of these mechanisms appears to be another of the roles of the catecholamines. It must be stressed that the release of catecholamines from the chromaffin tissue of the head kidney occurs in reponse to a fall in blood oxygen content, or a reduction in arterial blood P_{O_2}, and plasma catecholamine levels begin to rise before the onset of any marked anaerobiosis. Thus, the concentration of catecholamines in the blood may increase whilst pH is still elevated above the level seen under normoxic conditions.

The release of catecholamines results in a sharp drop in blood pH (by 0.3–0.5 units) to a value below that seen under normoxia. The decrease in pH is brought about by the initiation of an Na^+/H^+ exchange over the membrane of the red blood cells. Provided that there is no additional challenge due to the onset of a metabolic acidosis, this decrease in pH will be transient in nature. After a short period the blood pH will increase slightly to approach a new steady-state value, close to the normoxic level. Should anaerobiosis be initiated, and a metabolic acidosis result, the pH of the blood plasma will remain lower than normoxic values, but red blood cell intracellular pH will be defended. Thus, the activation of the Na^+/H^+ exchange mechanism, resulting from the mobilization of the catecholamines, can be considered to be a 'preparative' physiological response that will enable the intracellular pH of the red blood cells to be maintained should there be a metabolic acidosis. This, in turn, enhances the affinity of the haemoglobin for oxygen, and serves to offset the negative effects of metabolic acidosis on branchial oxygen transfer and the oxygen transport capacity of the blood.

An additional physiological response often observed when circulating levels of catecholamines increase is a contraction of the spleen. Contraction of the spleen may also be under nervous control via innervation from the sympathetic nervous system. In the majority of fish species the spleen is the most important of the haemopoietic organs (involved in the synthesis of red blood cells) and it may also be a major reservoir of red blood cells. At any given time the spleen may contain 15–25% of the body's red blood cells. Consequently, a contraction of the spleen, with an expulsion of the stored red blood cells into the circulatory system, would be expected to lead to an augmentation of the oxygen-carrying capacity of the blood. The release of red blood cells from the spleen appears to occur most readily under conditions of acute, severe, hypoxia, when the need for an augmented oxygen-carrying capacity would be expected to be most pressing.

It is, therefore, apparent that the onset of hypoxia results in the initiation of a series of adaptive physiological responses involving changes in gill ventilation and perfusion, in the oxygen transport system, and in the heart and cardiovascular system. These changes are invoked in an attempt to maintain the oxygen supply to the tissues under conditions of increasing environmental oxygen deficiency, but the physiological changes that occur with the onset of hypoxia lead to the efficiency with which oxygen is extracted from the water suffering a decline. This has the consequence that rates of oxygen uptake cannot be maintained indefinitely irrespective of the severity of the hypoxic challenge. In the common carp, for example, the rate of oxygen uptake can be maintained down to an external P_{O_2} of 25 mm Hg or below, but many other species of fish are unable to maintain their rates of oxygen uptake at such low external P_{O_2} levels.

15.6 DISSOLVED OXYGEN AND METABOLIC RATES

A variety of physiological responses are initiated when fish species are exposed to reductions in external oxygen levels, and these enable the fish to maintain rates of oxygen uptake when challenged by hypoxic conditions. Thus, the majority of fish species are able to maintain more or less constant rates of oxygen uptake under conditions of widely ranging external oxygen levels. These species can be said to be *regulators*, in that they can regulate their rates of oxygen consumption independently of environmental oxygen over a wide range. Rates of oxygen uptake are not inevitably found to be independent of external levels, and in *conforming* species there is a direct relationship between the external oxygen level and the rate of oxygen consumption displayed by the animal. Even in the regulating species, however, rates of oxygen consumption cannot be maintained independent of environmental levels indefinitely and, at some critical P_{O_2} (P_C), oxygen uptake will be limited by, and dependent upon, the ambient external level (Fig. 15.4).

The common carp, a sluggish, midwater or bottom-living freshwater species that often inhabits eutrophic habitats subjected to hypoxic conditions, appears to be able to regulate its rate of oxygen uptake over a very wide range of environmental oxygen conditions. Thus, in this species the rate of oxygen uptake may not become dependent upon external oxygen levels before a P_{O_2} of about 25 mm Hg is reached. Many other species are not as tolerant of oxygen-deficient conditions as carp, and rates

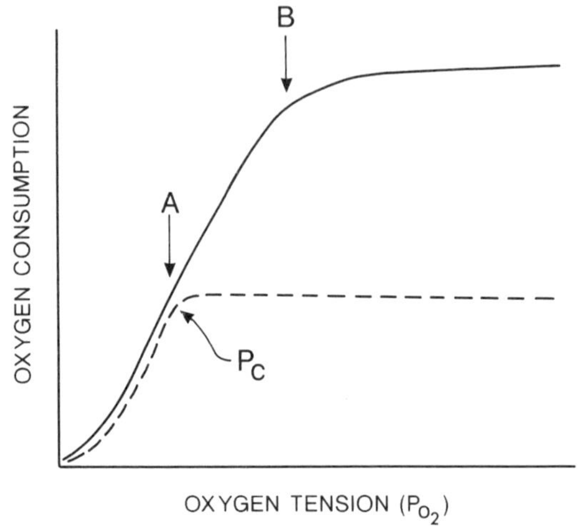

Fig. 15.4 The relationship between rates of oxygen consumption and oxygen tension (P_{O_2}) in a regulating species. As oxygen tension declines, oxygen consumption remains independent of tension until a critical tension (P_C) is reached. Below P_C the rate of oxygen consumption becomes directly dependent upon oxygen tension. Note that P_C is not a constant value, but is dependent upon the rate of oxygen consumption displayed by the fish. When rates of oxygen consumption are high, such as when fish are active or when they are held at high temperature, the P_C is higher (denoted by point B) than when rates of oxygen consumption are low (denoted by point A).

of oxygen consumption become dependent upon external oxygen level at higher P_{O_2}s. The critical external P_{O_2} (P_C) for several species appears to lie within the range 40–60 mm Hg, but the P_C of the cold-water salmonid species may be somewhat higher (80–100 mm Hg) (Table 15.2). Thus, the oxygen tension at which environmental oxygen availability becomes limiting for the maintenance of rates of uptake appears to differ between species, and the P_C may be related to the habitat and lifestyle of the species under study.

This division is, however, simplistic and may, to some extent, be misleading since the P_C is not a species-specific character and it will tend to vary depending upon test conditions. In other words, for any given species of fish, the oxygen tension (P_C) at which rates of oxygen consumption become dependent upon environmental oxygen level may be influenced by a range of different factors. In general it may be said that any factor that leads to an increase in the tissue oxygen demand of the fish will also lead to an increase in the P_C (Fig. 15.4). For example, water temperature, which is known to influence metabolic rate and oxygen uptake in fish, may also affect P_C. Thus, an increase in temperature will

Table 15.2 The relationship between habitat and lifestyle and the critical oxygen tensions (P_C) recorded for a range of freshwater and marine fish species

Species	Habitat	Temperature (°C)	Critical P_{O_2} (P_C) (mm Hg)
Common carp, *Cyprinus carpio*	Eurythermal, freshwater, mid/bottom dweller, sluggish	25	25
Bluegill sunfish, *Lepomis macrochirus*	Eurythermal, freshwater, midwater	13	60
Tilapia, *Oreochromis niloticus*	Tropical, freshwater	35	30
Brook trout, *Salvelinus fontinalis*	Cold water, freshwater, active	15	80
Rainbow trout, *Oncorhynchus mykiss*	Freshwater, cold water, active	12	90
Arctic charr, *Salvelinus alpinus*	Freshwater, cold water, active	10	100
Plaice, *Pleuronectes platessa*	Marine, benthic	10	45
Toadfish, *Opsanus tau*	Marine, benthic	22	29
Stargazer, *Genyagnus monopterygiust*	Marine, benthic	17	45
Lingcod, *Ophiodon elongatus*	Marine	10	45

often be found both to give rise to an increase in the rate of oxygen consumption of the fish and to result in a higher P_C (Table 15.3). Similarly, factors such as swimming activity and feeding level, which also influence tissue oxygen demand, will affect P_C. In other words, the P_C is usually found to be higher for actively swimming fish than for those at rest, and rates of oxygen consumption of well-fed fish become dependent upon environmental oxygen levels at higher tensions than do those of starved fish.

The critical oxygen tension is a measure of the environmental P_{O_2} at which the fish is no longer able to obtain sufficient oxygen from the water to meet tissue demands. In practice, the metabolic demand is not for oxygen *per se*, but for ATP-equivalents. Thus, as oxygen tensions fall below the critical level, and ATP production can no longer be met by aerobic metabolism, there must be a gradually increasing contribution from anaerobic metabolic pathways (Fig. 15.5). As a consequence of this, the P_C may closely coincide with the P_{O_2} at which there begins to be a marked increase in metabolic indictors of anaerobic metabolism (e.g. a rise in tissue lactate levels and an increase in the lactate:pyruvate ratio).

Table 15.3 The influence of temperature on rates of oxygen consumption and critical oxygen tensions (P_C) for a range of fish species

Species	Temperature (°C)	Critical P_{O_2} (P_c) (mm Hg)	Normoxic O_2 consumption (mg kg^{-1} h^{-1})
Largemouth bass, *Micropterus salmoides*	30	60	175
	25	50	100
	20	40	68
Tilapia, *Oreochromis niloticus*	35	30	125
	20	19	55
Spangled perch, *Leiopotherapon unicolor*	35	65	580
	30	47	345
	20	26	210
	10	19	100

Hypoxia and anaerobiosis

With increasing severity of hypoxia a greater and greater proportion of the ATP produced is generated via anaerobic metabolic pathways (Fig. 15.5). A major disadvantage with anaerobic metabolic pathways is that they are very costly in terms of fuel, so that as oxygen availability is reduced there will be a marked tendency for the depletion of glycogen reserves to increase (Pasteur or Crabtree effect). For example, the anaerobic metabolism of glycogen to lactate gives an ATP yield of 3 ATP for each glycosyl unit metabolized (2 ATP with glucose as substrate), whereas aerobic metabolism of the same quantity of respiratory substrate would give a yield of 37 ATP (36 ATP with glucose as substrate). Thus, very few species are capable of tolerating severely hypoxic conditions for prolonged periods of time, and in order for a fish to meet this challenge, it is clear that substantial reserves of fuels must be available.

In the crucian carp, *Carassius carassius*, a cyprinid species which may be exposed to severely hypoxic, or even anoxic, conditions whilst overwintering in ice-covered lakes, there occurs an accumulation of glycogen stores during the course of the late summer and autumn. In this species, which is extremely tolerant of hypoxia, the liver may increase in size from 2% to around 12–15% of total body weight, and glycogen may account for 30% of the dry weight of the liver by late autumn. Glycogen stores are also accumulated in the muscle tissue, and these reserves may account for 3–4% of the muscle on a dry-weight basis.

The glucose, or glycogen, to lactate anaerobic fermentation pathway is the most widely distributed and important of the anaerobic metabolic pathways in animal tissues, but the ATP yield is poor. The anaerobic

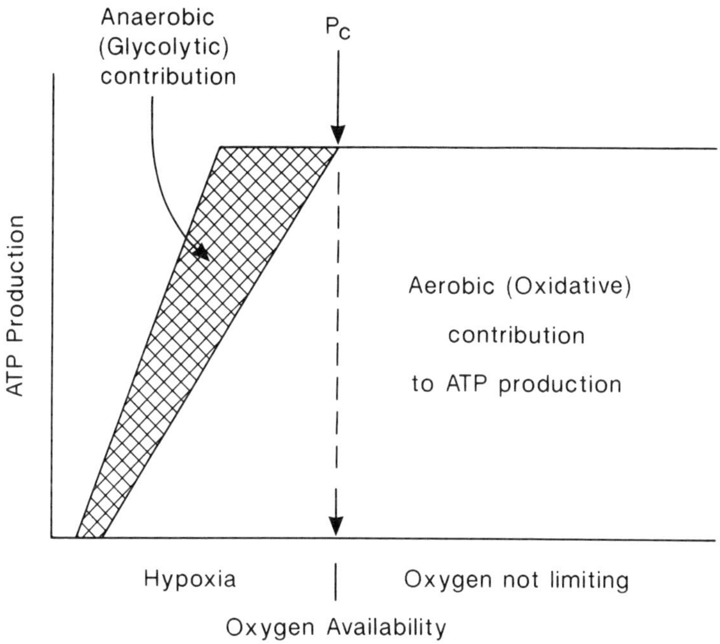

Fig. 15.5 Schematic representation illustrating the aerobic and anaerobic contributions to ATP production at different levels of oxygen availability. At oxygen tensions above the critical (P_C), the entire demand for, and production of, ATP can be met via aerobic (oxidative) metabolic processes. As oxygen tension falls below P_C, and the fish experience hypoxic conditions, production of ATP can be maintained, but the anaerobic (glycolytic) contribution increases.

metabolism of glucose to lactate is not the only fermentation pathway found in tissues, and some of the alternative pathways give a greater ATP yield for each unit of respiratory substrate metabolized (Fig. 15.6). For example, when the end products are succinate or propionate, the ATP yields are 4 and 6 ATP, respectively, for each unit of glucose metabolized. Whilst the fermentative formation of succinate and propionate from glycogen appear to be major pathways for the generation of ATP when many invertebrate species are exposed to hypoxic conditions, these pathways are of lesser importance in fish. The enzyme (phosphoenol-pyruvate carboxykinase, PEPCK) required for the catalysis of the carboxylation of phosphoenolpyruvate (PEP) to oxaloacetate forms the key branching point determining whether lactate or succinate is the end product of the anaerobic metabolism of glucose and glycogen. The enzyme is found in the liver of many fish species, but levels in other tissues are often

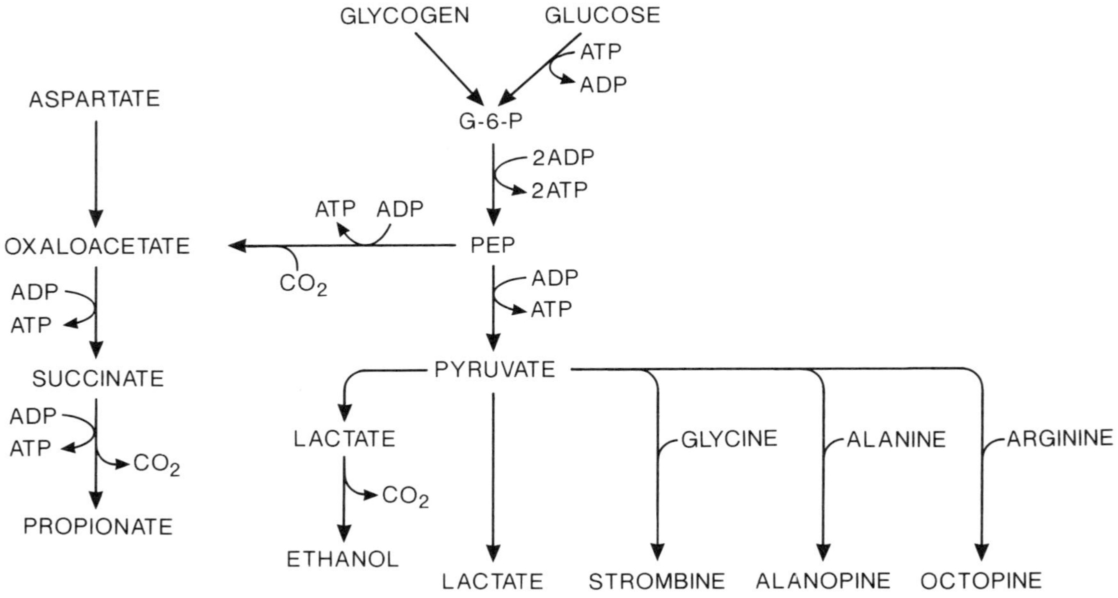

Fig. 15.6 Schematic representation of the major anaerobic metabolic pathways present in animal tissues. ADP, adenosine diphosphate; ATP, adenosine triphosphate; G-6-P, glucose-6-phosphate; PEP, phosphoenolpyruvate.

extremely low. Nevertheless, succinate has been shown to accumulate in the muscle tissue of fish experiencing hypoxic conditions, and it appears that the succinate arises not from the fermentation of glycogen, but from the use of amino acids as respiratory substrate (Fig. 15.6). For example, both aspartate and glutamate can be fermented to succinate or propionate to yield ATP. The ATP production resulting from these metabolic pathways seems, however, to be of little quantitative importance when compared with the amounts of ATP generated from the mobilization of glycogen, and subsequent fermentation to lactate.

The amino acids may not, however, always enter the fermentation pathways directly, and alternative routes of amino acid metabolism may become important when fish experience severe hypoxia. When fish are exposed to oxygen-deficient conditions, the rates at which nitrogenous excretory products, and particularly ammonia, are produced and excreted may be maintained at levels similar to those seen under normoxic conditions. This suggests that amino acid deamination continues during hypoxic exposure. The amino acid carbon-skeletons produced are used for gluconeogenesis, and the glucose produced then enters the lactate fermentation pathway.

Another disadvantage associated with anaerobic metabolism, in addition to the low ATP yield, is that it results in the accumulation of end products that have potentially deleterious effects on the cells and tissues. Thus, during anaerobic metabolism there is a gradual accumulation of organic anions (e.g. lactate) and protons (H^+). Lactate accumulation creates problems but these can, to some extent, be offset by the production of alternative end products, such as succinate, propionate or a range of opines (Fig. 15.6). Accumulation of lactate ions may, for example, give rise to osmotic problems leading to large fluid movements between extracellular and intracellular compartments. The substitution of imino acid opines (e.g. octopine, alanopine, strombine) for lactate as the end product of anaerobic metabolism provides one possible solution to the problem of osmotic imbalance. Opine formation occurs via the reductive condensation of pyruvate with an amino acid in the presence of an imino acid dehydrogenase enzyme as a catalyst.

The production of protons is, however, a more serious problem than lactate accumulation, and the coupling of glycolytic fermentation with the subsequent hydrolysis of ATP results in the generation of two protons (H^+) for each unit of glucose metabolized. Because hydrogen ion production and accumulation is generally considered to be the critical limiting factor in anaerobic metabolism, the ideal metabolic strategy would be to maximize the amount of ATP generated and turned over per acid equivalent produced. The succinate and propionate pathways result in a greater ATP turnover per H^+ than the lactate pathway, but all the anaerobic pathways of glycogen metabolism result in the production of acid equivalents (H^+).

Some buffering of the acid is possible, either via the bicarbonate buffering system or due to the presence of organic buffering agents (e.g. histidine-alanine, carnosine, anserine, ophidine) in the tissues. Buffering capacity may, however, be relatively limited, and there will eventually arise problems with acid–base balance. There would, therefore, appear to be considerable advantages to be gained by producing end products of anaerobic metabolism that did not lead to the generation of acid equivalents. The anaerobic fermentation of glycogen to ethanol is such a pathway, and is one that is found in a small number of hypoxia-tolerant fish species (e.g. crucian carp, goldfish and bitterling, *Rhodeus amarus*). In these species, the lactate resulting from the fermentation of glycogen is taken up by the muscle tissues where it is decarboxylated to ethanol and carbon dioxide (CO_2). Provided that the CO_2 can be removed from the system and excreted, the end products of this anaerobic metabolic pathway will not result in acidification and disturbance of the internal acid–base balance. Ethanol is a neutral, water-soluble product that will tend to diffuse readily to the water surrounding the fish. Thus, internal concentrations of ethanol will tend to remain at low levels, due to this passive diffusion out of the fish, and there will not be any inhibition of the fermentation pathway as a result of end-product accumulation. The excretion of ethanol does, however, represent a net loss of carbon, and the ethanol fermentation pathway can, therefore, be considered as being wasteful of respiratory substrate.

Acclimation to hypoxic conditions

The hypoxia tolerance of a fish will be dependent upon whether or not it has been previously exposed to low, but non-lethal, levels of dissolved oxygen. If a fish is kept in oxygen-deficient water for an extended period of time it will become more hypoxia tolerant. This improved tolerance is attributable to several changes that occur during long-term exposure to non-lethal hypoxia. Firstly, the efficiency with which the fish can extract oxygen from the water is improved following acclimation to hypoxic conditions. This improvement is possibly a result of increases in the oxygen-transporting capacity of the blood brought about by increased blood volume and increases in haemoglobin concentration. There may also be an increase in the tissue glycogen reserves and an elevation in the levels of several enzymes in the liver, leading to a higher gluconeogenic and anaerobic capacity following acclimation to hypoxia.

15.7 CARBON DIOXIDE AND HYPERCAPNIA

The fact that carbon dioxide is frequently used as an anaesthetic attests to the ability of this gas to disrupt the normal physiological functions of the fish. Whilst concentrations of about 500 $mg.l^{-1}$ are usually used, the anaesthetic effects of carbon dioxide may be observed at concentrations as low as 150–200 $mg.l^{-1}$.

When carbon dioxide dissolves and reacts with water, the resulting solution will equilibrate to contain a mixture of free carbon dioxide (CO_2), carbonic acid (H_2CO_3) and bicarbonate (HCO_3^-), with the concentration of each depending upon the pH of the solution: $CO_2 + H_2O = H_2CO_3 = H^+ + HCO_3^-$. Between pH 7 and 9, the equilibrium will be shifted towards bicarbonate, and below pH 6 free carbon dioxide will dominate.

The amount of free carbon dioxide in natural water bodies is quite variable, with the normal range being 1–6 $mg.l^{-1}$. Thus, the waters of lakes and rivers rarely contain more than 10 $mg.l^{-1}$ free CO_2, but total levels ($CO_2 + H_2CO_3 + HCO_3^-$) may reach 50–60 $mg.l^{-1}$ or more. A considerable amount of carbon dioxide may accumulate during the course of the autumn and winter, when photosynthesis is at a minimum, and levels may fall during the summer months as a result of the intensification of photosynthesis and plant growth. Thus, fish may be exposed to a relatively wide range of carbon dioxide levels and these may, on occasion, be sufficiently high to cause physiological disturbances.

Physiological responses to hypercapnia

An increase in the environmental P_{CO_2} will result in an elevation of plasma P_{CO_2}, and such elevations exert their prime effects through disturbances to the internal acid–base equilibrium. The immediate change observed following an elevation of plasma P_{CO_2} is a fall in blood pH. The physiological responses invoked by this change will serve to eventually restore the pH to values close to normal, despite continued hypercapnia. The $HCO_3^-:CO_2$ ratio is returned to the normal value, resulting in the restoration of pH and a large increase in plasma bicarbonate concentration. The bicarbonate that accumulates is not produced metabolically, nor does it appear to be derived from the calcareous, e.g. skeletal tissues of the body, so the most likely source is the environment.

Under normal circumstances the ion-exchange systems of the gill epithelium allow for the uptake of chloride ions in exchange for bicarbonate, but under hypercapnia this exchange system may be inhibited or reversed. Thus, modulation of the Cl^-/HCO_3^- exchange mechanism, leading to active uptake of bicarbonate during hypercania, would result in an accumulation of plamsa bicarbonate. This would eventually lead to a rise in blood pH. This is possible, despite hypercapnia and high environmental P_{CO_2}, because the excretion of the protons (H^+) that result from respiratory processes may be uncoupled from the excretion of CO_2 (Fig. 15.7). The excretion of the H^+ may occur via an active 'proton pump' and/or ion-exchange mechanisms at the gills, e.g. Na^+/H^+ exchange, or in the urine.

There is convincing evidence that active uptake of bicarbonate from the environment makes a substantial contribution to the restoration of acid–base balance under hypercapnic conditions. Bicarbonate uptake seems to be a common strategy for both freshwater and marine species. Marine species tend to be better able to meet the challenge of hypercapnia than are freshwater species, with plasma pH values of marine fish returning to normal within 10–12 h of hypercapnic exposure. Longer periods, of up to a day or more, appear to be required before the plasma pH of freshwater species returns to normal. These differences appear to be largely due to the differences in availability of bicarbonate in the two environments. The greater levels of HCO_3^- in seawater are suspected to enable marine species to more easily reverse the Cl^-/HCO_3^- ion-exchange mechanisms, than is possible for freshwater fish inhabiting relatively bicarbonate-poor waters.

Long-term exposure to elevated P_{CO_2} has been shown to result in both prolonged hypercapnic acidosis and a proliferation of 'chloride cells' in the channel catfish, *Ictalurus punctatus*. Since the 'chloride cells' are usually considered to be intimately involved in the processes of ion exchange and acid–base balance, these observations can be taken as providing further evidence that branchial ion-exchange mechanisms are involved in the restoration of plasma pH in fish exposed to a hypercapnic challenge.

15.8 GAS SUPERSATURATION

Gas supersaturation can occur when saturated water is suddenly heated, or when there is a reduction in pressure on gas that has been introduced into water under high pressure. Water that has plunged from a height, as in a waterfall, is potentially supersaturated, as is ground water drawn from deep wells. Air sucked into a water supply system via leaking pump junctions can also lead to supersaturation. Whilst gas supersaturation can occur in natural water bodies, the greatest attention has been directed towards the examination of the possible effects of gas supersaturation on fish in culture facilities.

One possible deleterious effect associated with gas supersaturation is gas bubble trauma, which has often been the cause of serious losses of young fish in commercial salmonid hatcheries. Theoretically, gas bubble trauma

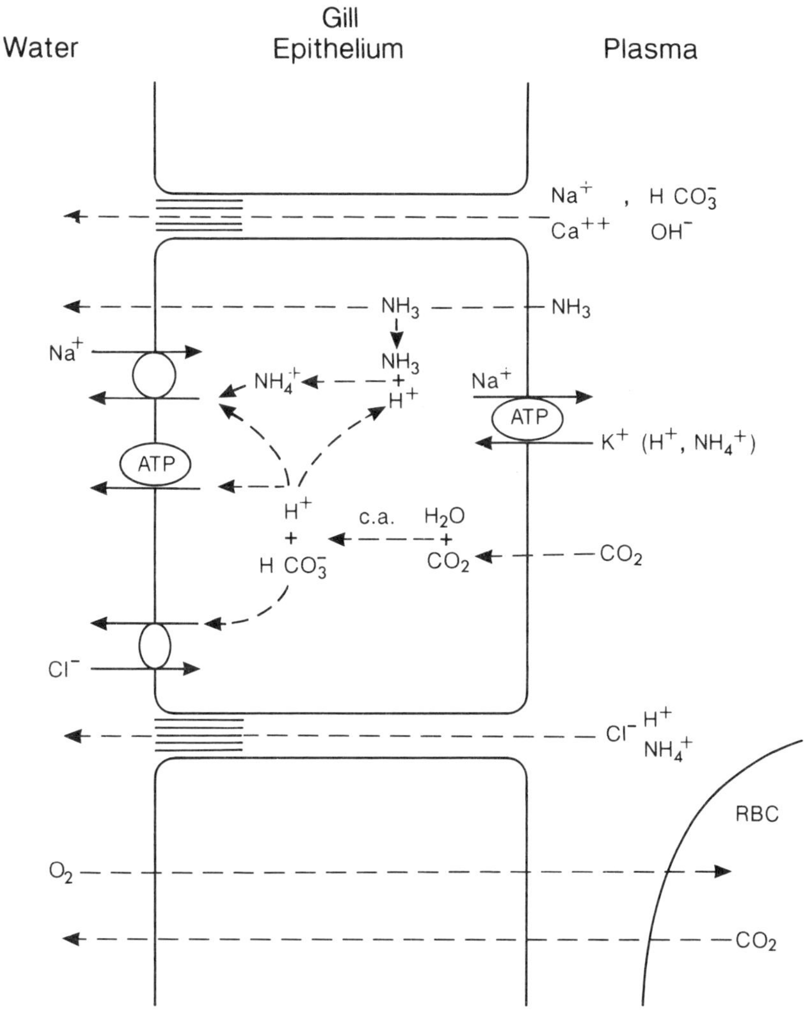

Fig. 15.7 Representation of exchange mechanisms in the gill of freshwater fish species. Carrier-mediated processes are indicated by solid lines and diffusional processes by dashed lines. RBC, red blood cell.

(or disease) can be caused by any supersaturated gas, but, in practice, the problem is almost invariably caused by supersaturation of the water with nitrogen. When the water is supersaturated with gas, the blood of the fish tends to become so as well, but excess levels of oxygen and carbon dioxide can be dealt with relatively effectively.

Oxygen supersaturation does not cause any problems for the fish until very high levels are reached, and excess carbon dioxide is an integral part of the bicarbonate buffering system, so moderate levels of hypercapnia are readily tolerated. Nitrogen, on the other hand, stays supersaturated in the blood, and any localized reduction of pressure on the gas can bring the nitrogen out of solution to form bubbles. Such bubbles, or emboli, can lodge in the blood vessels, can develop in the gills or under the skin, and may under certain circumstances lead to mortality.

All fish species are susceptible to gas bubble trauma, and long-term exposure to supersaturation levels as low as 103% has been shown to induce gas bubble trauma in the juveniles of some salmonid species. Threshold tolerances to nitrogen supersaturation may vary with species, but any saturation over 100% can be considered to pose a potential threat.

Chapter sixteen

Pollutants and toxic chemicals

16.1 INTRODUCTION

Pollutants and toxic chemicals can enter aquatic environments by several routes. These routes include direct precipitation, surface and subsurface water run-off, sewage discharges and industrial wastewater outfalls. There are many thousands of potentially toxic organic and inorganic compounds discharged into water bodies.

The major reason for carrying out toxicity tests is to determine which concentrations of a substance are harmful to aquatic organisms, and which are apparently without effect. The results of the tests provide the basis for the establishment of *water quality criteria.* These criteria are usually taken to refer to the highest concentrations of specific substances that are not expected to cause an appreciable negative effect on an aquatic system. There are, however, numerous potential problems inherent in the carrying out of toxicity tests and this makes the precise definition of water quality criteria extremely difficult. There are, for example, many factors that will have modifying effects upon the toxicity of a given substance, and this requires that the test conditions be strictly controlled and reported.

The characteristics of the water used in the test may, for example, have profound effects on the toxicities of a wide range of substances, so that the following physico-chemical characteristics must be known: pH, total hardness (calcium and magnesium contents), water conductivity and alkalinity (carbonate concentration), salinity, organic carbon content, temperature and dissolved oxygen.

Hydrogen ion concentration (pH) has a marked effect on the chemical form of a number of potentially toxic substances, and, consequently, pH can have major influences on toxicity. For some substances, such as ammonia, the un-ionized form is more toxic than the ionized form because it diffuses more readily across the gill membranes. The un-ionized form constitutes only a small fraction of the total ammonia at low pH, but the proportion of un-ionized ammonia increases at higher pH. Consequently, pH has a marked effect upon the toxicity of ammonia to fish. The ionization of hydrogen sulphide is also affected by pH, but in this case the un-ionized form predominates at low pH.

The toxicity of most metals is reduced as pH increases, because at high pH the metals bind to form hydroxide and carbonate complexes. This results in a reduction in the levels of the toxic metallic ion as the pH increases. In general, water hardness, alkalinity and pH increase or decrease together, so the effects of these factors on the toxicity of metals are often closely interwoven. The toxicity of metals may also be modified by the content of organic material in the water. A number of metals may bind or adsorb to organic matter and this will result in reduced toxicity of the metal to aquatic organisms.

There are interspecific differences in susceptibility to various toxicants, and, for a given species, there may be differences in vulnerability between different ontogenetic stages. In general, newly hatched larvae are usually found to be more sensitive to toxicants than are larger fish. This may be related to the fact that small individuals have higher weight-specific metabolic rates than do large. The overall health and nutritional status of the test fish may also influence the susceptibility to toxicants. Fish that are already weakened by other causes are expected to be much more sensitive to toxicants than are healthy individuals. Previous exposure and acclimation to toxicants can affect tolerance, with prior exposure possibly leading to a two-to-threefold increase in tolerance limits.

16.2 TOXICITY AND BIOACCUMULATION

When fish are exposed to potential toxicants the chemicals may both enter the body over, and cause damage to, the gill membranes. The toxicants may then affect the physiological functions of the fish in a variety of ways, but most commonly disturbances to iono- and osmoregulation are observed. These disturbances arise, primarily, due to damage to the gills which may, in turn, affect ion and respiratory gas exchange, acid–base balance and the excretion of waste products. The structural damage to the gills that results from exposure to toxicants may include dissociation of the lamellar epithelium from the underlying tissue, lamellar fusion, necrosis of pavement and 'chloride' cells, epithelial swelling and rupture, and infiltration of the tissue by leucocytes. Damage of this nature is almost invariably accompanied by a drop in plasma ionic concentrations in freshwater fish species. Falls in plasma ion concentrations have often been observed in freshwater fish following exposure to a variety of toxicants, including heavy metals, pesticides and chlorinated hydrocarbons.

There are several factors that contribute to the drop in the plasma levels of the various ions, but increased gill permeability seems to be of prime importance. An increase in the permeability of the gill lamellae to water and ions would lead to an increased efflux of ions from freshwater fish, and there would also be a concomitant enhancement of the osmotic uptake of water. An increase in water influx would, in turn, lead to increased urine production and thereby further exacerbate the loss of ions. In addition, any inhibition of active ion uptake by the cells of the gill epithelium would further contribute to the negative ion balance.

Exposure of fish to metals such as cadmium, copper and aluminium leads to an increase in gill permeability, probably by effects on membrane structure. The permeability of epithelial tissue is determined by the characteristics of the cellular membrane, particularly with respect to phospholipid composition, and by the nature of the junctions between the epithelial cells. A loss of bound calcium from the membranes and intercellular junctions leads to increased permeability, and it seems that many of the effects of heavy metals may be related to an interference with the binding of the calcium ions to the gill epithelial membranes. Thus, there may be competitive interactions between heavy metals and calcium for binding sites on the gill epithelia, and, consequently, heavy metal toxicity may be reduced if high levels of dissolved calcium can be maintained in the water.

Exposure of fish to water-borne cadmium, copper and aluminium may lead to decreased active uptake of ions despite the fact that morphological examination often reveals a proliferation of 'chloride' cells following exposure to these toxicants. On the other hand, levels of the ion-transporting ATPase enzymes may not display a similar increase, so it is open to question whether or not the morphological examinations present a true picture with regard to the tissue densities of 'actively transporting' cells. It is more likely that the increases in numbers of 'chloride' cells observed in morphological studies arise as a consequence of increased rates of cell death, necrosis and turnover resulting from exposure to toxicants. Consequently, direct measurements of Na^+,K^+ ATPase enzyme activity may give a more accurate reflection of the ionoregulatory capacity of the fish, and hence provide indications as to the ability of the fish to restore ionic balance when exposed to various types of water-borne toxicants.

In addition to entering the fish through the highly permeable gill epithelial membranes, toxicants may also be consumed along with the food and be absorbed from the gastrointestinal tract. These chemicals may then be deposited and stored in various tissues of the body, such that tissue concentrations continue to rise with prolonged exposure to the chemical. This *bioaccumulation* is not only dependent upon the rate of uptake of the chemical, but is also influenced by the mechanisms of tissue storage. A major determinant of the bioaccumulation of a toxicant under a given set of exposure conditions is the rate at which the chemical is metabolized, detoxified and cleared from the body.

Bioaccumulation of toxic compounds in fish has received attention because of concern about possible health hazards to humans related to the consumption of contaminated fish. Many of the substances that have been investigated from the human health viewpoint are organic compounds, such as pesticides and dioxins. These organic compounds are usually only sparingly soluble in water, but are readily soluble in lipids, so they become closely associated with the storage lipid depots of the fish.

Fish may also accumulate metals, such as mercury, cadmium and other heavy metals. Several instances of metal poisoning have been recorded in humans as a result of the consumption of contaminated fish. Perhaps the best known example of such heavy metal poisoning is that of the Minamata disease. During the course of the early 1950s there was reported to be a serious decline in the general health of people inhabiting the villages around Minamata Bay, Kyushu, Japan. A number of people were found to be suffering from neural disorders and there was a marked increase in women either aborting spontaneously, or giving birth to deformed children. The problems were found to be caused by mercury poisoning, and the human population had accumulated dangerous levels of the heavy metal via the diet.

The fish stocks in the bay were the major protein source for the local population and these fish were consumed in large quantities. Chemical analysis revealed that the fish were contaminated with mercury, and the source of the mercury was traced to industrial developments situated along the shores of the bay. Industrial wastewater, containing methyl mercury and other residues, was discharged into the bay and the toxicants entered the food chain via this route. Toxicants became deposited in the tissues of a wide range of invertebrates, were then accumulated by fish which fed on these prey, and finally entered the human population through the diet.

Mercury, unlike a number of other substances which are predominantly accumulated in the liver, is readily deposited in muscle. It is generally the muscle of the fish that is the edible portion. Thus, it can be seen that human health problems caused as a result of eating fish are more likely to be due to mercury poisoning than to other agents.

On the other hand, when molluscs and crustaceans are eaten by humans the entire soft body tissues are often consumed. The edible portions of these animals may, therefore, possibly contain a wide range of potential toxicants, since these organisms may readily accumulate such substances from waters polluted by industrial and other effluents.

16.3 DETOXIFICATION MECHANISMS

Not all toxic substances are accumulated in the tissues of the fish, and distributions may also differ from tissue to tissue. For example, whilst both mercury and cadmium may accumulate in fish tissues, other metals, such as zinc and copper, are usually deposited and stored much less readily. Fish have a number of mechanisms to prevent accumulation of many substances, and following exposure to potential toxicants a number of biochemical pathways are induced in order to detoxify and excrete the foreign compounds.

Transformation of potentially toxic compounds occurs in all of the body tissues, but the highest levels of the enzymes that metabolize toxicants are found in the liver. The transformation reactions usually involve an initial oxidation, reduction or hydrolysis of the toxicant, followed by a reaction in which the residues are transformed into water-soluble metabolites that can be readily excreted from the body. Oxidative reactions are the most important reactions during the initial stages.

One of the major enzyme systems involved in the transformation and detoxification of organic compounds, such as hydrocarbons and pesticides, is the cytochrome P-450 system. This system is not only involved in detoxification reactions, but is also implicated in the biotransformation and metabolism of endogenous compounds such as steroid hormones. There are known to be several P-450 isozymes that differ both in their substrate specificities and in the characteristics of synthesis and production induced by exposure to different organic compounds. Thus, enzymes of the P-450 system are inducible by a range of organic xenobiotics, including polychlorinated biphenyls (PCBs) and polyaromatic hydrocarbons (PAHs). The inductive response involves an increase in the rate of gene transcription, which results in increased levels of messenger RNA, and this is followed by increased synthesis of P-450 proteins. Whilst the highest levels of P-450 system enzymes are usually found in the liver, quite high levels may also be found in the kidneys, gills and gastrointestinal tract of fish species, with all of these organs playing an important role in the uptake, biotransformation and excretion of organic xenobiotics.

The biotransformation of organic xenobiotics occurs in two stages. The first stage involves a metabolic transformation in which an oxygen atom is introduced into the lipophilic organic substrate by the cytochrome P-450 enzyme system, and in the second stage the water solubility of the oxidized xenobiotic is increased via conjugation to glucuronic acid, sulphate or glutathione. The conjugated products are generally both less toxic and more easily excreted than the untransformed compounds.

Detoxification of inorganic compounds, such as heavy metals, occurs via processes different from those involved in the biotransformation of organic xenobiotics. Exposure of fish to heavy metals such as cadmium and copper leads to the synthesis of metallothioneins in liver, kidney and gill tissues. Metallothioneins are low-molecular-weight proteins that can bind to metals. The binding properties of the proteins vary, but the various metallothioneins generally have the lowest affinity for zinc and a higher affinity for cadmium and copper.

Metallothioneins are normal tissue components, where they are thought to play a major role in the regulation of the intracellular metabolism of zinc and copper. These two metals are essential co-factors for a number of enzymes. However, because exposure to heavy metals induces the synthesis of metallothioneins they are also thought to play an important part in protecting the cell from damage by these toxicants. The proteins may serve a protective function by sequestering the free ions of the heavy metals. This will prevent the metal ions from binding to, and inhibiting, cellular enzymes and other functionally important proteins.

The exposure of fish to low concentrations of metals has been found both to give increased tolerance to subsequent exposure and to lead to the induction of metallothionein synthesis. The implication is that the metallothionein proteins provide added protection against the toxic actions of heavy metals if the fish should be exposed to these agents at a later date. Exposure to heavy metals may, however, result in harmful effects being observed before the binding sites of the metallothionein proteins have become fully saturated with heavy metal ions. Thus, the expectation that the toxic actions of heavy metals should only become evident when the binding capacity of the metallothioneins has been exceeded and the metal ions begin to bind to other proteins is simplistic.

There is also some doubt as to the level of protection provided by the metallothioneins. Whilst exposure to heavy metals will generally lead to a several-fold increase in the levels of these metal-binding proteins in the tissues, the proteins may only have the capacity to bind a small proportion of the body load of the toxicant. In addition, since the metallothioneins display different affinities for the various heavy metals, the pattern of metal binding to these proteins may not provide an accurate reflection of either body loading or environmental concentrations.

Lipid peroxidation, which results in the destruction of cell membranes and disruption of membrane-bound enzyme activities, is thought to be one of the main mechanisms by which toxicants exert injurious effects. Lipid peroxidation is a chemical process that results in the oxidative deterioration of polyunsaturated fatty acids, and thereby leads to destructive changes in the lipids of the cell membranes (Chapter 4). Malondialdehyde is produced as a breakdown product of lipid peroxidation, and exposure of fish to a range of potentially toxic organic and inorganic substances (e.g. PCBs, carbon tetrachloride, heavy metals) has been shown to stimulate lipid peroxidation and malondialdehyde production.

The lipid peroxidation chain reaction is initiated by the removal of a hydrogen atom from a polyunsaturated fatty acid. This results in the formation of a lipid radical. Lipid radicals react very rapidly with oxygen to form hydroperoxide radicals, and these, in turn, may remove hydrogen atoms from fatty acids in other lipid molecules. This results in the formation of a molecule of hydroperoxide and a new lipid radical, which continues the chain reaction. The hydroperoxide can be degraded to give a variety of products, including malondialdehyde.

Protection of the cell membrane lipids against peroxidation can be provided by two broad classes of antioxidants, the *preventive* and the *chain-breaking* antioxidants. Preventive antioxidants reduce the rate at which lipid radicals are formed. Thus, the preventive antioxidants act by hindering the initiation of the lipid peroxidation chain reaction. The most important preventive antioxidants are enzymes (e.g. catalase, glutathione peroxidase) that are capable of reducing hydroperoxide radicals.

The chain-breaking antioxidants owe their antioxidant activity to their ability to bind with and trap hydroperoxide radicals. Ascorbic acid (vitamin C) is an important water-soluble chain-breaking antioxidant, and the most important lipid-soluble antioxidants are the tocopherols (vitamin E and related compounds). By itself, ascorbic acid is a relatively ineffective antioxidant. There appears, however, to be a synergistic interaction between ascorbic acid and the tocopherols. This results in a greater protection than would be predicted on the basis of the sum of the protective effects imparted by the different compounds acting alone. Thus, ascorbic acid is proposed to play a role in the regeneration ('repair') of the tocopherols. The restoration of the antioxidant function to the tocopherols may involve the reduction, by ascorbic acid, of the tocopheroxyl radicals formed when the tocopherol traps hydroperoxides.

Exposure of fish to toxicants might be expected to lead to marked changes in tissue concentrations of the preventive and chain-breaking antioxidants. For example, tissue levels of ascorbic acid can be seriously depleted following exposure to a range of organic and inorganic toxicants. Thus, the dietary requirement for ascorbic acid may be higher in fish living in polluted environments than in those inhabiting unpolluted water bodies.

16.4 FRESHWATER ACIDIFICATION

As water bodies become more acidified, the fish fauna becomes depleted, both because of a reduction in species diversity and because of a decline in the size of the populations that inhabit the acidic waters. When the pH falls below 4, almost no fish species will be found inhabiting the water body.

Falls in pH, which could result from the periodic discharge of acidic industrial wastes, can lead to respiratory distress, disturbance of the iono- and osmoregulatory performance, and problems with the acid–base balance of the body fluids. The severity of these disturbances appears to be related to the extent of the acidification of the water body and to the calcium ion concentration of the water.

Respiratory distress may arise as a result of reduced oxygen diffusion over the gill membranes caused by an increase in the thickness of the mucous layer covering the secondary lamellae. An increase in mucus secretion is a typical response to acute exposure to acid conditions, and following longer-term exposure to low pH there may be a hyperplasia of mucus cells of the gill epithelia. Longer-term exposure to acid conditions may also result in there being changes in both numbers and morphology of the epithelial 'chloride' cells. The increase in 'chloride' cell numbers is most pronounced in fish held in acidified ion-poor water. These changes in both numbers and morphology of the 'chloride' cells are suspected to result in changes in gill ionic absorption capacity and thereby influence, and restore, ionoregulation and acid–base balance.

Acute exposure to acidified water results in an inhibition of hydrogen ion excretion and a reduction in ion uptake. For example, calcium uptake was reduced by about 40% in carp exposed to water of pH 5.2, in comparison with control fish held in water of pH 7.2. The reduction in net uptake of calcium appeared to be related to a reduced influx of calcium at low pH, with calcium efflux being little affected by the fall in pH from 7.2 to 5.2.

Reductions in the net uptake of ions in acidified water can result in a gradual reduction in plasma ion levels, and the inhibition of hydrogen ion excretion may give rise to acidosis if the fish are in water of low calcium content. The reduction of the ionic content of the plasma may arise as a result of a reduced influx, as with calcium, or both a reduced influx and an increased efflux of the affected ions, as appears to be the case with sodium.

Following exposure to acid conditions, there is a substantial increase in the permeability of epithelial surfaces to water and ions, and one of the major factors affecting epithelial permeability appears to be the degree of calcium binding. Acidification of the water reduces binding of calcium ions, epithelial permeability increases, and, in freshwater species, there is an increased water influx and an increased ionic efflux. The influx of water leads to an increase in urine production which, in turn, results in further loss of ions. Thus, there is a fall in the osmotic concentration of the plasma and, in addition, the blood is acidified owing to the influx of hydrogen ions. These disturbances are accompanied by an increase in haematocrit due to an increase in the number of circulating red blood cells, an increase in red blood cell volume and a decrease in the total blood volume owing to a reduction in plasma volume.

Whilst exposure to acid conditions may lead to a fall in blood pH of no more than a few tenths of a pH unit, the physiological consequences of the pH changes are potentially quite large. A lowering of blood pH may result in a fall of the pH in the red blood cells, which in turn affects the affinity and capacity of haemoglobin for oxygen. Affinity may be reduced due to the Bohr effect and reductions in capacity occur due to the Root shift. The net result is that the haemoglobin will remain relatively unsaturated with oxygen. This can, to some extent, be counteracted by the release of catecholamines, leading to a restoration of intracellular pH within the red blood cells (Chapters 8, 15). Nevertheless, the end result of severe acidification may be a reduction in the overall capacity of the blood to transport oxygen to the tissues.

This reduction in oxygen-carrying capacity may impose limits on the maximum aerobic capacity of the fish in acid waters, and fish exposed to acid conditions have significantly lower maximum sustained (aerobic) swimming speeds than do fish held in waters of slightly higher pH. The problems associated with acidosis are most marked for fish swimming at moderately high speed, and the maintenance of oxygen delivery to the tissues is a much less serious problem for fish at rest.

Except in very extreme cases of acidification, the buffering capacity of the fish blood appears to be able to cope with the potential respiratory problems created by increased hydrogen ion influx. Thus, if acute exposure to acidified water results in the death of the fish the most probable cause is osmotic and ionic imbalance.

Whilst industrial wastewater outfalls may lead to changes in the pH of rivers and lakes, there is equal concern about the long-term chronic acidification of water bodies that has occurred as the result of increased industrial atmospheric emissions. Industries (e.g. electric power plants, smelters and steel mills) that use large quantities of fossil fuels (coal and oil) will produce waste emissions that contain considerable amounts of sulphuric oxides, and the exhaust emissions from road vehicles contain quantities of nitric oxides. Both of these types of gaseous emission will enter the atmosphere and can later fall as acid precipitation.

'Acid rain' will fall in areas located downwind of large industrial and urban complexes, and it is this acid precipitation with its content of sulphuric and nitric acids that causes the most severe acidification of surface waters. The acid rain water may enter streams and lakes directly, as surface run-off, or after percolation through the soil and subsoil.

The safe range of environmental pH for the survival of fish is approximately 5–9, but productivity is maximized in waters of pH 6.5–8.5. The pH of seawater is generally in the range 7.8–8.2, and the large buffering ability of seawater means that the effects of acid precipitation on pH are negligible. In the oceans, lakes and rivers, carbonate–bicarbonate represents the primary buffering system. Following an input of acid, hydrogen ions will combine with bicarbonate to form carbonic acid, a weak acid, which dissociates incompletely to give carbon dioxide and water.

The buffering capacity of fresh waters is dependent upon the carbonate–bicarbonate system, and buffering capacity is, therefore, related to the carbonate content of the water. Since freshwater bodies vary markedly in carbonate and ionic contents, buffering capacities are also quite variable. Consequently, the pH of fresh waters covers a wider range than that of oceanic waters, and the effects of acid precipitation can be much more marked in rivers and lakes than in the sea. Acidification will be most dramatic in water bodies that lie over, or receive run-off from, bedrock that is poor in carbonate and other soluble minerals that can contribute to buffering capacity.

When acid precipitation falls on land it may reach larger water bodies following percolation through the soil and subsoil. Interactions between the acidic precipitation and the organic and inorganic constituents of the soil can exert a significant influence on the composition of the drainage water, and thereby have profound effects on lakes and streams.

Soils will contain hydrogen ions bound to both organic matter and the cation-exchange sites of the soil clay minerals. Soil water will contain additional hydrogen ions, and the sum of the ions in solution and those bound to organic and inorganic matter represents the total acidity. The hydrogen ion concentration (pH) of the soil water will be influenced by the relative amounts of base cations (calcium, magnesium, sodium and potassium), aluminium cations and hydrogen ions on the soil cationic exchange sites. The stability of the soil pH will be determined by the proportion of the cation exchange capacity occupied by base cations. Hydrogen ion input in the form of acid precipitation may lead to displacement of base cations from the exchange sites, followed by leaching of these cations from the soil in the drainage water.

If the base cations are not replaced via inputs from base-rich water, erosion of bedrock, or the decomposition of biological materials, the consequence will be a gradual depletion of these cations. Calcium is the most mobile cation present and will be the first to be leached out and depleted. Generally, the greater the amount of water passing through the soil, and the greater the acid input, the greater will be the extent of leaching. As calcium becomes depleted, other ions, such as aluminium, are leached out of the soil in increasing quantities. Thus, the effects of acid precipitation on the mobilization of soil aluminium in drainage waters are likely to be most marked in areas of high rainfall and calcium-poor soils. Aluminium is extremely toxic to fish at a water pH of 5, so the entry of aluminium into water bodies already affected by acid precipitation is an additional hazard likely to result in fish mortality.

Exposure of fish to aluminium in acid water (pH 5–5.3) can cause severe disruptions of the ion-exchange mechanisms, resulting in reduced net uptake of ions such as calcium and sodium by freshwater fish species. These reductions in net ion uptake may be due to the effects of the aluminium on influx and/or ion-efflux mechanisms. For example, the reduced net uptake of calcium by carp exposed to aluminium at low pH appears to be due more to an inhibition of influx than to marked increases in calcium ion efflux, although some increase in efflux is observed. By contrast, in the same species, exposure to aluminium under acidic conditions does not seem to affect sodium influx to any great extent. There is, however, a net loss of sodium brought about by the marked increase in ion efflux that follows exposure to aluminium.

The observation that sodium influx is not affected by exposure to aluminium suggests that the inhibition of calcium ion influx is not the result of general damage to the ion-transporting cells, but is due to specific effects on calcium uptake and transport mechanisms. Calcium enters the cells via channels in the apical membrane, and is then transported through the cell to the basolateral membrane where it is extruded into the blood via Ca^{2+}-ATPase activity, and it is possible that aluminium has inhibitory effects on one or more of these processes.

Exposure to aluminium results in increased effluxes of both calcium and sodium, possibly via effects on the junctional complexes that form the barrier for paracellular movement of ions. High concentrations may influence the functional integrity of these complexes, thereby enabling increased ionic effluxes to occur.

Aluminium toxicity to fish and other aquatic organisms is much reduced as pH is increased, so attempts have been made to counteract the effects of acid rain by the liming of water bodies in order to increase the buffering capacity and thereby increase and stabilize pH. Remedial action of this type has been taken in areas where there have been marked declines in resident populations of acid-sensitive fish species such as salmonids. In some cases, attempts at environmental restoration have included combinations of liming and fish restocking programmes.

16.5 AMMONIA AND OTHER NITROGENOUS WASTES

Most of the nitrogenous end products of metabolism excreted by aquatic organisms originate from protein catabolism, with deamination of the amino acids ingested in the food being the major nitrogen source. In freshwater fish, approximately 80% of the total nitrogen excreted will be in the form of ammonia, with urea, amino acids, amino oxide derivatives, creatine, creatinine and uric acid making up the remainder. Ammonia is favoured as the principal product of nitrogenous excretion in the majority of aquatic organisms because the synthesis of ammonia is energetically more efficient than that of alternative excretory products. Ammonia is, however, relatively toxic, and when the ammonia concentration in the water increases, this can have negative consequences for growth and survival of the fish.

The excretion of the nitrogenous end products of metabolism from fish living in their natural environments is only rarely likely to have a direct negative effect on the fish. Under normal circumstances, the excretory products will either be rapidly carried away and diluted by flowing water, or will be rapidly oxidized to harmless products by other organisms in the environment. However, in an unnatural environment, such as an intensive culture unit, where fish are held at high stocking densities and are provided with plentiful supplies of food, the rates of production and excretion of nitrogenous waste products may exceed the rates at which they are eliminated from the system. Thus, unless care is taken to ensure that rates of water throughput and replenishment are adequate, there may be a gradual accumulation of nitrogenous excretory products, eventually resulting in these compounds reaching toxic levels.

Ammonia and ammonium

Of the nitrogenous end products excreted by fish, ammonia is generally considered to be the most toxic. Ammonia is, however, present in water in two forms – an un-ionized portion (NH_3) and an ionized form (NH_4^+) – and the term 'ammonia' is often used when referring to the sum of the un-ionized ammonia and ionized ammonium forms. In the discussion that follows ammonia/ammonium will be used when referring to 'total ammonia', and ammonia and ammonium when referring to the un-ionized and ionized forms, respectively.

The major proportion of ammonia/ammonium excreted by fish is cleared from the blood in the gills and passes over the gill epithelium into the surrounding water. The main mechanisms by which excretion occurs are thought to be by passive outward diffusion of both ammonia and ammonium down their respective diffusion gradients, and via a carrier-linked ionic exchange involving sodium and ammonium. Outward diffusion of ammonia is undoubtedly the most important of the nitrogenous excretory mechanisms in the gills (Chapter 7).

Ammonia is a gas that is highly soluble in water, and in aqueous solutions, un-ionized ammonia exists in equilibrium with the ammonium ion (NH_4^+) and the hydroxide ion (OH^-): $NH_3 + H_2O = NH_4^+ + OH^-$. The proportions of ammonia and ammonium are dependent primarily upon water pH and temperature, and to a lesser extent upon salinity and pressure. For example, a pH increase from 7 to 8 (within the temperature range 0–30 °C) results in an approximately tenfold increase in the concentration of ammonia (NH_3), and a temperature increase of 5 °C (in the range 0–30 °C) at pH 7 will result in the concentration of ammonia increasing by 40–50%. Thus, within the range of conditions prevalent in aquatic environments inhabited by fish, there can be several-fold differences in the proportions of ammonia/ammonium found in the un-ionized and ionized forms, with high water pH and high temperatures leading to increased formation of un-ionized ammonia.

When the ammonia concentration of the surrounding water rises, the diffusion gradient for ammonia, from the blood to the water, over the gill epithelium is reduced. This will lead to a slowing of the rate of outward diffusion, causing an elevation in blood ammonia and, possibly, ultimately a toxic reaction. One way in which the fish could reduce the risk of ammonia toxicity would be to increase the conversion of ammonia to urea, a much less toxic excretory product.

Many teleost species appear either to lack some of the enzymes necessary for the synthesis of urea from ammonia, or to have the enzymes present in the tissues but at very low levels. Thus, it is open to question whether the conversion of ammonia to urea can provide a major detoxification pathway in the majority of teleost species. Converting ammonia to urea is not the only possible way of reducing ammonia toxicity. Glutamine is less toxic than ammonia and could be used as an alternative nitrogenous end product under conditions where there is a risk of ammonia toxicity. Relatively high levels of glutamine synthesase activity have been found in the tissues, particularly the brain, of a number of fish species, so it is clear that the possibility exists that ammonia can be detoxified via the production of glutamine. Nevertheless, under conditions where the ambient ammonia concentration in the environment rises either too high or too quickly, blood ammonia levels will also become markedly elevated and toxicity effects will become manifest.

Ammonia/ammonium toxicity

Toxicity manifests itself chiefly as neurological disturbances, with acute responses including hyperventilation, erratic swimming, convulsions, coma, and ultimately death. There may also be histopathological changes in gill structure, with a general thickening of the epithelial membrane and associated hyperplasia. The structure of the secondary lamellae may be disrupted, leading to a reduction in the surface area over which gaseous exchange can occur effectively. Changes in gill structure may occur as part of the acute response, but are more likely to be seen as a result of chronic exposure. Chronic toxicity symptoms may include hyperplasia of the gill epithelium and fusion of the secondary lamellae. Long-term exposure may also result in kidney dysfunction and damage to the tissues of the liver.

It has commonly been believed that ammonia is the toxic form of ammonia/ammonium, probably because of the ease with which the un-ionized ammonia can diffuse across the gill epithelium. Ammonium may, however, also be toxic and toxicity may, therefore, be more dependent upon the total ammonia/ammonium level than on ammonia *per se*. Toxicity is, therefore, not limited to the ammonia moiety, and, as the environmental pH changes, the different forms appear to show varying degrees of toxicity. For example, high concentrations of ionized ammonium in water of low pH can produce toxic effects, even though the levels of ammonia may be well below lethal levels. It is undoubtedly an over-simplification to attribute toxicity effects solely to ammonia at high pH, and ammonium at low pH, since there will probably always be a contribution from both forms whenever the ammonia/ammonium concentration is sufficiently high to produce toxicity.

Much of the work on ammonia/ammonium toxicity to fishes has been carried out with the aim of finding 'safe' levels that can be recommended for intensive culture systems. The values for safe levels obtained in the different studies are, however, extremely variable, possibly as a result of interactions with other factors that could contribute to the development of toxicity symptoms (e.g. levels of dissolved oxygen and carbon dioxide, pH, temperature, other toxicants). There is, therefore, no universally recognized safe standard, although a concentration of 0.0125 mg ammonia per l has been widely cited as being the maximum safe concentration in fish culture facilities. In some cases, however, exposure to ammonia concentrations well above this level has not been shown to result in either toxicity symptoms or a depression in rates of feeding and growth. Other results, however, point to toxicity effects at ammonia concentrations below 0.0125 $mg.l^{-1}$, and permissible safe levels have been suggested as being as low as 0.002 mg ammonia per l and/or 1 mg ammonia/ammonium per l.

Nitrification

In natural waters, some of the ammonia/ammonium excreted by aquatic organisms will be oxidized, first to nitrite, and then to nitrate. There may also be some decomposition of urea, this being carried out by ureolytic bacteria such as *Bacillus pasteurii.* The urea is hydrolysed to ammonia and carbon dioxide, and the ammonia may then be oxidized to nitrite and nitrate. Thus, in nature, ammonia/ammonium and urea are transformed to nitrate under aerobic conditions, but this occurs via oxidation to nitrite as an intermediate step.

Nitrite may be toxic to fish, and some negative effects have been observed following long-term exposure to concentrations as low as 0.1 $mg.l^{-1}$. In well-oxygenated natural water bodies nitrite concentrations are typically less than 0.005 $mg.l^{-1}$, but under some circumstances nitrite levels may become substantially higher. For example, nitrite may accumulate in the bottom waters of deep lakes during the periods of the year when stratification occurs. When stratification breaks down and the waters become well mixed, the nitrite may appear in surface waters in concentrations that are toxic to fish.

Industrial, and other human, activities can give rise to increases in nitrite concentrations in aquatic systems. Nitrite concentrations of 50 $mg.l^{-1}$ and higher have been recorded in waters receiving effluents from industrial complexes producing metals, dyes and celluloids. Sewage effluents can also contain large amounts of nitrite, and nitrite may occasionally increase to toxic levels in fish culture systems that rely on recirculated water.

The oxidation of ammonia/ammonium to nitrate by aerobic bacteria is known as *nitrification.* Both heterotrophic and chemoautotrophic bacteria may play a role in the nitrification process, but it is generally assumed that the chemoautotrophs are the most important. Thus, for general purposes, nitrification can be considered as being the oxidation of ammonia/ammonium to nitrate in a two-step process by aerobic, chemoautotrophic bacteria. The two major groups of bacteria involved are *Nitrosomas* spp., which oxidize ammonia/ammonium to nitrite, and *Nitrobacter* spp., which oxidize nitrite to nitrate. Other genera of bacteria may be involved in the oxidation processes, but to a much lesser degree than the nitrobacters and nitrosomonads.

For the nitrification process to be activated quickly, and proceed effectively, certain requirements must be met. Conditions must be aerobic, and solid surfaces are required for growth of both nitrobacters and nitrosomonads. Both types of bacteria appear to require phosphate, magnesium and iron for growth, and, in addition, sodium, calcium and copper appear to be necessary to ensure adequate growth of nitrosomonads. Growth of both nitrobacters and nitrosomonads can occur within the temperature range 5–30 °C, but the pH optimum (6.2–7.0) for *Nitrobacter* spp. appears to be somewhat lower than that (7.2–8.0) of the nitrosomonads. It should also be clear that the *Nitrobacter* spp. are, by and large, dependent upon the nitrosomonads for supplies of substrate (i.e. nitrite) for oxidation.

The nitrification process can be inhibited by high levels of un-ionized ammonia (NH_3). Ammonia, however, inhibits the action of nitrobacters at much lower concentrations than those at which it inhibits nitrosomonads. Thus, when levels of ammonia are relatively high, there may be incomplete nitrification, in that ammonia/ammonium is oxidized to nitrite, but the oxidation of the nitrite to nitrate is inhibited. Under these circumstances, there is the risk that nitrite could accumulate and eventually reach toxic concentrations.

Since the proportions of ammonia and ammonium present in aqueous solutions are markedly affected by pH, conditions under which high concentrations of ammonia/ammonium co-occur with high water pH will increase the risk of incomplete nitrification and nitrite accumulation. Other factors that affect or inhibit the nitrification bacteria differently could also lead to the accumulation of nitrite.

Nitrite toxicity and detoxification

Nitrite is taken up actively by most freshwater fish species, and nitrite concentrations in the blood may reach more than ten times the concentrations in the surrounding water. Nitrite levels do not, however, increase in the blood of all species, and some centrarchids have been shown to be capable of maintaining plasma nitrite concentrations below those in the environment. The susceptibility of different fish species to nitrite toxicity appears to be related to the extent to which they either concentrate nitrite in or exclude nitrite from the plasma. Fish such as salmonids and ictalurids that readily concentrate nitrite in the plasma are particularly sensitive, whereas the centrarchids, which exclude nitrite, are much less so.

The process by which nitrite is taken up by the fish appears to be similar to that of chloride uptake, and it has been suggested that nitrite and chloride may compete for the same transport systems in the gill epithelia. Thus, nitrite uptake and toxicity appear to be markedly reduced in the presence of adequate concentrations of chloride ions. Other monovalent anions may also have an ameliorative effect, with bromide appearing to be highly effective in reducing nitrite toxicity. Bicarbonate and nitrate, whilst having some positive effect, are not nearly so effective as chloride and bromide. Divalent and trivalent anions seem to have very little effect on nitrite toxicity.

Nitrite toxicity is associated with uptake over the gill epithelium and accumulation in the plasma. From the plasma the nitrite enters the red blood cells, where it oxidizes the iron in the haemoglobin, resulting in the formation of methaemoglobin. Methaemoglobin does not have the capacity to reversibly bind oxygen, so an increase in methaemoglobin can lead to a decrease in the oxygen-carrying capacity of the blood. One sign of a high methaemoglobin content is that the blood takes on a brownish, rather than red, colour.

A certain amount of methaemoglobin is formed spontaneously, even in the absence of nitrite, so fish blood will always be found to contain a measurable amount of methaemoglobin and ratios of 10 methaemoglobin : 90 haemoglobin are not particularly uncommon. Methaemoglobin can be reconverted to haemoglobin by a reductase enzyme found in the red blood cells, and this restoration process usually occurs at a steady rate. When nitrite is present, the ultimate level of methaemoglobin present in the blood is a result of the balance between rates of formation and reconversion to haemoglobin by the reductase. When the ratios of methaemoglobin to haemoglobin are below 50:50 there is usually little or no mortality, so fish are generally quite resistant to death from methaemoglobinaemia. When the proportion of methaemoglobin increases to over 50%, however, the oxygen-carrying capacity of the blood may become severely reduced and the fish may become torpid. Because nitrite toxicity is probably related to methaemoglobinaemia, and reduced oxygen-carrying capacity of the blood, any factors that affect oxygen availability or demand would be expected to result in changes in the levels of nitrite having negative effects upon the fish. Thus lower concentrations of nitrite would be expected to give rise to toxicity symptoms under conditions either where levels of available oxygen were low, or where the tissue oxygen demands of the fish were high.

Toxic effects of nitrite on fish have also been observed in the absence of pronounced methaemoglobinaemia, but other causal mechanisms of toxicity remain obscure.

Chapter seventeen

Environmental stressors

17.1 INTRODUCTION

The physiological systems of fish can be challenged or 'stressed' by a wide range of biological, chemical and physical factors. The term *stressor* can be used to describe these factors if the challenge is sufficiently severe to elicit a compensatory physiological response. Thus, an environmental challenge (stressor) that is severe enough to cause stress is one that requires a compensating response to be made. Acclimation can be said to have occurred if the compensatory stress response enables a restoration of physiological variables to within limits that do not compromise survival.

Sublethal stressors elicit a complex of changes that can, in the long term, potentially lead to impairment of growth, reproductive performance, or disease resistance. Chronic exposure to these sublethal stressors can, therefore, result in reduced reproductive success or decreased survival of individuals, which may have serious consequences for the survival of entire populations. Thus, the success of fish and fish populations in acclimating to environmental changes depends upon the abilities of individuals to carry out and maintain successful compensatory responses when exposed to these stressors.

Exposure of fish to environmental stressors induces a characteristic series of endocrine, and other, responses that are analagous to the 'general adaptation syndrome' (GAS) seen in mammals. Under this model the stress responses may be divided into three categories or stages:

1. *Alarm* A series of physiological changes are initiated and homeostatic control systems begin to compensate for the disturbance.
2. *Resistance* The compensatory processes successfully restore physiological balance, but the costs involved ultimately reduce performance capacity.
3. *Exhaustion* The duration or severity of the disturbances caused by exposure to the stressors exceeds biological tolerance limits. The compensa-

tory changes become maladaptive, and there may be adverse and deleterious effects.

Within this general framework, it is usual to consider the *stress response* in terms of *primary, secondary* and *tertiary* changes that involve successively higher levels of biological organization, starting at the level of the endocrine system and concluding with changes displayed at the whole-animal level.

17.2 PRIMARY STRESS RESPONSE – THE ENDOCRINE SYSTEM

Adverse stimuli (stressors) are detected by external or internal sense organs, which, in turn, relay the information to the hypothalamus, the integrative centre of the brain. Afferent neural pathways run in the sympathetic nervous system from the hypothalamus to the chromaffin tissue of the head kidney (pronephros), and direct stimulation of this tissue leads to the release of catecholamines. The catecholamines are released rapidly from the chromaffin tissue following the exposure of a fish to a stressor, and plasma levels of catecholamines may be elevated by over an order of magnitude within the course of a few minutes. Some elevation may persist for several hours. The liberation of catecholamines into the bloodstream leads to them being widely distributed in the body, where they initiate a series of secondary effects on other organ systems, such as the cardiovascular system (Chapters 8, 15). These rapid changes are the *primary stress response.*

Other neurons within the hypothalamus run to the vicinity of the cells in the anterior of the pituitary gland (adenohypophysis). A neuropeptide is transported from the hypothalamus to the pituitary within this system, and release of the neuropeptide stimulates production and secretion of adrenocorticotropic hormone (ACTH). There may be alternative sources of peptides that stimulate ACTH release, and collectively these peptides are known as corticotropin releasing factors (CRFs). The CRFs stimulate the release of ACTH from the corticotrops that form a sheet of cells at the interface between the rostral pars distalis of the adenohypophysis and the neurohypophysis in the pituitary gland. The ACTH is released into the blood and travels via the systemic circulation to the interrenal cells of the head kidney, where it stimulates the production and release of corticosteroid hormones.

In the majority of teleost fish species, cortisol is the major corticosteroid hormone, although other corticosteroids, e.g. corticosterone, 11-deoxycorticosterone, and 11-deoxycortisol, are produced in the interrenal tissue. Plasma levels of ACTH may be elevated within the space of a few minutes following exposure to an environmental stressor, and this is followed by an elevation in circulating levels of cortisol. The release of cortisol is controlled by a feedback system, with high levels of circulating cortisol having a negative feedback on ACTH production and release by the pituitary. The cortisol may also have a feedback action on the hypothalamus, and thereby influence the release of the neuropeptide CRF. Since there are close links between the hypothalamus and the release of ACTH and cortisol, acting as a cascade, it has become usual to refer to this series of interactions as being governed by the hypothalamic–pituitary–interrenal axis (HPI axis).

In salmonids, exposure of the fish to an acute stress, such as handling, may lead to a rapid rise of plasma cortisol from a basal level of under 5 $ng.ml^{-1}$ to a value within the range 40–200 $ng.ml^{-1}$. This elevation in plasma cortisol may persist for several hours, but, provided the exposure to the stressor has been of short duration (acute stress), the plasma cortisol concentration will usually have returned to baseline levels within 24 h. When a fish is subjected to either repeated, or chronic (i.e. continuous), exposure to the stressor, plasma cortisol levels may remain elevated for periods of several days, or even weeks. However, even in chronically stressed fish, plasma levels of cortisol will usually eventually return to baseline values, but in these fish the kinetics of cortisol secretion and plasma clearance may be markedly different from those of unstressed individuals. There is also evidence to suggest that chronic exposure to environmental stressors leads to a down-regulation (i.e. reduction in number, or sensitivity) of cortisol receptors in the tissues of these fish, and this would be expected to have consequences for the responses shown in the event of exposure to additional stressors.

A major target organ for corticosteroid action in fish appears to be the liver. Cortisol enters the cells and binds to nuclear receptors. This results in gene activation, with the production of a series of enzymes which have a range of metabolic effects.

17.3 SECONDARY RESPONSES – TISSUE ALTERATION

The release of catecholamines and cortisol triggers a broad suite of biochemical and physiological changes known collectively as *secondary stress responses.* The metabolic effects may include hyperglycaemia, hyperlacticaemia, depletion of tissue glycogen reserves, lipolysis and inhibition of protein synthesis. There may also be increased catabolism of muscle protein, and alterations in the plasma levels of amino acids, free fatty acids and cholesterol. The tissue changes may also include a depletion of ascorbic acid (vitamin C) reserves, particularly those of the interrenal tissue.

Catecholamines are thought to cause the initial elevation in plasma glucose levels by mobilizing the glycogen reserves of the liver (glycogenolysis), whereas the corticosteroids may contribute to the maintenance of hyperglycaemia via the stimulation of gluconeogenesis from amino acids. The synthesis of glucose from amino acids requires deamination, and liver levels of aminotransferase enzymes may be increased following the administration of cortisol. Increased deamination of amino acids may also result in rates of nitrogenous excretion being elevated in fish treated with cortisol. In addition to maintaining hyperglycaemia, the increase in synthesis of glucose from amino acids could, in the longer term, also result in the restoration of liver glycogen levels.

Osmotic and ionic disturbances may occur as the result of diuresis and the loss of electrolytes from the blood. There may also be changes in haematology, with erythrocytic (red blood cell) changes and a reduction in numbers of white blood cells (leucopenia). The catecholamines have marked influences on the cardiovascular system (Chapters 8, 15), leading to changes in patterns of blood flow and increased gill perfusion, and alterations in the oxygen transport capacity of the blood, whereas the corticosteroids are known to stimulate the ion-transporting mechanisms present in the gills and kidney. These effects are of particular importance in the period of recovery from stress, when the fish attempts to maintain oxygen supplies to the tissues, and to regain osmotic and ionic equilibrium.

17.4 TERTIARY RESPONSES – LONGER-TERM EFFECTS

The secondary stress responses, which are induced by the rapid endocrine changes (primary responses), result in the rapid mobilization of energy reserves. Thus, the secondary stress responses are believed to have evolved as adaptive mechanisms that enable the animal to meet the increased energy demands imposed by exposure to environmental stressors. Typically the biochemical and physiological changes persist for only a few hours or days, following acute exposure to the stressor. They do not, therefore, result in any serious deleterious effects to the animal. In contrast, chronic exposure to stressors can induce a number of pathological changes, and can lead to reductions in reproductive success, depression of growth rates, and decreased disease resistance. These changes are known as the *tertiary stress effects or responses.*

Exposure of fish to many different types of environmental stressors may result in increased susceptibility to a wide range of common pathogens, e.g. viruses, bacteria, fungi and protozoans. This stress-induced increase in susceptibility to disease organisms can be mimicked by the administration of physiological doses of corticosteroids to otherwise unstressed fish. The close association between circulating levels of corticosteroids and susceptibility to disease appears to be mediated via the effects of cortisol on the immune system, because high circulating levels of cortisol have been shown to result in immunosuppression in a range of fish species. Reductions in disease resistance do not, however, seem to be directly dependent upon plasma cortisol concentrations being maintained at very high levels, because prolonged maintenance of cortisol concentrations as low as 10–15 $ng.ml^{-1}$ seem to be capable of producing this effect in a number of salmonid species.

Exposure to environmental stressors, such as acid waters and sublethal levels of chemical toxicants, has been shown to reduce the reproductive success of a range of fish species, and this seems, also, to be mediated via the corticosteroids. Both exposure to stressors, and the elevation of plasma cortisol levels due to cortisol implantation, have been shown to lead to suppression of the pituitary–gonadal axis, resulting in a reduction in circulating levels of sex hormones in salmonids. In males, chronic exposure to stressors may result in reduced pituitary gonadotropin (GTH) content, reductions in plasma levels of both testosterone and 11-ketotestosterone, decreased testis size and reduced sperm counts compared with unstressed fish. In the female fish, the reduction in the pituitary GTH content may be accompanied by a reduction in circulating oestradiol and a drop in the plasma level of the yolk precursor, vitellogenin. The overall result of this may be that vitellogenesis is impeded, ovarian and egg size reduced, and there is a significant delay in the time of ovulation compared with unstressed fish.

The ultimate effects of stress on reproductive success relate to the influences on egg mortality and fry survival. Such effects have been demonstrated in rainbow trout where there is clear evidence that exposure to environmental stressors during the period of vitellogenesis and gonadal development can lead to a significant reduction in survival of the offspring (60% vs. 85% survival to hatch for eggs of stressed and unstressed fish, respectively).

One of the most noticeable responses of fish following exposure to any form of environmental stressor is a cessation of feeding. This, accompanied by the catabolic effects of the catecholamines and corticosteroids on the energy reserves stored in the body tissues, must result in reduced growth in stressed fish. Cortisol probably exerts an inhibitory effect upon tissue protein synthesis, but the exact ways in which this inhibition and the suppression of growth are mediated remain to be elucidated.

Nucleic acid ratios (RNA:DNA) in the body tissues of fish have been extensively used as indicators of somatic growth. The quantity of RNA is assumed to give a direct reflection of anabolic activity, whilst the concentration of DNA is conservative and represents a standard reference for the number of cells in a given tissue. The RNA:DNA ratios of muscle and liver tissues are markedly influenced by nutritional manipulations, such as starvation and restricted feeding, and hypophysectomy may reduce, and growth hormone replacement therapy restore, tissue RNA: DNA ratios. Thus, the RNA:DNA ratio and/or tissue RNA concentrations may be relatively sensitive indicators of the growth status of the fish. RNA:DNA ratios have been found to be depressed in fish exposed to various types of environmental stressors. For example, exposure of fish to sublethal doses of a range of chemical toxicants, including heavy metals, cyanide, and organic compounds, has been shown to lead to both a fall in tissue RNA:DNA ratio and a marked reduction in growth. The possible links between the HPI axis and the anabolic events occurring in the tissues remain, however, to be elucidated, as do the possible interactions with other endocrine mediators of growth, such as the gonadal steroids, thyroid and growth hormones.

Thus, when fish are exposed to environmental stressors a hierarchy of responses is initiated, and if the stress is severe or long-lasting, succeedingly higher levels of biological organization become affected. The cumulative effects of prolonged exposure to environmental stressors at sublethal levels may affect disease resistance, growth and reproductive success of individuals, which, in turn may reduce recruitment to successive life stages and, eventually, cause population numbers to decline.

Chapter eighteen

Adaptations to the environment – concluding summary

Fish display a wide range of adaptations to meet the challenges imposed by the changes that can occur within the aquatic environment. The main purpose of the preceding chapters has been to describe some of the behavioural, physiological and biochemical adaptations that may occur when fish are exposed to an environmental stress. These may range from simple avoidance reactions, in which the fish swim away from the source of the stressing agent, to the initiation of complex suites of physiological changes requiring the integration and coordination of input signals from a variety of receptor systems.

It has not been the intention to present a detailed catalogue of water quality criteria nor to give a comprehensive list of the tolerance limits of fish for potentially harmful stressors. It has, however, been deemed appropriate to present some information about how tolerance limits and preferences may be determined experimentally. To what extent the experimentally determined tolerance limits have practical application under natural conditions is open to question because fish will rarely be faced by the challenge of a single stressor in isolation. For example, chemical effluents will seldom contain a single toxicant, but will be 'chemical cocktails' of a mixture of potentially harmful substances. Whilst each substance may be present in the mixture at a sublethal concentration, it is possible that interactions between the different components of the mixture may produce deleterious effects resulting in mortality. Similarly, industrial effluents pumped into rivers and other water bodies may differ in temperature from that of the receiving body, with the result that such effluents could impose the simultaneous challenges of 'acute' temperature change, reduced oxygen availability and increased concentrations of chemical toxicants.

Multiple-toxicant, or stressor, experiments are rarely performed, but it is clear that such interactions do exist. For example, the pH of the water may have significant effects upon the oxygen tension (P_C) at which oxygen uptake becomes dependent upon availability. The P_C of carp held at 15 °C, for example, appears to increase with acidification of the water, from about 22 mm Hg at pH 8 to almost 50 mm Hg at pH 4.

A major problem with multiple-stressor studies is that the numbers of possible interactions of this type are myriad. Thus, whilst the multiple-factor approach deserves more attention and exploration it may be extremely difficult to decide which of the possible interactions are likely to produce the most dramatic effects and, therefore, warrant closest attention.

PART THREE: FURTHER READING

Books

Adams, S.M. (ed.)(1990) *Biological Indicators of Stress in Fish* (*Am. Fish. Soc. Symp.* 8), American Fisheries Society, Bethesda, MD.

Alabaster, J.S. and Lloyd, R. (eds)(1980) *Water Quality Criteria for Freshwater Fish*, Butterworths, London.

Cairns, V.W., Hodson, P.V. and Nriagu, J.O. (eds)(1984) *Contaminant Effects on Fisheries*, Wiley, New York.

Cossins, A.R. and Bowler, K. (1987) *Temperature Biology of Animals*, Chapman & Hall, London.

Hoar, W.S. and Randall, D.J. (eds)(1970) *Fish Physiology*, Vol. IV, Academic Press, London.

Hoar, W.S. and Randall, D.J. (eds)(1971) *Fish Physiology*, Vol. VI, Academic Press, London.

Hoar, W.S. and Randall, D.J. (eds)(1984) *Fish Physiology*, Vol. XA, Academic Press, London.

Mason, B.J. (1992) *Acid Rain: Its Causes and its Effects on Inland Waters*, Clarendon Press, Oxford.

Morris, R., Taylor, E.W., Brown, D.J.A. and Brown, J.A. (eds)(1989) *Acid Toxicity and Aquatic Animals*, Cambridge University Press, Cambridge.

Pickering, A.D. (ed.)(1981) *Stress and Fish*, Academic Press, London.

Rand, G.M. and Petrocelli, S.R. (eds)(1985) *Fundamentals of Aquatic Toxicology*, Hemisphere, Washington, DC.

Rankin, J.C. and Jensen, F.B. (eds)(1993) *Fish Ecophysiology*, Chapman & Hall, London.

Satchell, G.H. (1991) *Physiology and Form of Fish Circulation*, Cambridge University Press, Cambridge.

Review articles and papers

Andersson, T. and Förlin, L. (1992) Regulation of the cytochrome P450 enzyme system in fish. *Aquat. Toxicol.*, **24**, 1–20.

Beitinger, T.L. (1990) Behavioral reactions for the assessment of stress in fishes. *J. Great Lakes Res.*, **16**, 495–528.

Beitinger, T.L. and McCauley, R.W. (1990) Whole-animal physiological processes for the assessment of stress in fishes. *J. Great Lakes Res.*, **16**, 542–75.

Bols, N.C., Mosser, D.D. and Steels, G.B. (1992) Temperature studies and recent advances with fish cells *in vitro*. *Comp. Biochem. Physiol.*, **103A**, 1–14.

van der Boon, J., van den Thillart, G.E.E.J.M. and Addink, A.D.F. (1991) The effects of cortisol administration on intermediary metabolism in teleost fish. *Comp. Biochem. Physiol.*, **100A**, 47–53.

Buhler, D.R. and Williams, D.E. (1988) The role of biotransformation in the toxicity of chemicals. *Aquat. Toxicol.*, **11**, 19–28.

De Vries, A.L. (1988) The role of antifreeze glycopeptides and peptides in the freezing avoidance of Antarctic fishes. *Comp. Biochem. Physiol.*, **90B**, 611–21.

Eddy, F.B. and Williams, E.M. (1987) Nitrite and freshwater fish. *Chem. Ecol.*, **3**, 1–38.

Goss, G.G., Perry, S.F., Wood, C.M. and Laurent, P. (1992) Mechanisms of ion and acid–base regulation at the gills of freshwater fish. *J. Exp. Zool.*, **263**, 143–59.

Hazel, J.R. and Williams, E.E. (1990) The role of alterations in membrane lipid composition in enabling physiological adaptation of organisms to their physical environment. *Progr. Lipid Res.*, **29**, 167–227.

Hodson, P.V. (1988) The effect of metal metabolism on uptake, disposition and toxicity in fish. *Aquat. Toxicol.*, **11**, 3–18.

Jobling, M. (1981) Temperature tolerance and the final preferendum – rapid methods for the assessment of optimum growth temperatures. *J. Fish Biol.*, **19**, 439–55.

Kramer, D.L. (1987) Dissolved oxygen and fish behaviour. *Env. Biol. Fishes*, **18**, 81–92.

Lewis, W.M., jun. and Morris, D.P. (1986) Toxicity of nitrite to fish: A review. *Trans. Am. Fish. Soc.*, **115**, 183–95.

Liem, K.F. (1989) Respiratory gas bladders in Teleosts: Functional conservatism and morphological diversity. *Am. Zool.*, **29**, 333–52.

Motais, R., Borgese, F., Fievet, B. and Garcia-Romeu, F. (1992) Regulation of Na^+/H^+ exchange and pH in erythrocytes of fish. *Comp. Biochem. Physiol.*, **102A**, 597–602.

Perry, S.F. (1986) Carbon dioxide excretion in fishes. *Can. J. Zool.*, **64**, 565–72.

Perry, S.F. and Wood, C.M. (1989) Control and coordination of gas transfer in fishes. *Can. J. Zool.*, **67**, 2961–70.

Pickering, A.D. (1992) Rainbow trout husbandry: management of the stress response. *Aquaculture*, **100**, 125–39.

Pickering, A.D. and Pottinger, T.G. (1989) Stress responses and disease resistance in salmonid fish: Effects of chronic elevation of plasma cortisol. *Fish Physiol. Biochem.*, **7**, 253–8.

Smatresk, N.J. (1988) Control of the respiratory mode in air-breathing fishes. *Can. J. Zool.*, **66**, 144–51.

Thomas, S. and Perry, S.F. (1992) Control and consequences of adrenergic activation of red blood cell Na^+/H^+ exchange on blood oxygen and carbon dioxide transport in fish. *J. Exp. Zool.*, **263**, 160–75.

van den Thillart, G. and van Waarde, A. (1985) Teleosts in hypoxia: Aspects of anaerobic metabolism. *Molec. Physiol.*, **8**, 393–409.
van Waarde, A. (1991) Alcoholic fermentation in multicellular organisms. *Physiol. Zool.*, **64**, 895–920.
Wendelaar Bonga, S.E. and Lock, R.A.C. (1992) Toxicants and osmoregulation in fish. *Neth. J. Zool.*, **42**, 478–93.

Author index

Adams, S.M. 292
Addink, A.D.F. 293
Adelman, I.R. 165
Alabaster, J.S. 292
Anderson, D.P. 88
Andersson, T. 293
Arnesen, A.M. 206

Baardvik, B.M. 206
Bagenal, T. 205
Barker, G. 206
Barton, B.A. 88
Beamish, F.W.H. 206
Bedford, M.R. 88
Beitinger, T.L. 293
Beitz, D.C. 206
Bell, J.G. 88
Blaxter, K. 205
Blazer, V.S. 88
Bols, N.C. 293
Borgese, F. 293
Boujard, T. 205
Bowler, K. 205, 292
Brafield, A.E. 205
Brett, J.R. 205–6
Bromage, N. 206
Brown, D.J.A. 292
Brown, J.A. 292
Brown, M.E. 89
Buhler, D.R. 293

Cairns, V.W. 292
Calow, P. 205
Camire, A. 88
Camire, M.E. 88
Campbell, G.L. 88
Christiansen, J.S. 177, 178, 206
Cossins, A.R. 205, 292
Cowey, C.B. 88
Davies, B. 206
Davies, P.S. 140
Davison, W. 206
De Vries, A.L. 293
Duncan, A. 205
Duston, J. 206

Eddy, F.B. 293
Elliott, J.M. 101
El-Sayed, A-F.M. 88
Evans, L.V. 88

Fievet, B. 293
Forlin, L. 293
Forster, R.P. 206

Garcia-Romeu, F. 293
Gerking, S.D. 205
Gill, H.S. 205
Goldstein, L. 206
Goolish, E.M. 206
Goss, G.G. 293
Greene, D.H.S. 88
Grodzinski, W. 205

Halver, J.E. 88
Hara, T.J. 88
Hardy, R.W. 89
Hawkins, A.J.S. 206
Hazel, J.R. 293
Henderson, R.J. 88
Hilton, J.W. 89
Hoar, W.S. 205, 292
Hodston, P.V. 292–3
Houlihan, D.F. 206
Huntingford, F.A. 205

Iwama, G.K. 88

Jensen, F.B. 205, 292
Jobling, M. 140, 177, 178, 195, 197, 206, 293
Jones, J. 206
Jørgensen, E.H. 178, 206

Kamler, E. 205
Kelly, J.M. 206
Klekowski, R.Z. 205
Kramer, D.L. 293
Krumhar, K. 88

Lands, W.E.M. 89
Laurent, P. 293
Leatherland, J.F. 206
Lewis, W.M., jun. 293
Liem, K.F. 293
Llewellyn, M.J. 205
Lloyd, R. 292
Lobley, G.E. 206
Lock, R.A.C. 293

McBride, B.W. 206
McCauley, R.W. 293
Mackie, A.M. 88
Mason, B.J. 292
Miglavs, I. 197
Millward, D.J. 206
Mommsen, T.P. 206
Morris, D.P. 293
Morris, R. 292
Mosser, D.D. 293
Motais, R. 293
Moyle, P.B. 205

Nriagu, J.O. 292

Pedersen, T. 195
Perry, S.F. 293
Petrocelli, S.R. 292
Pickering, A.D. 89, 292–3
Pottinger, T.G. 293

Rand, G.M. 292
Randall, C. 206
Randall, D.J. 205–6, 292
Rankin, J.C. 205, 292
Reeds, P.J. 206

Sandnes, K. 89
Sardesai, V.M. 89
Satchell, G.H. 292
Schreck, C.B. 205
Selivonchick, D.P. 88
Shearer, K.D. 89
Smatresk, N.J. 293
Springate, J. 206
Steels, G.B. 293

Taylor, E.W. 292
Teshima, S-I. 88
Thomas, S. 293
Thorpe, J.E. 205
Thrush, M. 206
Tocher, D.R. 88
Torrissen, O.J. 89
Tytler, P. 205

van den Thillart, G.E.E.J.M. 293–4
van der Boon, J. 293
van Waarde, A. 293

Walsh, P.J. 206
Weatherley, A.H. 205
Wendelaar Bonga, S.E. 293
Williams, D.E. 293
Williams, E.E. 293
Williams, E.M. 293
Wilson, R.P. 88
Woiwode, J.G. 165
Wood, C.M. 293
Wootton, R.J. 205
Wright, P.A. 206

Species index

SCIENTIFIC NAMES

Actinopterygii 242
Amia calva *239*, 242, 244
Ammodytes marinus *62*
Amphipnous cuchia 242
Anabas spp. *239*, 242
Ancistrus chagresi *239*, 244
Anguilla anguilla 182–4, *239*, 241
Anguilla rostrata 14
Arapaima gigas *239*

Boreogadus saida *62*
Brachiopterygii 242
Brachydanio rerio 103

Carassius auratus *118*, 214
Carassius carassius 258, 262
Channa striatus 184
Chanos chanos 64, 182
Clarias spp. *239*
Clupea harengus *58*
Colossoma mitrei 175
Ctenopharyngodon idella 64
Cyprinus carpio *36*, 54–5, 182, 221, 226, *257*

Dicentrarchus labrax *100*, 138
Dipneusti 240, 242

Electrophus electricus *239*, 242
Eremophilus mutisii *239*, 242
Esox lucius 139
Euphausiids *14*, 73

Fundulus heteroclitus 215

Gadus morhua *128*, 138, 195, 215, 229, 248–9
Gadus ogac 229
Genyagnus monopterygiust *257*
Gymnotus carapo 243

Hemitripterus americanus 229
Heteropneustes fossilis 184
Hoplerythrinus unitaenitus 243
Hoplosternum spp. *239*
Hypopomus brevirostris 241
Hypopomus spp. *239*
Hypostomus plecostomus 244

Ictalurus nebulosus 214, 216
Ictalurus punctatus 35, *54*, *108*, 164, 167–8, 189, 264

Leiopotherapon unicolor *258*
Lepidosiren paradoxa *239*, 243
Lepisosteus osseus 242, 244
Lepisosteus spp. *239*, 242
Lepomis cyanellus 166, 176, 223, *226*
Lepomis macrochirus *118*, *257*
Limanda limanda *128*

Macropodus spp. 242
Macrozoarces americanus 229
Mallotus villosus *14*, *62*
Melanogrammus aeglefinus *128*, 193
Microgadus tomcod *229*
Micromesistius poutassou *62*
Micropterus salmoides *100*, 139, *258*
Misgurnus spp. *239*
Morone chrysops 157, 164–5, 181
Morone saxatilis 157, 164–5, 181, 224, *226*
Myoxocephalus scorpius 229

Neoceratodus forsteri *239*, 243

Oncorhynchus kisutch 167
Oncorhynchus mykiss *37–8*, *54*, *61*, *65*, *100–1*, 102, *109*, *118*, 139, 152, *157*, 167, 175, 193–4, 200–1, 219, 223, 249, 257, 289
Oncorhynchus nerka *100*, *128*, 155, *157*
Oncorhynchus tshawytsca *36*, 161
Oncorhynchus gorbuscha *157*
Ophiocephalus spp. *239*
Ophiocephalus striatus 242
Ophiodon elongatus *257*
Opsanus tau *257*
Oreochromis mossambicus 182
Oreochromis niloticus *36*, *54*, *108*, 182, *257–8*
Osmerus mordax 228

Pagrus major 40
Paralichthys olivaceus 182
Perca fluviatilis *14*
Periophthalmus spp. *239*, 241
Piaractus mesopotamicus 175, 237–8
Pimephales promelas 236
Platichthys flesus *128*
Plecoglossus altivelis 67
Pleuronectes platessa *100*, 140, *257*
Poecilia reticulata 175, 193, 237
Protopterus annectens 247
Protopterus spp. *239*, 243
Pseudopleuronectes americanus 193, 228–9

Rhodeus amarus 262

Salmo salar 103, 152, 173, 175, *178*, 190, 197
Salmo trutta *100–1*, *108*, 139, *157*
Salvelinus alpinus *100*, 152, *157*, 163, 174, 177–8, 197, *257*
Salvelinus fontinalis 193, 235, *257*
Scomber scombrus *62*
Scyliorhinus canicula 250
Sprattus sprattus *62*
Synbranchus marmoratus 241

Tilapia aurea 182
Tilapia nilotica 182
Trichogaster spp. 242
Trisopterus esmarkii *62*

COMMON NAMES

African lungfish 247
American eel 14
Anchovy 61
Arctic charr *100*, 152, *157*, 163, 174, 177–8, 197, *257*
Argentine 60
Asiatic catfish 184
Atlantic salmon 103, 152, 173, 175, *178*, 190, 197
Australian lungfish 243, 246
Ayu 67

Banded knife-fish 243
Bichirs 242
Bitterling 262
Blue whiting *62*
Bluegill sunfish 118, *257*
Bowfin 242, 244, 246
Brisling *62*
Brook trout 193, 235, *257*
Brown bullhead 214, 216
Brown trout *100–1*, *108*, 139, *157*

Capelin 60
Carp *36*, *54–5*, 182–4, 221, *226*, 255, *257*
Catfish *239*, 244
Channel catfish 35–*6*, *54–5*, 64, 80, 83–4, *108*, 164, 167–8, 189, 264
Characids 237
Chinook salmon *36*, *54*, 161
Cichlids 108
Clupeids 68
Cod *128*, 138, 200, 215, 217, 219
Coho salmon 167
Columbian catfish *239*, 242
Crucian carp 258, 262

Crustaceans 271
Cyprinids 31, 35, 68, 87, 108, 157, 191, 196

Dab 128

Eel 182–4, 241
Electric eel *239*, 242

Fathead minnow 236
Flounder 128

Gadoids 60, 68, *108*, 196
Gar 242, 244
Gobiids 108
Goldfish *118*, 214, 217, 221, 223, 262
Grass carp 64
Green sunfish 166, 176, 223, *226*
Guppy 175, 193, 237

Haddock *128*, 193
Herring 58

Japanese flounder 182

Knifefish 241

Largemouth bass *100*, 139, *258*
Lingcod *257*
Lungfish 240, 242, 244

Mackerel *62*
Milkfish 64, 182
Molluscs 271
Mudskipper 241
Mummichog 215

Norway pout *62*
Notothenids 229

Ocean pout 229

Pacu 175, 237–8
Perch 14
Percids 31, *108*
Pike 139
Pink salmon *157*
Plaice *100*, 140, *257*
Pleuronectids 108
Polar cod *62*
Pomacentrids 108

Rainbow smelt 228
Rainbow trout *37–8*, 59, *61*, *65*, *100–1*, 102, *109 118*, 139, 152, *157*, 167, 175, 193–4, 200–1, 219, 223, 249, 257, 289
Reedfish 242
Rock cod 229

Salmonids 31, 35, 68, *72*, 80, 84, 87–8, *108*, 152, 157, 161, 168, 170, 173, 176–7, 183–4, 191, 196, 201, 264, 289
Sandeel 62
Sea bass *100*, 138
Sea bream 40, 72
Sea raven 229
Serrasalmids 237–8
Short-horned sculpin 229
Snake-head 184, 242
Sockeye salmon *100*, *128*, 155, 157
South American catfish 244
South American lungfish 243
Spangled perch 258
Spotted dogfish 250
Stargazer 257
Stichaeids 108
Striped bass 157, 164–5, 181, 224, *226*
Swamp eel 241–2

Tilapia 31, 35, *36*, 67–8, 72, 87, *108*, 182–4, *257–8*
Toadfish 257
Tomcod 229
Trahira 243

Winter flounder 193, 228–9

Zebrafish 103

Subject index

Page numbers in **bold** represent figures and those in *italic* represent tables

Absorption efficiency (AE) 104–5
  determination 105–7
  factors influencing 107–9
  feed ingredients *108*
  fish species *108*
Acclimation
  and hypoxia tolerance 262
  resistance 209
  to temperature change 125–6, **127**, 219–30
Acetyl CoA (acetyl coenzyme A) 26–**9**
Acid rain 275–6
Acid–base balance
  in acidified water 116, 274
  in anaerobiosis 253–4
  and hypercapnia 263–4
  and strenuous exercise 137
Acidification, freshwater 274–7
ACTH (adrenocorticotrophic hormone) **84**, 286–7
Acute response to temperature **126**, **127**
Acute thermal preferenda 216
Aerial respiration 247–8
Aflatoxins 59
Aggression 153, 173, 176, 178
Air breathing
  facultative 239
  obligate 239
Air breathing fish 238–9
  receptor systems 244–8
Air breathing organ (ABO) *239*, 241–4
  circulatory system **240**
  mechanoreceptors 247–8
Aldose, structure **6**
Algae 4, 9–10, 73
  and oxygen 234
Alginates, as binders 58, 60
Alginic acid 9
Allergens 65
Aluminium
  and freshwater acidification 276–7
  toxicity 269, 277
Aluminosilicates 59
Amino acids
  chemical structures **23**
  in citric acid cycle **28**
  classification 21–2
  deamination 109–10, **112–13**, 278, 288
  dietary interactions 77–9
  essential 31–2, 37, 141
  interactions 36
  metabolism in hypoxia 261
  nonessential 32
  requirements 31–6, 37
  urea formation 119
Ammonia
  diffusion 111
  excretion 95, 111, 114, 278–9
  nitrification 110, 281–2
  toxicity 268, 280
Amylopectin, chemical structure **8–9**
Amylose, chemical structure **8**
Anabolic steroids 164, 184–7
Anaerobic metabolic pathways 257–9, **260**
Androgens 19, 184
Animal by-product meals, feed ingredient *61*
Anoxia 234
Antibiotics, feed ingredient 74–5

Antifreeze glycoproteins (AFGPs) 229–30
Antifreeze peptides (AFPs) 229–30
Antimicrobial agents, feed ingredient 73–4
Antinutrients, in soybean meal 64
Antioxidants
feed ingredient 62, 72
and lipid peroxidation 81–2, 273
Aquatic surface respiration (ASR) 236–7, 238
Ascorbic acid
chemical structure **43**
and disease 83–4, 86
and heavy metals 80
and lipid peroxidation 70, 81, 273
requirements 42–3
and stress 83–4, 287
Assimilation efficiency (AsE) 105
ATP (adenosine triphosphate) 24–5, 27–8, 131–2, 135–6
generation 224
hydrolysis 261
production 257–9, **260**
and protein degradation 143–4
Atresia 194
Attractants, feed ingredient 74

Bacterial fermentation 63
Bacterial kidney disease (BKD) 84–6
Bentonites 59
Beta-carotene 49
chemical structure **50**
Beta-oxidation spiral 27–8, **29**
Binding agents
as feed ingredient 57
and feed production 58–61
Bioaccumulation, and toxicity 268–71
Biomass production 175
Biotin 45
chemical structure **47**
Blood acidosis
in acidified water 274–5
in hypoxia 254–5
in strenuous swimming 132–3, 134
Body size effect 171–**2**
Bohr effect 275
Bradycardia 249–50, 253
Broken line response 33

Cadmium, toxicity 272
Calcium 76, 80, 276–7
and acidified water 274
and heavy metals 269
and phosphorus 76
requirement 53–4
transport 277
water content 267
Canola meal, feed ingredient 64, 76
Capacitance coefficient 233
Carbohydrates
as binders 57–8
chemical structures 5–10
classification 5
digestibility 58, 75
metabolism 87
Carbonic anhydrase 134
Cardiac output
in hypoxia 250
in strenuous exercise 134, 138
Cardio-respiratory synchrony 250
Carotenoids 19, 72–3
Catecholamines
and hypoxia 251, 253–5
and red blood cells 275
and strenuous exercise 132–3, 134
and stress 83, 287–9
Cellular maintenance functions 143
Cellulose 106
chemical structures **8**
Chemical composition
amino acid analyses 66
fatty acid analyses 69
fish species *62*
foodstuffs 3–4, 5–21, 57–76
prey organisms *58*
Chemoreceptors, role in hypoxia 245–6
Chitin 8–9, 106
Chloride cells
in acidified water 274

Chloride cells *contd*
  and heavy metals 269
  and hypercapnia 264
  in ion-poor water 54, 274
  and toxicants exposure 262
Cholecalciferol 51, 79
  chemical structure **52**
Cholesterol 287
  chemical structure **18–19**
Choline
  chemical structure 43
  requirements 44
Chromic oxide 106
Citric acid cycle 24–**5**
  amino acids, entry points **28**
  biochemical changes **27**
Clonidine 182
CoA (coenzyme A), chemical structure **48**
Cobalt, requirement *54*
Compensation
  complete 220
  inverse 220
  partial 220
Condition factor 195
Copper
  requirement *54*
  toxicity 80, 269, 272
Cori cycle 135
Corn gluten meal 85–6
Corn meal 59
  digestibility *108*
Corticosteroid hormones 19
Corticotropin releasing factors (CRFs) 286–7
Cortisol 17–**18**, 19, 83, 86, 286–9
Cottonseed meal 59, 85–6
Creatine 111, 114
Creatine phosphate 131–2
Critical oxygen tension (P_c) 253, *256–8*, 292
  see also Oxygen tension
Critical swimming speed 130–1
Critical thermal maximum (CTM) **214**, 216
Crustaceans *4*, 271
Cyanocobalamin 49, 79
  chemical structure **46**
Cytochrome P-450 system 271

Deficiency symptoms 82–3
Detoxification
  enzymes 271–2
  metallothioneins 272
  nitrite 283
Dietary requirement determination **33**
Dietary requirement studies 35
Digestibility
  carbohydrates 58, 75
  determination methods 105–*7*
  food nutrients 104
  markers 106
  proteins *61*, 105
Diseases
  and nutrition 67
  and stress 288–9
Dominance hierarchies 101, 153
  size-related 171
  and sustained exercise 178
Dosage-mortality curves **211**–12
Dose-response curves 33
  analysis 34–**5**
Drugs, feed ingredient 74–5

Eggs 187–**8**, 191
Eicosanoids 17, 67–8
Energy balance equation 155, 169
Energy budget equation 93–7
Energy costs 121
Energy metabolism 22–9
Enkaphalin analogue 182
Enzyme activity ratio 225
Enzymes
  adaptations to temperature 221, 225–30
  ATPases 143
  and detoxification 63, 71, 75–8
  glycolytic 24, 226
  oxidative 24
  in pet foods 76
  vitamin destruction 69
Ergocalciferol 51, 79
  chemical structure **52**

Ethanol 262
Excretion
  definition 99
  losses 104–9
Fatty acids
  beta-oxidation spiral **29**
  chemical structure **11**–12
  classification *13*
  elongation and desaturation 38–**9**, 40
  environmental compositions *14*
  and environments 12
  essential 41, 66–7, 87
  in farmed fish 69
  phospholipid profile 15
  requirements 37–41
  and salinity 15–16
  species differences 87
  and thermal acclimation 223
Fecundity 193–4
Feed contaminants 59
Feed types
  coatings 71
  compressed pellets 57, 59, 194
  digestibility 107
  extruded pellets 57, 60
  formulated 57
  moist pellets 60–1, 64
  protein sources 61–6
  wet feeds 60–1
Feeding
  factors influencing 99–104
  frequency 102–3
  restricted 152–4, **172**, 194–**6**
  to excess 150–1
  to satiation *100*, 151–2
Feeding behaviour 100–4
Final temperature preferendum 216–17
Fish oil, feed ingredient 61–*2*
Fish silage
  feed ingredient 60
  production 63–4
Fishmeal 80
  digestibility *108*
  feed ingredient 60–*1*, *62*, 64
  production 62
Flavin adenine dinucleotide (FAD) 44
Flavin mononucleotide (FMN) 44
Folic acid 48, 79
  chemical structure **46**
Food availability 145
Food conversion, and exercise **177**
Food intake
  and body weight *100*
  and competition 102–3, 171, 173–4
  and daylength 104, 163–**5**, 166
  and distribution 173, 178
  and exercise *178*
  metabolic rate increase 139–**40**, 141
  and nutritional status **149**, **197**–201
  and stress 289
  and temperature 100–*1*, **156**–7, **159–60**
Foodstuffs, chemical composition 3–*4*, 5–21
Freezing avoidance 227–30
Fumarate 26, **28**

Galactans 9
Gas
  absorption coefficient 232
  capacitance coefficient 233
  coefficient of solubility 232
Gas bladder, and air breathing **240**, 242–4
Gas supersaturation 264–6
General adaptation syndrome (GAS) 285
Gill
  blood flow resistance 134
  epithelium and ammonia 280
  epithelium and toxicants 268-9
  gas-exchange in hypoxia 137, 219, 249–51
  ion-exchange mechanisms **265**
  lamellar recruitment 132, 251
  and nitrogenous excretion 110–11, 114
  oxygen diffusion and tension 167
  and oxygen tension 23–4

Gill *cont*
perfusion in hypoxia 249–51
protein turnover 178
secondary lamellae 241, **252**
ventilation
in hypoxia 137, 239–41, 249–51
inhibition 248
in strenuous swimming 132–5
Gluconeogenesis 25, 261–2, 288
Glucose 258–9
derivatives **7**
metabolism 6–**7**, 8–9
Glucuronic acid **7**
Glutamine 279
Glycans 6, 9
Glycerol 10
chemical structure **11**
and temperature 228, 230
Glycinin 65
Glycogen 8–9, 25, 132, 135–**6**, 224–5, 258–9
Glycogenolysis 288
Glycolysis 24
Glycolytic enzymes, and thermal acclimation *226*
Glycolytic pathway **26**
Gonadal steroids 19
Gonadosomatic index (GSI) 190
Growth
and behavioural factors 171, 173–6
and body size 147, 169–71
compensatory 194–201
hormonal manipulation 179–89
optimum temperature 158–7
photoperiodic effects 163–**5**, 166
relative 170, **172**
reproductive 190–4
restriction 153
RNA:DNA ratio 181, 289–90
and salinity 155, 160–**2**, 163
selection programmes 189–90
and social interactions 171, 173–4, 176
somatic 190–4
stocking density effects 174–6
and sustained exercise 176–*8*, 179
temperature effects **156**–*7*, **158–60**
Growth hormone 164, 180–2
Growth rate
absolute 148
equation 150
essential fatty acids 67
instantaneous 148
specific 148, 170
Growth responses 33–4
Growth studies 32
Growth-ration curves, interrelationships 171–**2**
Growth-rations curve 148–**9**, 157–**8**, 160–1, **162**–3
Growth-rations relationships 153

Haemoglobin **133**–4, 245
Heart disease, high fat diet 67–8
Heart rate
in hypoxia 249
and strenuous exercise 134
Heat-shock proteins 227
Heavy metals
bioaccumulation 270
and gill structure 269
toxicity 269–72
Hering–Breuer reflex 247
Highly unsaturated fatty acid, *see* HUFA
Homeoviscous adaptation 222–3
HPI axis 83–**4**, 287, 290
HUFA 12, 16–17, 37–8, 40, 51, 69, 88
formation of long-chained **39**
human health 67–8
and thermal acclimation 223–4
Hydroperoxide radicals 81
Hypercapnia 263–4
Hyperphagia 197, 200
Hypothalamic–pituitary–interrenal axis, *see* HPI axis
Hypothermia 217
behavioural 219
Hypoxia
and acclimation 262
and anaerobiosis 244–6, 253–4, 258–62
avoidance reactions 217, 219, 235

and cardiovascular system 239–41, 253–5
and gill ventilation 137

Immune system
immuno-suppressive effects of stress 289
potentiators 86
stimulation 86
Immune-inflammatory disease 68
Incipient lethal level 212, 213–**14**, 215
Industrial atmospheric emmisions 275
Ingestion, *see* Food intake
Insulin 183–4
Iodine, requirement *54*
Ionic balance 54–5
Iron 80–1
and immune system 85
and lipid peroxidation 81
requirement *54*
Isosmotic hypothesis 161

Ketose, structure **6**
Krill *14*, 73

Lactate 24, 131–2, 135–**6**, 253, 261
Lactate: pyruvate ratio 257
Lactic acid 63
Laminarin 10
LD_{50} (median lethal dose) **210–11**, 212
Lethal limits 209
Leukotrienes 67
Lipid-soluble vitamins 49–53
Lipids
chemical structures 10–**11**, 12–20
deposition 147
feed ingredients 59–60, 66–9
oxidation 95
peroxidation 81–**2**, 272–3
requirements 37–41
synthesis 142
in temperature acclimation 222
Liposomes 86
Lipoxygenase 17
LT_{50} (median lethal time) **210**–11
LT-process 62

Macronutrients and disease 85–6
Magnesium
absorption 80
requirement *54*–6
water content 267
Maillard reaction 66
Manganese
absorption 80
requirement *54*
Mechanoreceptors, role in hypoxia 246–7
Membrane structure
and fluidity 222
phospholipids 15–17
and temperature changes 222–4
and toxicants 269
Mercury, and human health 270
Metabolic costs 94
Metabolic intensities 124
Metabolic rate
active 121–*2*, 123–8
and body size 124, 169–71
conforming 255
and dissolved oxygen 94–6, 166–8, 255–*8*
fasting 122–3, 142–3
and feeding 139–42, 199
field metabolic rates 137–9
increase after food intake 139–**40**, 141
nutritional history effects 142–5
regulators 255
reorganization in thermal acclimation 224–5
resting 122
routine 122
and salinity 161
and sustained activity 176–9
swimming activity 128–37, **136**
and temperature 125–*8*, 160
terminology definitions *122*
and thermal acclimation 220–1
Metabolism
aerobic 96, 121

Metabolism *contd*
  anaerobic 96, 261
    in burst swimming 130–1
    iin hypoxia 253–5, 258–62
  as heat loss 121
  products 94
  standard 123
  and stress 287–8
Metallothioneins 272
Methaemoglobin 283
Methionine 64–5, 78
Methyltestosterone 184–**8**
Minerals
  bioavailability 53–6, 76, 80
  dietary overfortification 76
  and disease 85
  requirements 53–*4*, 55–6
Minimal metabolism, *see* Metabolic rate, fasting
Mitochondria volume increase 221
Mitochondrial enzymes, and thermal acclimation *226*
Molluscs 4, 271
Muscle
  depletion during reproduction 193
  enzymes and temperature 225–30
  hormonal control of protein synthesis **179**
  hypertrophy 177
  metabolism
    activity estimation 138
    in strenuous exercise 131–2
  protein turnover 178
  repletion on refeeding 198
  structural changes and temperature 221–2
Myo-inositol
  chemical structure **43**
  requirements 43–4

Niacin 45, 48
  chemical structure **46**
Nitrification 281–2
Nitrogenous excretion 105
  ammonia 95, 110, 114, 278–9
  and body size *118*
  endogenous 118
  exogenous 116
  products 109–19
  and temperature *118*
  urea 114–**15**, 116–**17**
Nutrient classes 3–21
Nutrient concentration, dose-response data **34**
Nutrient interactions
  amino acid 36, 77–9
  mineral 80
  vitamin 79–80
Nutrition
  and disease 82–6
  and metabolic reorganization 224–5

Oestradiol 19
Oestrogens 19
Oilseed meals, feed ingredient 64
Oocytes 191, 193–4
Optomotor response 129
Ornithine-urea cycle 114, 116–**17**
Overcompensation 20
Oxidative phosphorylation 24
Oxycalorific coefficient 95–*6*, 121
Oxygen
  dissolved 233–6
  and food intake 101
Oxygen consumption, *see* Metabolism; Metabolic rate
Oxygen debt 135
Oxygen extraction, in normoxia and hypoxia 237–8
Oxygen receptors 235, 244–5, 248
  stimulation 249
Oxygen saturation 167–**8**
Oxygen tension 231, 233–4
  and consumption **256**
  habitat and lifestyle *257*
  see also Critical oxygen tension (P_c)

PAHs (polycyclic aromatic hydrocarbons) 271
Pantothenic acid 45
  chemical structure **47**
Parasites, drugs in feed 75
Pasteur/Crabtree effect 258

PCBs (polychlorinated biphenyls) 271, 273
Peptide linkage formation **20**
Pharyngeal lungs 242
Pheromones 171
Phosphatidyl choline, chemical structure **11**
Phospholipids 68
  in cell membranes 37, 222–3
  chemical structure **11**, 15
  classification 15
Phosphorus 76
  dietary 80
  requirements 53–*4*
Photostimulation 164–**5**, 166
Phytate 76
Phytic acid 64, 80
Phytoplankton *4*
Pigments, feed ingredient 72–3
Pituitary 164, 180
Pituitary gonadotrophin (GTH) 289
Planes of nutrition 142
Plant oils, feed ingredient 64
Polyunsaturated fatty acid, *see* PUFA
Potassium 276
Poultry by-products, digestibility *108*
Progesterone 19
Prostacyclins 67
Prostaglandins 17, 67
Protein concentrates 65
Protein metabolism 141–2
Protein synthesis 143–4, 177–8, 189
Proteins
  chemical structure 20–2
  digestibility 105
  in feeds 61–6
  nutritional requirements 31–7
Protons (H^+) exchange
  gill epithelium 134
  gill excretion 264
  red blood cells **133**–4
PUFA 12, 15–*16*, 40
  lipid destruction **82**
  and thermal acclimation 223–4
Purine nucleotide cycle 79, 110, **112–13**
Pyridoxine
  chemical structure **47**
  and disease 84–5
  requirement 45

Q_{10} 126–8

Ration
  maintenance 148–**9**
  maximum **149**, **159**, **162**
  optimum **149**, **159**
Red blood cells, in acidified water 275
Repartitioning agents 187–9
Reproduction
  resource allocation 191–2
  stress 190
  and stress 288
Resistance acclimation 209
Respiration, *see* Metabolism; Metabolic rate
Respiratory quotient (RQ) 95–*6*, 121
Respiratory substrates 87
  and amino acids **260**
  ATP yields 24–**5**, 135, 224
  energy content 119
  lipid role 66–7
  oxycalorific coefficients 94–*6*
Respirometers **129**
Respirometry 123
Retinol 49
  chemical structure **50**
  and disease 84
Rhodopsin 49
Riboflavin 44
  chemical structure **47**
Ribosomes 144
Rice bran, feed ingredient 76
RNA 144, 271
RNA:DNA
  growth indicator 181
  and stress 289–90
Root shift 132, 275
RQ, *see* Respiratory quotient

Saturated fatty acid 10
Seaweeds 9–10
Selenium
  and lipid peroxidation 81

Selenium *contd*
requirement *54*
Sensitivity analysis 20
Size-effect studies 169–71
Social interactions
and food intake 102–3, 171, 173–4
and growth 153, 171, 173–4, 176
Sodium pump 143
Soybean meal 80
digestibility *65*, *108*
feed ingredient 64
preparation 65
Space effect 176
Sphingomyelin 15, 222
Spleen 254
Squid *14*, 69, 74
as feed ingredient 60
Starch
chemical structures **8**
in feeds 58, 60
Starvation 144, 152, 195–6
Steroid hormones
anabolic 164, 184–7
chemical structures 19–20
metabolism 20
see also under hormone name
sex reversal 184–**6**
stress response 83, 289
Stress
and catecholamines 84, 289
and corticosteroids 84
and disease 288–9
and growth 289–90
and immune system 289
metabolic effects 287
and reproduction 288–9
Stress responses 19, 83, 285–6
endocrine system 286–7
tertiary 288–90
tissue alteration 287–8
Stressor, definition 285
Stroke volume 138
Swimming activity
burst swimming 131
critical swimming speed 130–1
and growth 176–7, *178*–9
metabolic rate 128–37
prolonged 130
sustained 130, 131–2, 176–9
and temperature 220–3

Telemetry 137
Temperature
compensation 220
and digestibility *109*
and dissolved oxygen 217
effects on growth **156**–*7*, **158–60**
and food intake 100–*1*
and metabolic rate 125–*8*, 140, 160
metabolic reorganization 224–5
optimum for growth 156–*7*
and oxygen tension 256, *258*
preference studies **218**
and swimming activity 220
tissue adaptations 221–4
tolerance 158–**9**, 213–19
tolerance polygon 213–**14**, 215
Temperature acclimation 125–**6**, 219–30
and cell membranes 222–4
and enzymatic adaptations 225–30
and swimming performance 222
Temperature coefficients 125–**7**
Terpenoids 17
Territorial behaviour 173–4
Testosterone 19, 184
in stress 289
Thermal guilds 219
Thermal hysteresis 229
Thermal tolerance zone 213
Thiamin 44
chemical structure **47**
deficiency 69
Thromboxanes 67–8
Thyroid hormones 164, 182–3
Tocopherols
and disease 51, 84–6
and lipid peroxidation 51, 81, 273
requirement 51
Tolerance limits 291
Tolerance limits measurements **210**

Toxicity
  ammonia 268, 280
  avoidance 248
  metals 268
  nitrite 282–3
Toxicity tests 267
Transport systems, amino acids 77–9
Triacylglycerol 68–9
  chemical structure **11**
TSH (thyroid stimulating hormone) 183

Unsaturated fatty acids 12
Urea 111, 114, 116, 278–9
  production **117**
Urea pathway **115**
Uric acid 95
Uricolysis 114–**15**

Ventilation frequency 138
Ventilation volume 250
Vitamin A 49–51
  chemical structure **50**
  and disease 84–5
Vitamin C, *see* Ascorbic acid
Vitamin D 51, 79–80
  chemical structure **53**
Vitamin E, *see* Tocopherols
Vitamin K 51, 53
  chemical structure **53**
Vitamins
  and disease 69, 83–5
  interactions 79
  lipid-soluble 49–53
  requirements 41–*2*, 43–56
  stability and destruction 69–*70*, 71
  water-soluble 41–9
Vitellogenesis 191, 194, 289
Vitellogenin 19, 191, 289

Water quality criteria 267
Water-soluble B-complex vitamins, chemical structures **46–7**
Water-soluble macrovitamins, chemical structures **43**
Wax ester 10

X-radiography 150–1
Xenobiotics 271–2

Yeast, feed ingredient *61*, 73
Yolk production 191

Zinc 76, 80, 272
  requirement *54*
Zone of resistance **211**, **214**–15
Zone of tolerance **211**, **214**
Zooplankton *4*, 73
  chemical composition *58*